D1035208

Ionospheric Tomography

Springer
Berlin
Heidelberg
New York
Hong Kong
London
Milan
Paris
Tokyo

Physics and Astronomy ONLINE LIBRARY

http://www.springer.de/phys/

Physics of Earth and Space Environments

http://www.springer.de/phys/books/ese/

The series *Physics of Earth and Space Environments* is devoted to monograph texts dealing with all aspects of atmospheric, hydrospheric and space science research and advanced teaching. The presentations will be both qualitative as well as quantitative, with strong emphasis on the underlying (geo)physical sciences. Of particular interest are

- contributions which relate fundamental research in the aforementioned fields to present and developing environmental issues viewed broadly
- concise accounts of newly emerging important topics that are embedded in a broader framework in order to provide quick but readable access of new material to a larger audience

The books forming this collection will be of importance for graduate students and active researchers alike.

V. Kunitsyn E. Tereshchenko

Ionospheric Tomography

With 88 Figures, Including 16 Color Plates

Springer

Professor Dr. Viacheslav E. Kunitsyn
M. Lomonosov Moscow State University
Physics Faculty, Atmospheric Physics Dept.
Moscow 119992
Russian Federation

Dr. Evgeny D. Tereshchenko
Polar Geophysical Institute
Khalturina Str. 15
Murmansk 183010
Russian Federation

Library of Congress Cataloging-in-Publication Data
Kunitsyn, V.E. (Viacheslav Evgen'evich)
Ionospheric tomography / V. Kunistyn, E. Tereshchenko.
p. cm. – (Physics of earth ans space environments, ISSN 1610-1677)
Includes biblographical references and index.
ISBN 3-540-00404-1 (acid-free paper)
1. Ionosphere–Remote sensing. 2. Tomography–Scientific applications.
I. Tereshchenko, E. D. II. Title. III. Series.
QC881.2.I6K83 2003 551.51'45'0287–dc21 2003041206

ISSN 1610-1677

ISBN 3-540-00404-1 Springer-Verlag Berlin Heidelberg New York

Springer-Verlag Berlin Heidelberg New York
a member of BertelsmannSpringer Science+Business Media GmbH

Printed in Germany

Typesetting: LE-TEX Jelonek, Schmidt & Vöckler GbR
Cover design: Erich Kirchner, Heidelberg

Printed on acid-free paper 54/3141/YL - 5 4 3 2 1 0

Preface

The monograph is devoted to a new branch of remote sounding of the ionosphere – ionospheric tomography. Adoption of tomographic methods seems to be an inevitable stage of the evolution of almost all diagnostic systems. Advanced techniques for remote sensing and progressive means for data processing open the possibility of reconstructing the spatial structure of the medium on the base of tomography. In this book, mainly the problems of satellite radio tomography of the ionosphere are discussed, and only one subsection is alloted to optical ionospheric tomography. Modern radio sounding techniques make it possible by means of satellite facilities to probe the ionosphere within a wide range of varying positions of transmitting-receiving systems and to apply tomographic methods. In this connection, in recent years, active tomographic investigations of the ionosphere have been done.

The purpose of this monograph is to set forth the tomographic methods developed for recovering the 2-D and 3-D structure of the ionosphere and to discuss experimental implementation of these methods. The topic of discussion is reconstruction of electron density distribution and effective collision frequency. The structure of the ionosphere is known to be quite complex; along with a quasi-stratified background with large scale variations in electron density, there are also local irregularities of various scale sizes, including turbulent volumes. In accordance with this, the problems of reconstructing an inhomogeneous ionosphere should be divided into statistical and deterministic ones. The latter, in their turn, can be split into problems of diffraction radio tomography and ray radio tomography when diffraction effects are inessential.

In this monograph, a theory is considered for solving inverse problems of diffraction tomography for weak and strong irregularities (with a stratified background ionosphere taken into account) based on small angle scattering data, including a solution obtained by a holographic approach. Ray radio tomography methods are discussed, including the phase-difference tomography technique. A theory is developed for tomographic reconstruction of the spatial distribution of statistical parameters of a randomly inhomogeneous ionosphere. A technique is elaborated for radio tomography experiments based on the use of signals from navigation satellites (150/400 MHz) that opens prospects for wide scope investigation of the structure of an irregular ionosphere. Schemes are presented of developed measuring radio tomographic systems. By diffraction tomography, the 2-D structure of kilometer-scale irregularities is recovered from data measured at a receiving chain transverse to the satellite path. The tomographic reconstruction is based on methods of holographic field recording. An example is shown of tomographic reconstruction of the spatial distribution of electron density fluctuations in a randomly inhomogeneous ionosphere. Examples are given to illustrate experimental reconstruction of ionospheric sections by ray radio tomography. Examples of the sections of near-equatorial, middle, and subauroral ionosphere are shown, in particular, the sections of the ionospheric trough and equatorial anomaly are portrayed.

We are very grateful to our colleagues in cooperation with whom ionospheric radio tomography was developed: E.S. Andreeva, B.Z. Khudukon, Y.A. Melnichenko.

We also highly appreciate valuable discussions and cooperation with L.D. Bakhrah, A. Brekke, J. Foster, S. Franke, T. Hagfors, J. Klobuchar, Y.A. Kravtsov, D.S. Lukin, M. Lehtinen, V.V. Migulin[1], T. Nygrén, and K.C. Yeh.

Finally our sincere thanks to M. Kozlova and J. Shapovalova for their help in preparing of the manuscript.

Moscow, Murmansk,
March 2003

Viacheslav Kunitsyn
Evgeny Tereshchenko

[1] Academician V.V. Migulin passed away in September 2002.

Contents

List of Symbols

Latin

A amplitude
$\hat{A}$ matrix of RT problems
A_q projection of the correlation function of complex potential
B correlation function
$\boldsymbol{B}_0$ geomagnetic field
c speed of light
e electron charge
$\boldsymbol{E}$ electric field, electromagnetic field
$\mathrm{E_s}$ sporadic E layer
$\boldsymbol{E}_\mathrm{s}$ scattered field
E_o ordinary wave
E_e extraordinary wave
E_H holographically reconstructed field
f frequency
f, f_i probing frequency
f_N plasma frequency
f_H hyro frequency
F hypergeometric function, also spectrum of irregularities
$\hat{F}$ Fresnel operator
$F_T(\omega)$ Fourier transform of function $F(t)$ over interval T
g group path
$\hat{G}$ Green's function
h height
$\boldsymbol{H}$ magnetic field (intensity)
I linear integrals
$\hat{I}$ unit diagonal tensor
$\boldsymbol{J}$ scattering current
J_1 Bessel function
k wave number
k, k_0 free space wave number
k_B Boltzmann's constant
$\mathcal{K}$ correlation coefficient
K projection of correlation coefficient
L thickness of ionospheric layer, also transfer matrix
L_0 outer scale of irregularities
m electron mass
$\boldsymbol{m}$ vector of measurements
m_i ion mass
n, n_i refractive index
N, N_0 electron concentration (electron density)
N_n, N_n0 neutral molecules concentration
δN electron density fluctuations

$\hat{P}$ projection operator
$\boldsymbol{P}_{\mathrm{s}}$ power spectrum of the scattered field
$P_{\boldsymbol{\varepsilon}}(\boldsymbol{\varepsilon})$ probability density of $\boldsymbol{\varepsilon}$
$P(\boldsymbol{m}/\boldsymbol{x})$ conditional probability density of $\boldsymbol{m}$ for given $\boldsymbol{x}$
$P(\boldsymbol{\varepsilon})$ multidimensional density distribution
$P(\boldsymbol{x})$ a priori probability density
$P(\boldsymbol{x}/\boldsymbol{m})$ a posteriori probability
q complex potential
$\boldsymbol{q}$ scattering vector
q_z projection of complex potential
$\hat{Q}$ Fisher information matrix
r_{e} classical electron radius
R the Earth's radius
$\hat{R}$ resolution matrix
R_{F} Fresnel radius
t time
T temperature
T_{i} ion temperature
z longitudinal coordinate (coordinate in the direction of wave propagation)
X, Y, Z coefficients in equations of magnetoionic theory
Z_{u}, Z_{d} upper and lower boundaries of the ionospheric layer

Greek

α elongation of irregularities along the geomagnetic field
β elongation of irregularities in a plane perpendicular to the geomagnetic field vector, also elevation angle of a satellite
Γ gamma function
$\Gamma_{1,1}$ first coherence function
$\Gamma_{2,0}$ second coherence function
δ delta function
δ_{ij} Cronecker symbol
δ resolution of a recording system, also a measure of numerical simulation error
$\hat{\varepsilon}$ dielectric permittivity
$\boldsymbol{\varepsilon}$ vector of measurement errors
ζ reduced distance
θ the angle between the geomagnetic field vector and the direction of wave propagation
Θ aperture angle
λ wavelength
λ_{D} Debye length
μ_0 magnetic permeability
ν collision frequency
∇ differentiation operator

$\boldsymbol{\rho}$ transverse coordinate
$\hat{\sigma}$ conductivity tensor
σ rms (root-mean-square) deviation
σ_{χ}^2 variance of logarithmic relative amplitude
σ_V scatter cross section
σ^2 variance of electron density fluctuations
Σ coefficient in equations of magnetoionic theory
$\hat{\Sigma}$ matrix of errors τ horizontal distance measured across the Earth's surface
Φ phase, also spectrum of irregularities
χ amplitude level (logarithmic relative amplitude)
$\omega_{\mathrm{p}} = 2\pi f_{\mathrm{N}}$ plasma frequency (angular)
Ω Doppler frequency
ψ orientation angle of cross-field anisotropy of irregularities

Introduction

Investigation of the structure of ionospheric irregularities is a very important topic for many practical problems, both in geophysical research (in studying the structure of the ionosphere and the physics of processes running there) and in various radio applications associated with radio wave propagation (since the ionosphere as a medium of propagation may cause a noticeable effect on the operation of various navigation, location, and communication systems). Among various sounding techniques, the major ones are remote sensing methods that provide real-time information, which is of key significance due to the inconstancy of the ionosphere. Of special interest are radio probing methods that use well-developed equipment and advanced measurement techniques. During their more than 70-year history, these methods proved to be most contributory to the investigation of the ionospheric structure – the main body of facts has been obtained by just radio sounding methods. Optical methods for ionosphere diagnostics have also been successfully used for many years.

There is an extensive literature on radio sounding of the ionosphere and on the methods for analysis of radio wave propagation in the ionospheric plasma. In this monograph, we did not aspire to present an exhaustive analysis of a wide range of existing methods for remote optical and radio sounding of the ionosphere; here, only a brief outline of the methods suitable for monitoring the ionosphere is given. The aim of this work is to describe the tomographic methods developed for reconstructing the 2-D and 3-D structure of the ionosphere along with their experimental implementation. Non-model approaches to the problems of reconstruction will be considered. The number of works concerning the reconstruction of spatial irregularities is not very big now. The bibliographic references and the analysis of these works will be given below. Hereinafter, the reconstruction of irregularities from remote radio sounding data will be understood as recovering the structure $N(r,t)$ of electron density and effective collision frequency $\nu(r,t)$ as a function of three spatial variables and time. Together with the data on the correlation function descriptive of electron density fluctuations, information about the three mentioned functions is basic in forecasting the parameters of ionospheric radio signals. Naturally, it is not always that such a general problem can be stated and solved; in several cases, it is quite enough to reconstruct 2-D sections of the spatial structure of stationary irregularities.

Inherent in many fields of science and techniques is the problem of reconstructing the medium's structure from the radiation that passes through. In other words, it is necessary to "image" the object by recording the scattered probing waves. In the 1960s, from the success in optical holography that made it possible to "discern" 3D objects, similar investigations also started in other scientific branches. Special terms appeared such as seismic, acoustic, microwave, and radio holography and other types of holography. In the 1970s and 1980s, again not without impressive achievements in X-ray tomography, the terminology and the trend of remote sensing investigations have undergone changes. Such terms as acoustic, seismic, optical, microwave, and other tomographies were more often used at that time. In theoretical works associated with remote sounding of media and reconstruction of irregularities, the term "inverse problems" (IP) is mostly used.

There is no point in strictly classifying progress in any scientific field; however, we should make certain demarcations and explanations of terminology, so as not to be reproached with inventing unnecessary new terms in the further discussion. Let us remember that the Greek "holography" means "complete record." Therefore, it makes sense to understand radio holography as a method of "complete" field recording and further reconstruction of the field from the measured data, as has become already customary in optical and other frequency ranges. Inverse problems of reconstructing irregularities in the medium from a measured field are no longer holographic. But if the structure of the medium is recovered directly from a holographically reconstructed field, it is quite reasonable to keep the term "holographic approach to the inverse problem." The word "tomography" stems from the Greek $\tau o \mu o \sigma$ – section, layer. When certain projections or sections of an inhomogeneous object (or its transform, for example, the Fourier transform) are known from remote sounding data and a problem is set up of reconstructing the irregular structure of this object, it is expedient to consider the problem as tomographic. At present, the mathematical fundamentals of tomography are used as related to integral geometry, where it is required to reconstruct the object from data available in the form of integrals over manifolds of less dimensionality. Therefore the term "tomography" is usually understood not in the narrow original sense of the word (as a layer-by-layer investigation of the structure of inhomogeneous objects), but in a wider sense, as a recording of sections or projections of the object and subsequent reconstruction of the object's structure from these data. Note that just the projections of the object are integrals of different kinds over manifolds of less dimensionality. For instance, a problem can be stated such as recovering the 3-D structure of an object from a set of its 2-D projections or reconstructing a 2-D object from a set of 1-D projections the latter problem leads to a classical Radon transform.

The range of problems considered in this monograph concerning remote radio sounding of the ionosphere by artificial earth satellites includes reconstruction of an irregular ionosphere from sets of various data of the tomographic type (projections and sections), which give good grounds for using the term "radio tomography of the ionosphere."

The inverse problems (IP) of reconstructing the irregular structure of the ionosphere should be divided into deterministic and statistical ones. For the deterministic IP, the structure of some large-scale irregularities or group of irregularities is to be reconstructed. But if a great number of irregularities occupies some volume in the space, it is pointless to reconstruct a particular realization of continuously varying irregularities; here, it makes sense to state the problem of reconstructing the statistical parameters of irregularities such as the correlation function of electron density fluctuations and so on.

Thus, in our view, the following terminology is suitable for problems of the tomography of the ionosphere or another medium. First of all, tomographic problems can be divided according to the type of the sounding radiation used: radio tomography, optical tomography, etc. Then, tomographic problems split into deterministic and statistical ones. When the tomographic problem is to reconstruct the statistical properties of a medium from the statistical parameters of the sounding waves, it is expedient to use the term "statistical tomography." Due to the specificity of the mathematical techniques used, deterministic problems can be divided into those for diffraction and ray tomography when diffraction effects are negligible (Fig. 1).

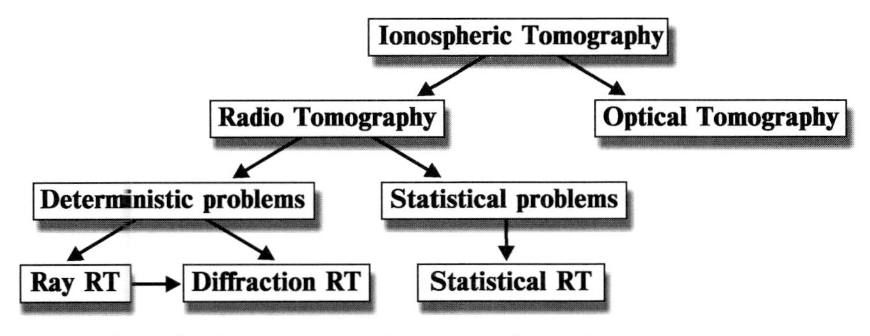

Fig. 1. Classification of ionospheric tomography problems

Let us single out some principal peculiarities of the problems of ionospheric tomography, or IP of reconstructing the irregular ionosphere. The dimensions of feasible receiving-transmitting systems are much less than the distance to the reconstructed irregularities of about a few hundred km, that is, the aperture angles are small. Since it is very difficult and expensive to build the receiving-transmitting systems with a great number of receivers, the necessity to synthesize an aperture along one of the coordinates becomes evident. It is possible to synthesize the aperture, for instance, by using a transmitter onboard the moving satellite. We would like to stress that it is just satellite radio sounding that makes it possible to obtain in practice the sets of various tomographic data and to realize radio tomography of an irregular ionosphere. As the aperture angles are small, it is pertinent to state the problem of recovering the structure of the ionosphere with a resolution noticeably exceeding the wavelength. Therefore, we will consider here the reconstruction of large, compared to the wavelength, irregularities. Note that the sizes of irregularities can be either greater or less than the Fresnel radius, so that in some cases the diffraction effect should be taken into account.

The scheme of an experiment on satellite radio tomography is shown in Fig. 2. The satellite with an onboard transmitter moves around Earth; receiving systems are located at Earth's surface. The receiving system can contain one receiver, a chain of receivers, or a set of receiving chains, depending on the particular problem. For ray RT, it is necessary to have a series of receivers spaced about a few hundred kilometers apart installed along the satellite path. For diffraction RT, km-long and longer receiving chains transverse to the satellite path are required. Statistical RT problems can be solved by measurements at a few receivers along the satellite path or at a few transverse chains, depending on the particular statement of the problem.

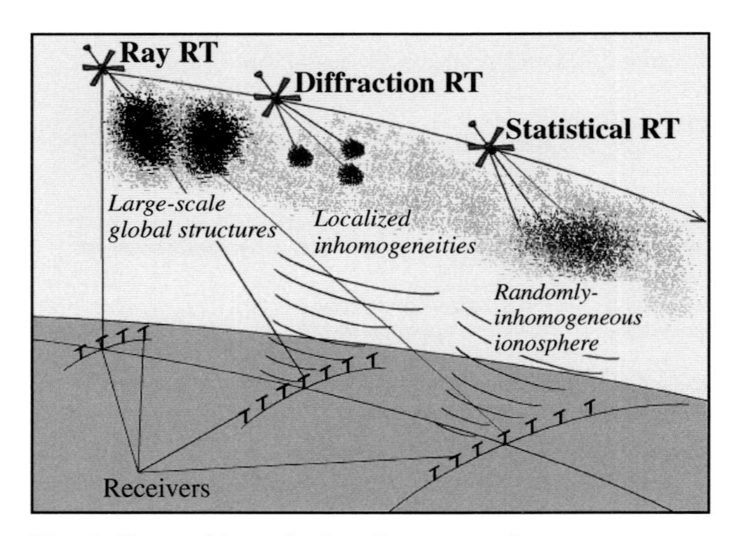

Fig. 2. Types of ionospheric radio tomography

1. Ionosphere: Structure and Basic Methods for Radio Sounding and Monitoring

1.1 An Outline of Ionospheric Structure

The term "ionosphere" for defining the ionized shells encircling Earth was introduced by V. Watt in 1929. The hypothesis that a conducting region should exist in the upper atmosphere was first advanced at the end of the nineteenth century first by B. Stewart and then by A. Schuster [Schuster, 1889] to explain diurnal variations of the geomagnetic field. However, this hypothesis became widespread only after Heaviside and Kennely explained in 1902 the propagation of radio waves over large distances. The existence of ionized regions in the upper atmosphere was directly proved in the 1920s. Of greatest importance was the invention of the pulse ionospheric station by G. Breit and M. Tuve [Breit and Tuve, 1926], who were the first to carry out radio probing of the ionosphere. In Russia, the first results in this field were obtained by M.V. Shuleykin [Shuleykin, 1923] who inferred from studying the operation of broadcasting stations that a radio wave exists that comes from above to Earth's surface. He calculated the reflection height as 260 km.

In the earliest investigations, information about the structure of the ionized atmospheric region used to be obtained by ionospheric stations where electron density was determined from the top frequencies of radio waves transmitted vertically upward and reflected from the ionized atmospheric region. These measurements seemed to detect several layers in the ionosphere. First, the E layer was thus revealed, called so after reflection of an electric (E) field. Further, the existence of the D and F layers was established. These layers, located lower and higher than the E layer, were named in alphabetical order in correspondence with the E layer. Although modern investigation showed that the idea of distinctly separated ionospheric layers was not correct, this classification is still in use in the literature right up to now and relates mostly to the heights of ionospheric processes.

Figure 1.1 portrays the height variations of the vertical profile of the electron density [Brekke, 1997]. The figure gives only an approximate distribution of electron density. Actual values of the electron density may differ significantly from the curve shown, since they depend on a lot of parameters such as the location of the observation site, local time, season, processes in the Sun, solar wind parameters, and others. The planetary distribution

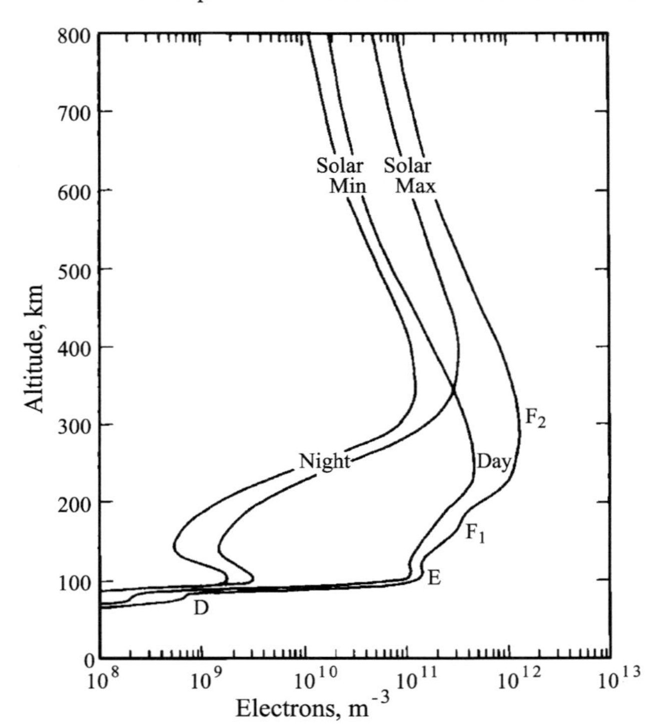

Fig. 1.1. Typical midlatitude ionospheric electron density profiles for sunspot maximum and minimum conditions in daytime and nighttime. The different altitude regions in the ionosphere are labeled with appropriate nomenclature

of the electron density calculated for spring equinox conditions is shown in Fig. 1.2.

As seen from this figure, despite the fact that the electron density distribution is simulated for the spring equinox and is practically symmetrical about the equator, it still shows a rather complicated structure. When considering the ionosphere, it is used to distinguish between the equatorial, midlatitude, and high-latitude ionosphere. Sometimes the high-latitude ionosphere is subdivided into the auroral and polar cap ionosphere, and also a separate region of the main ionospheric trough located equatorward of the auroral zone is usually singled out. Such a division is based on physical grounds and associated with the structure of the geomagnetic field.

The high-latitude ionosphere differs significantly from the midlatitude region in its structure and changeability since it is linked with the magnetospheric regions where the structure of the geomagnetic field is not quasi-dipolar. Physical processes that take place in these regions are highly fickle and complex since in the high latitudes, the effects of the solar wind, electric fields, and particle precipitation are most strongly pronounced. The charged

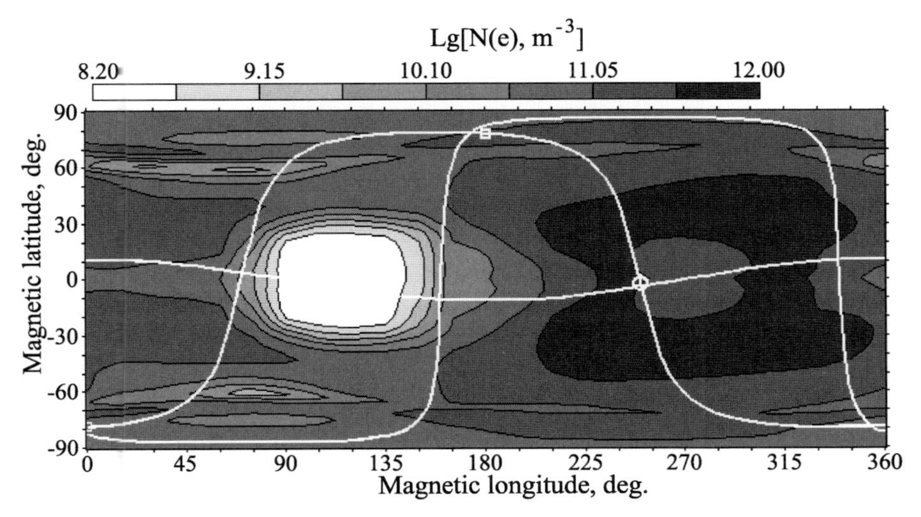

Fig. 1.2 Planetary distribution of the electron density calculated from the Global Upper Atmosphere Model [Namgaladze et al., 1988, 1996] at 24:00 UT on March 24, 1987 at 300 km height. The circle shows the position of the sub-Sun point located practically at the geographical equator displayed by the white nearly horizontal line; the square is the geographic North Pole. The line connecting the sub-Sun point with the geographic North Pole is a projection of the midday–midnight meridian in geomagnetic coordinates; the line departing from the latter by 90° indicates the location of the morning–evening meridian

particles intrude into the ionosphere in the auroral oval and cusp regions. Particle intrusion is permanent [Akasofu, 1974], but at low activity, the intensity of the invading flow is comparatively low and, most important, sharply confined in space. With increasing activity, the intensity of the flow grows, and the area covered by precipitating particles expands. The width and the size of the auroral oval increase noticeably. As the oval widens, it moves away from the pole down to lower latitudes. The behavior of the electric field alters similarly to the changes in magnetospheric disturbance. Therefore, any change in the magnetosphere causes a response in the high-latitude ionosphere. The features typical of the structure of the high-latitude ionosphere are shown in Fig. 1.2. So, an enhancement of electron density is seen at 70–85° produced by particle precipitation in the auroral region. The large-scale decrease of electron density by a factor of 2 observed at the equator of the auroral region is called the main ionospheric trough. The region of depleted electron density poleward of the auroral region is an auroral cavity, or high-latitude trough.

Middle latitudes are the area to which magnetospheric volume of closed field lines is projected. Therefore, except for periods of solar flares, the changes in electron density in this latitudinal interval, generally, depend on short-wave solar radiation and thus have a comparatively regular structure.

The equatorial ionosphere also has its specific features. The existence of crossed meridional magnetic and zonal electric fields causes intensive upward electrodynamic plasma flow from F region heights up to 500–600 km wherefrom the charged particles spread along the field lines to both both South and North Hemispheres reaching the top-height F-layer ionosphere at about 20° latitude. Hence, in daytime conditions, minimum electron density takes place close to the geomagnetic equator, and increased density appears on both sides of the equator (north and south of it, Fig. 1.2).

Together with the above discussed global changes in ionospheric electron density, smaller scale irregularities are also encountered in the ionosphere. Those with scale sizes from a few centimeters up to several kilometers, we will call small scale irregularities. They are produced by different kind instabilities occurring in the ionospheric plasma [Keskinen and Ossakow, 1983]. There are two main maxima in the global distribution of intensities of small scale irregularities, the auroral and equatorial maxima [Aarons, 1982]. Besides these regions, irregularities are also observed in the main ionospheric trough as well as in the polar cap region.

1.2 Basic Methods for Radio Sounding and Monitoring the Ionosphere

Incoherent scattering (IS) is a highly informative method for investigating the ionosphere. It is based on the effect of scattering of the electromagnetic field in a plasma. In a weak scatter approximation at small frequency shifts, the scattered field is defined by fluctuations of electron density. The theory developed which relates spatiotemporal fluctuations with physical conditions in a quasi-uniform medium makes it possible to evaluate basic ionospheric parameters such as electron concentration, electron and ion temperature, ion composition, ion-neutral collision frequency, and electron and ion drift velocity in the presence of an electric field. A lot of these parameters were unavailable from former measurements, and the development of the IS method allows us to obtain a more complete picture of ionospheric processes. The use of VHF and UHF waves allows us to settle several problems typical of ionosonds and opens the possibility of measuring a lot of parameters simultaneously in different ionospheric regions with better spatial resolution.

Nowadays, the IS method is successfully used in the USA, England, France, Northern Scandinavia, and on Svalbard.

The problem of the scattering of electromagnetic (EM) waves in a plasma may be formulated as follows. Consider a plasma volume V, illuminated by EM wave $\boldsymbol{E}_0(\boldsymbol{r},t)$ emitted in the direction of unit vector $\boldsymbol{n}_i{}' = (\boldsymbol{r}_0 - \boldsymbol{r}')/|\boldsymbol{r}_0 - \boldsymbol{r}'|$ from a source at position $\boldsymbol{r}_0$ (Fig. 1.3).

It is implied here that the mean frequency of incident wave ω_0 is much higher than the electron plasma frequency, electron gyro frequency, and electron collision frequency. Therefore, the bulk of the EM wave energy passes

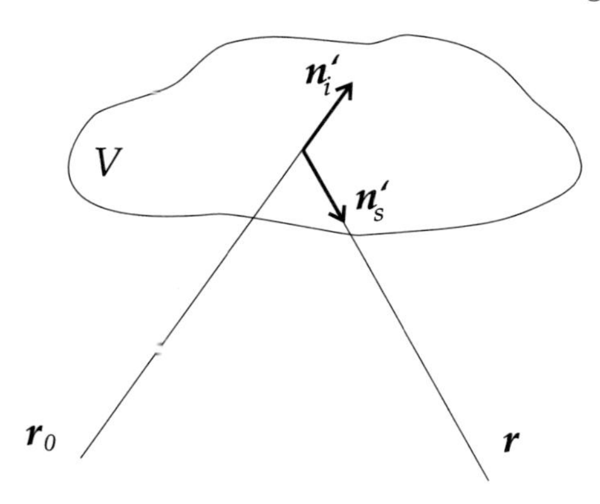

Fig. 1.3. Geometry of scattering

through the plasma, and damping caused by scattering and absorption is negligible small. This means that the initial field has the same intensity at each point of plasma volume V. Let us also assume that the initial incident field does not exceed some critical limit so that it does not change the state of the plasma.

Let us seek the scattered field $\boldsymbol{E}_{\mathrm{s}}(\boldsymbol{r},t)$ at distance $|\boldsymbol{r}'-\boldsymbol{r}|$ from the point of scattering in the direction of unit vector $\boldsymbol{n}_{\mathrm{s}}'=(\boldsymbol{r}-\boldsymbol{r}')/|\boldsymbol{r}-\boldsymbol{r}'|$. In our calculations, we consider small relative fluctuations of the plasma parameters and assume that the scattered field is small compared with the incident field; therefore the scattered EM field can be calculated by a Borne approximation. A combined approach can be used in solving the problem when the propagation of the scattered wave is described in terms of macroscopic equations, the same as for an incident wave, whereas in solving for scattering current and its dependence upon plasma parameters, a deeper physical consideration will be used [Suni et al., 1989].

Let us assume that electric field $\boldsymbol{E}(\boldsymbol{r},t)$ in a fluctuating medium obeys the uniform wave equation,

$$\mathrm{rot\ rot}\ \boldsymbol{E}+\frac{1}{c^2}\frac{\partial^2}{\partial t^2}\boldsymbol{E}+\mu_0\frac{\partial}{\partial t}\hat{\sigma}\,\boldsymbol{E}=0\,, \tag{1.1}$$

where $\mu_0=4\pi\cdot10^{-7}$ H/m.

Conductivity $\hat{\sigma}$ specifies the properties of the scattering volume and depends upon the parameters of the medium:

$$\hat{\sigma}=\hat{\sigma}(s_i)\,,\qquad i=0,1,\dots\,,$$

where s_i symbolically denotes various parameters of the plasma, for example, s_0 – electron density, s_1 – ion density, etc.

Taking into account the fluctuating nature of the medium, each parameter can be described as a sum of its mean value $\langle s_i\rangle$ and deviation δs_i:

$$s_i=\langle s_i\rangle+\delta s_i\,,$$

where angular brackets denote an ensemble average, and thus,

$$\hat{\sigma} = \hat{\sigma}\left(\langle s_i \rangle + \delta s_i\right) .$$

If a HF signal propagates in the ionosphere, the conductivity of a plasma may be well described by a "cold" plasma approximation [Krall and Trivelpiece, 1973] when the conductivity is assumed to depend only on electron density. Thus assuming $\hat{\sigma} = \hat{\sigma}\left(N + \delta N\right)$ and taking into account the smallness of δN compared with mean value N, one can expand $\hat{\sigma}$ into a Taylor series. Retaining only linear terms in the expansion, we obtain

$$\hat{\sigma} = \hat{\sigma}\left(N\right) + \left.\frac{\partial \hat{\sigma}}{\partial N}\right|_{\delta N = 0} \delta N .$$

Hence, the equation for the field in a plasma can be written as

$$\text{rot rot}\, \boldsymbol{E} + \frac{1}{c^2}\frac{\partial^2}{\partial t^2}\boldsymbol{E} + \mu_0 \frac{\partial}{\partial t}\hat{\sigma}(N)\boldsymbol{E} = -\mu_0 \frac{\partial}{\partial t}\left(\left.\frac{\partial}{\partial N}\hat{\sigma}\right|_{\delta N = 0} \delta N \boldsymbol{E}\right) .$$

The right-hand part of the equation is the result of interaction between the wave and the irregularities of the medium; therefore, it can be considered as some fictive source.

Let the scattering be weak and substitute incident wave $\boldsymbol{E}_0$ instead of $\boldsymbol{E}$ in the right-hand part of the above equation. Defining the scattering current as

$$\boldsymbol{J}\left(\boldsymbol{r}, t\right) = \left.\frac{\partial \hat{\sigma}}{\partial N}\right|_{\delta N = 0} \delta N\, \boldsymbol{E}_0 ,$$

one obtains the following expression for the scattered field:

$$\text{rot rot}\, \boldsymbol{E}_{\text{s}} + \frac{1}{c^2}\frac{\partial^2}{\partial t^2}\boldsymbol{E}_{\text{s}} + \mu_0 \frac{\partial}{\partial t}\left.\hat{\sigma}\right|_{\delta N = 0} \boldsymbol{E} = -\mu_0 \frac{\partial \boldsymbol{J}}{\partial t} . \tag{1.2}$$

Let us apply the Fourier transform in time domain $F(\omega) = \int \mathrm{d}t\, F(t) \exp\left(\mathrm{i}\,\omega t\right)$ to the equation for a scattered field and use the approximation of a "cold" stationary nonmagnetized plasma for conductivity

$$\hat{\sigma}\left(\omega_1\right) = \mathrm{i}\,\frac{e^2 N}{m\omega_1}\,\hat{I} ,$$

where $\hat{I}$ is a unit diagonal tensor, m is electron mass, and e is electron charge. Thus we obtain

$$\text{rot rot}\, \boldsymbol{E}_{\text{s}}\left(\boldsymbol{r}, \omega_1\right) - \frac{\omega_1^2}{c^2}\left(1 - \frac{4\pi r_{\text{e}} N}{k_1^2}\right)\boldsymbol{E}_{\text{s}}\left(\boldsymbol{r}, \omega_1\right) = \mathrm{i}\,\omega_1 \mu_0 \boldsymbol{J}\left(\boldsymbol{r}, \omega_1\right) , \tag{1.3}$$

where $k_1 = \omega_1/c$, $r_{\text{e}} = \mu_0 e^2/4\pi m = 2.82 \cdot 10^{-15}$ m – classical electron radius, and

$$\boldsymbol{J}(\boldsymbol{r}, \omega_1) = \mathrm{i}\,\frac{4\pi r_{\text{e}}}{\mu_0 \omega_1}\int \mathrm{d}t\, \delta N(\boldsymbol{r}, t)\boldsymbol{E}_0(\boldsymbol{r}, t)\exp(\mathrm{i}\,\omega_1 t) .$$

Since we consider the scattering of HF waves, the term $4\pi r_{\text{e}} N/k_1^2$ may be neglected in (1.3) as it is of the order of the squared ratio of plasma frequency to the frequency of the scattered wave.

Using Green's function $\hat{G}(\boldsymbol{r},\boldsymbol{r}')$, one can write the solution of (1.3) as

$$\boldsymbol{E}_\mathrm{s}(\boldsymbol{r},\omega_1) = \mathrm{i}\,\omega_1\mu_0 \int \mathrm{d}\boldsymbol{r}'\, \hat{G}(\boldsymbol{r},\boldsymbol{r}')\boldsymbol{J}(\boldsymbol{r}',\omega_1)\,.$$

Let us neglect the terms that vanish when probing the far-field region by quasi-monochromatic waves, which is equivalent to replacing the exact expression for $\hat{G}(\boldsymbol{r},\boldsymbol{r}')$ by its approximation [Suni et al., 1989]

$$\hat{G}(\boldsymbol{r},\boldsymbol{r}') \simeq -\frac{1}{4\pi}\left[\hat{I} - \frac{(\boldsymbol{r}-\boldsymbol{r}')(\boldsymbol{r}-\boldsymbol{r}')}{|\boldsymbol{r}-\boldsymbol{r}'|^2}\right]\frac{\exp\left(\mathrm{i}\,\frac{\omega_1}{c}|\boldsymbol{r}-\boldsymbol{r}'|\right)}{|\boldsymbol{r}-\boldsymbol{r}'|}\,. \tag{1.4}$$

Therefore, one obtains

$$\boldsymbol{E}_\mathrm{s}(\boldsymbol{r},\omega_1) = r_\mathrm{e}\iint \mathrm{d}\boldsymbol{r}'\,\mathrm{d}t\;\delta N(\boldsymbol{r}',t)\,\{\boldsymbol{n}_\mathrm{s}{}'\times[\boldsymbol{n}_\mathrm{s}{}'\times\boldsymbol{E}_0(\boldsymbol{r}',t)]\}$$
$$\times\frac{\exp\left(\mathrm{i}\,\omega_1 t + \mathrm{i}\,\frac{\omega_1}{c}|\boldsymbol{r}-\boldsymbol{r}'|\right)}{|\boldsymbol{r}-\boldsymbol{r}'|}\,. \tag{1.5}$$

For further calculations, let us specify the shape of the probing signal. Assume that the scattering volume is in the far-field region of a transmitting antenna. Set the initial field as

$$\boldsymbol{E}_0(\boldsymbol{r}',t) = \mathcal{E}_0\boldsymbol{l}_0\frac{\exp\left(-\mathrm{i}\,\omega_0 t + \mathrm{i}\,k_0|\boldsymbol{r}-\boldsymbol{r}'|\right)}{|\boldsymbol{r}-\boldsymbol{r}'|}\,,$$

which corresponds to a spherical wave with an amplitude $\mathcal{E}_0$, unit polarization vector $\boldsymbol{l}_0$ and wave number $k_0 = \frac{\omega_0}{c}$. Then, the Fourier component of the scattered field will be

$$\boldsymbol{E}_\mathrm{s}(\boldsymbol{r},\omega_1) = r_\mathrm{e}\mathcal{E}_0\int \mathrm{d}\boldsymbol{r}'\;\delta N(\boldsymbol{r}',\omega_1-\omega_0)\,[\boldsymbol{n}_\mathrm{s}{}'\times(\boldsymbol{n}_\mathrm{s}{}'\times\boldsymbol{l}_0)]$$
$$\times\frac{\exp\left(\mathrm{i}\,\frac{\omega_1}{c}|\boldsymbol{r}-\boldsymbol{r}'| + \mathrm{i}\,\frac{\omega_0}{c}|\boldsymbol{r}_0-\boldsymbol{r}'|\right)}{|\boldsymbol{r}-\boldsymbol{r}'||\boldsymbol{r}_0-\boldsymbol{r}'|}\,. \tag{1.6}$$

In further computations, a Fraunhofer approximation is usually applied [Hagfors, 1977], and $\boldsymbol{E}_\mathrm{s}(\boldsymbol{r},\omega_1)$ is obtained in the form of a spatiotemporal Fourier spectrum of δN. Note that in real experiments, neither the receiving nor the transmitting antenna is in a Fraunhofer zone. Indeed, should the, e.g., receiving antenna be in the Fraunhofer zone of scattering volume V with plane size L, condition $|\boldsymbol{r}_0-\boldsymbol{r}'| \gg kL^2$ will be satisfied. The size of the scattering volume is a product of the spacing $\Delta\Theta$ and the width of the radiation pattern $|\boldsymbol{r}_0-\boldsymbol{r}'|$, i.e., $L = |\boldsymbol{r}_0-\boldsymbol{r}'|\Delta\Theta$. Therefore, the necessary condition will be $|\boldsymbol{r}_0-\boldsymbol{r}'| \ll \frac{1}{k(\Delta\Theta)^2}$. Substituting typical values of the IS radar parameters $k\sim 1\ \mathrm{m}^{-1}$, $\Delta\Theta \approx 10^{-2}$ rad, we find that $|\boldsymbol{r}_0-\boldsymbol{r}'| \ll 10$ km. Obviously, this is infeasible in ionospheric investigations.

Nevertheless, a suitable expression for the power spectrum of a scattered radio signal can be obtained that will coincide formally with the result derived from a Fraunhofer approximation. To obtain this expression, an assumption should be made that the correlation length of electron density fluctuations δN in a plasma (the Debye length) is much shorter than any other scale sizes. This assumption is quite natural as the Debye length in the ionosphere is of the order of a few centimeters [Ginzburg and Rukhadze, 1975].

Let $F_T(\omega)$ be the Fourier transform of the function $F(t)$ over an interval T. Then, the power spectrum is defined as

$$\frac{1}{T}\langle|F_T(\omega)|^2\rangle .$$

Taking into account (1.6), one can find that the power spectrum of the scattered field $\boldsymbol{E}_\mathrm{s}$ is

$$\begin{aligned}\boldsymbol{P}_\mathrm{s}(\boldsymbol{r},\omega_1) &= \frac{r_\mathrm{e}^2\mathcal{E}_0^2}{T}\iint \mathrm{d}\boldsymbol{r}'\,\mathrm{d}\boldsymbol{r}''\,\langle\delta N(\boldsymbol{r}',\omega_1-\omega_0)\delta N(\boldsymbol{r}'',\omega_1-\omega_0)\rangle\\ &\times[\boldsymbol{n}_\mathrm{s}'\times(\boldsymbol{n}_\mathrm{s}'\times\boldsymbol{l}_0)]\,[\boldsymbol{n}_\mathrm{s}''\times(\boldsymbol{n}_\mathrm{s}''\times\boldsymbol{l}_0)] \quad (1.7)\\ &\times\frac{\exp\left[\mathrm{i}\frac{\omega_1}{c}\left(|\boldsymbol{r}-\boldsymbol{r}'|-|\boldsymbol{r}-\boldsymbol{r}''|\right)+\mathrm{i}\frac{\omega_0}{c}\left(|\boldsymbol{r}_0-\boldsymbol{r}'|-|\boldsymbol{r}_0-\boldsymbol{r}''|\right)\right]}{|\boldsymbol{r}-\boldsymbol{r}'||\boldsymbol{r}-\boldsymbol{r}''||\boldsymbol{r}_0-\boldsymbol{r}'||\boldsymbol{r}_0-\boldsymbol{r}''|} .\end{aligned}$$

Let us convert from variables $\boldsymbol{r}'$ and $\boldsymbol{r}''$ to variables $\boldsymbol{R}_+ = (\boldsymbol{r}' + \boldsymbol{r}'')/2$ and $\boldsymbol{R}_- = (\boldsymbol{r}''-\boldsymbol{r}')$. The Jacobian of this conversion is unity. In a statistically quasi-uniform medium, the spatial correlation of density fluctuations depends weakly on $\boldsymbol{R}_+$ but changes considerably at $\boldsymbol{R}_-$ scales of the order of the Debye length approaching zero at $|\boldsymbol{R}_-| > \lambda_\mathrm{D}$. Making use of this fact, one can neglect the difference between $\boldsymbol{r}'$ and $\boldsymbol{r}''$ in (1.7) in the factors not contained in the exponent.

Now consider the exponents. A Taylor expansion of, e.g., $|\boldsymbol{r}-\boldsymbol{r}'|$ in terms of $\boldsymbol{R}_-$, gives

$$\begin{aligned}|\boldsymbol{r}-\boldsymbol{r}'| &= \sqrt{\left(\boldsymbol{r}-\boldsymbol{R}_+ + \frac{\boldsymbol{R}_-}{2}\right)^2} \simeq |\boldsymbol{r}-\boldsymbol{R}_+| - \boldsymbol{n}_\mathrm{s}\frac{\boldsymbol{R}_-}{2}\\ &+\frac{1}{2}|\boldsymbol{r}-\boldsymbol{R}_+|\left(\frac{(\boldsymbol{R}_-)_\perp}{2|\boldsymbol{r}-\boldsymbol{R}_+|}\right)^2 + O\left(\frac{|\boldsymbol{R}_-|^3}{|\boldsymbol{r}-\boldsymbol{R}_+|^2}\right) .\end{aligned}$$

Taylor expansions of other components can be written in analogy with the above expression.

One can retain only the first three terms of the series provided that

$$\frac{k_0\lambda_\mathrm{D}^3}{|\boldsymbol{r}_0-\boldsymbol{R}_+|^2} \ll \pi , \qquad \frac{k_1\lambda_\mathrm{D}^3}{|\boldsymbol{r}_0-\boldsymbol{R}_+|^2} \ll \pi ,$$

which is obviously satisfied in the ionosphere experiments.

The next fact to be taken into account is that narrow antenna beams with a main lobe width of the order of 1° are used in the experiments. This means that integration over $\boldsymbol{R}_-$ is carried out in a domain where the vectors $\boldsymbol{n}_i'$ and $\boldsymbol{n}_\mathrm{s}'$ keep the same direction with an accuracy of 1°. Introducing a scattering vector independent of $\boldsymbol{R}_-$

$$\boldsymbol{q} = k_1 \boldsymbol{n}_{\mathrm{s}} - k_0 \boldsymbol{n}_i \equiv \boldsymbol{k}_1 - \boldsymbol{k}_0 \,,$$

that corresponds, e.g., to the center of scattering volume, one can transform (1.7) into

$$\boldsymbol{P}_{\mathrm{s}}(\boldsymbol{r}, \omega_1) = \frac{r_{\mathrm{e}}^2 \mathcal{E}_0^2}{T} \frac{\sin^2 \chi}{|\boldsymbol{r} - \boldsymbol{R}|^2 |\boldsymbol{r}_0 - \boldsymbol{R}|^2} \iint \mathrm{d}\boldsymbol{R}' \, \mathrm{d}(\boldsymbol{r}'' - \boldsymbol{r}') \qquad (1.8)$$
$$\times \langle \delta N(\boldsymbol{r}', \omega_1 - \omega_0) \delta N(\boldsymbol{r}'', \omega_1 - \omega_0) \rangle \exp\left[\mathrm{i}\, \boldsymbol{q}(\boldsymbol{r}'' - \boldsymbol{r}')\right] ,$$

with the notation,

$$\sin^2 \chi = \left| [\boldsymbol{n}_{\mathrm{s}}{}' \times (\boldsymbol{n}_{\mathrm{s}}{}' \times \boldsymbol{l}_0)] \right|^2 .$$

The subscript $_+$ in $\boldsymbol{R}_+$ is omitted in (1.8) as it will have no use in further analysis. Since there is a spatial Fourier transform under the integral in (1.8), one obtains the following expression for the scattering cross section σ_V per unit frequency and per unit plasma volume:

$$\sigma_V = 4\pi r_{\mathrm{e}}^2 \sin^2 \chi \frac{1}{T \cdot V} \langle |\delta N(\boldsymbol{k}_1 - \boldsymbol{k}_0, \omega_1 - \omega_0)|^2 \rangle \,. \qquad (1.9)$$

It is seen that a specific feature of EM wave scatter by plasma irregularities is a selectivity of scattering: the intensity of the scattered field is proportional to the spatial spectrum of the electron density fluctuations with only one actually contributing harmonic of wave number q.

Note that in the derivations, the spectral shape has not been specified except for the assumption that the correlation length of electron density fluctuations (Debye length) is short. Therefore the above results can be used as basics for the investigation of the spectral characteristics of quasi-stationary irregularities in the ionosphere and the magnetosphere as well as solar plasma disturbances, regardless of the nature of the irregularities which could be quite different. In particular, an "incoherent" scattering spectrum is determined by thermal fluctuations in a plasma.

The calculations of the spectrum of electron density fluctuations are based on the kinetic approach in the description of plasma processes [Ginzburg and Rukhadze, 1975]. These calculations are quite lengthy, although linearization of the problem is used in calculating the kinetic parameters. Therefore, we will give here only qualitative estimates of some features of the spectrum of electron density fluctuations in an ionospheric plasma.

Theoretical analysis [Suni et al., 1989; Sheffield, 1975] shows that it is important to draw a distinction between two different cases: $|\boldsymbol{q}|\lambda_{\mathrm{D}} \gg 1$ and $|\boldsymbol{q}|\lambda_{\mathrm{D}} \ll 1$. If $|\boldsymbol{q}|\lambda_{\mathrm{D}} \gg 1$, the EM wave is scattered by free thermal electrons, whereas ions do not contribute significantly to the scattering. Continuous thermal motion of charged particles within the scattering plasma volume causes a Doppler shift of the frequency of the scattered radio wave. Since free electrons are moving at a thermal speed, the spectrum of a scattered radio wave is purely Gaussian, and its half-width is proportional to the electron temperature. Otherwise $|\boldsymbol{q}|\lambda_{\mathrm{D}} \ll 1$, i.e., the spectral structure changes significantly if lower frequency waves are scattered. In this case, the contribution from ions becomes important and, furthermore, the spectral shape becomes strongly frequency-dependent. The latter reflects the fact that in addition to chaotic density fluctuations, there are also fluctuations in the plasma associated with specific electromagnetic oscillations that scatter the radio waves.

Figure 1.4 sketches the spectrum of the scattered radiation as a function of frequency shift $\omega_1 - \omega_0$ for the above limiting cases of long (*solid lines*) and short wavelengths (*dashed line*). For negative $\omega_1 - \omega_0$ values, the spectral function is not shown since it is symmetrical about a zero frequency shift when there is no directed motion of charged particles in a plasma. The dashed line displays scattering by free electrons. The solid lines depict scattering not only by plasma charged particles but by ion-acoustic Ω_s and plasma Ω_p oscillations as well. Note that Ω_p is related to plasma frequency ω_p by the formula

$$\Omega_p = \omega_p \sqrt{1 + 3|\boldsymbol{q}|^2 \lambda_D^2} = \sqrt{\frac{Ne^2 c^2 \mu_0}{m}} \sqrt{1 + 3|\boldsymbol{q}|^2 \lambda_D^2} \,,$$

and $\Omega_s = |\boldsymbol{q}| \sqrt{\frac{k_B (T_e + 3T_i)}{m_i}}$, where k_B is Boltzmann's constant and m_i and T_i are ion mass and ion temperature, respectively.

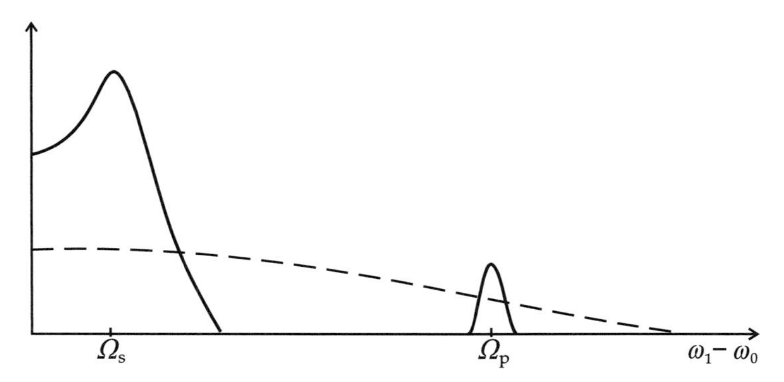

Fig. 1.4. Shapes of scattering spectra for different $|\boldsymbol{q}|\lambda_D$: $|\boldsymbol{q}|\lambda_D < 1$ (*solid line*) and $|\boldsymbol{q}|\lambda_D > 1$ (*dashed line*)

The multiparametric dependence of the scattering spectrum at $|\boldsymbol{q}|\lambda_D < 1$ is successfully used for determination of the electrodynamic characteristics of the ionosphere [Baron, 1986].

The main parameters that can be obtained from the ion line are electron density, electron and ion temperatures, ion masses, and ion velocity. Note that electron density is proportional to the received power of radiation from the scattering volume. In addition to the main physical parameters that are available directly from IS measurements, a series of secondary parameters can be calculated such as electric field, conductivity tensor, electric current, neutral wind velocity, etc. [Brekke, 1997].

Ionospheric radio probing by dynasonds has a prominent place among other methods for sounding the ionosphere, beginning from pioneer works by Breit and Tuve [Breit and Tuve, 1926], who provided a wealth of information on the dynamics and physics of the ionosphere. The wavelengths typically used in dynasonds exceed the Debye length significantly. The incident wave interacts with particles involved in Debye screening, that is, here we are dealing with "collective," or "coherent" scattering [Sheffield, 1975]. In this case, it is expedient to consider reflection of radio waves from the ionosphere as well as scattering from the irregularities in terms of the dielectric permittivity formalism. Then, instead of (1.1), it is convenient to use a similar equation with complex dielectric permittivity ε:

$$\mathrm{rot}\,\mathrm{rot}\,\boldsymbol{E} + \frac{1}{c^2}\frac{\partial^2}{\partial t^2}\,\hat{\varepsilon}\boldsymbol{E} = 0\,. \qquad (1.10)$$

This equation stems from the Maxwell equations and material equations. In the general case, the permittivity of the plasma even in the absence of the magnetic field is a tensor $\hat{\varepsilon}$, due to spatial dispersion. However, in either natural or artificial ionospheric irregularities, spatial dispersion is negligible, since the thermal velocity of electrons in ionospheric plasma inhomogeneities is rather less than the light velocity, which means that the "cold" plasma approximation is valid with high accuracy. Similarly, typical rates of diffusion, convection, and other transport processes, as well as the speed of the transmitting/receiving systems, do not exceed 10 km per hour ($v/c < 3 \cdot 10^{-4}$); therefore a quasi-stationary approximation is valid allowing for the "slow" time dependences of the parameters of the medium and fields in the form of parametric dependences. With this approach, such features as Doppler frequency shift, time variations in group delays are taken into account with the accuracy of relativistic effects.

Next, we will consider monochromatic sounding and proceed to the equations for complex amplitudes E of corresponding harmonics [$\boldsymbol{E} \to \boldsymbol{E}\exp(-\mathrm{i}\omega t)$]. Hereinafter, for the sake of brevity, we will call the complex amplitudes of monochromatic components $\boldsymbol{E}(r,t)$ with "slow" times as "field." Taking into account the foregoing notes, within the scope of a quasi-stationary approximation for "cold" plasma conditions, the Maxwell and material equations yield the equation for the field:

$$\Delta\boldsymbol{E} + \frac{\omega^2}{c^2}\hat{\varepsilon}\boldsymbol{E} - \mathrm{grad}\,\mathrm{div}\,\boldsymbol{E} = 0\,, \qquad (1.11)$$

where $\hat{\varepsilon}$ is a dielectric permittivity tensor. Analysis of radio wave propagation in an inhomogeneous magnetoactive plasma with a tensor dependence of $\hat{\varepsilon}(\boldsymbol{r})$ is quite a difficult problem.

In a stratified plasma, (1.10) simplifies. The case of a stratified plasma is of practical interest, since the ionosphere at large-scale distances can often be deemed a quasi-stratified medium. The process of radio wave propagation in a stratified ionospheric plasma is described in terms of classical magnetoionic theory [Ratcliffe, 1959; Budden, 1961; Davies, 1969]. Components (x, y) of the electric field vector in vertical sounding (along the z-axis) of a stratified ionosphere are described by a system of equations of the form

$$\frac{\mathrm{d}^2\boldsymbol{E}}{\mathrm{d}z^2} + k^2\hat{\varepsilon}(z)\,\boldsymbol{E} = 0\,, \tag{1.12}$$

where $\hat{\varepsilon}$ is a 2×2 matrix,

$$\hat{\varepsilon}(z) = \begin{pmatrix} 1 - a_1 & -ib \\ ib & 1 - a_2 \end{pmatrix}, \tag{1.13}$$

$$a_1 = \frac{X}{1 + \mathrm{i}Z}\left[1 + \frac{Y_\mathrm{L}^2}{(1 + \mathrm{i}Z)\Sigma}\right], \quad a_2 = \frac{X}{\Sigma}, \quad b = \frac{XY_\mathrm{L}}{(1 + \mathrm{i}Z)\Sigma},$$

$$X = \frac{Ne^2}{\varepsilon_0 m\omega^2}, \quad \boldsymbol{Y} = \frac{e\boldsymbol{B}_0}{m\omega}, \quad Z = \frac{\nu}{\omega},$$

$$\Sigma = 1 - \frac{Y_\mathrm{T}^2}{1 - X + \mathrm{i}Z} + \mathrm{i}Z - \frac{Y_\mathrm{L}^2}{1 + \mathrm{i}Z},$$

where $\boldsymbol{B}_0$ is the geomagnetic field and Y_L and Y_T correspond to the geomagnetic field components in and perpendicular to the direction of propagation of the probing radio wave.

Thus, vertical propagation of radio waves in a stratified plasma is expressed by a system of two coupled second-order differential equations (1.12). In the absence of absorption ($Z = 0$), coefficients a_1, a_2, and b have a $1/x$ singularity because Σ becomes zero at $X = 1 - \dfrac{Y_\mathrm{T}^2}{1 - Y_\mathrm{L}^2}$. From the field components E_x, E_y, one can pass to amplitudes of ordinary E_o and extraordinary E_e waves:

$$\begin{pmatrix} E_y \\ E_z \end{pmatrix} = \mathbf{M}\begin{pmatrix} E_\mathrm{o} \\ E_\mathrm{e} \end{pmatrix}, \quad \mathbf{M} = \begin{pmatrix} \dfrac{1}{\sqrt{1 - \mathtt{R}^2}} & \dfrac{\mathtt{R}}{\sqrt{1 - \mathtt{R}^2}} \\ \dfrac{\mathtt{R}}{\sqrt{1 - \mathtt{R}^2}} & \dfrac{1}{\sqrt{1 - \mathtt{R}^2}} \end{pmatrix}, \tag{1.14}$$

where $\mathtt{R}$ is the polarization of an ordinary wave:

$$\mathtt{R} = \frac{\mathrm{i}}{2Y_\mathrm{L}}\left(\frac{Y_\mathrm{T}^2}{1 - X + \mathrm{i}Z} - \sqrt{\frac{Y_\mathrm{T}^4}{(1 - X + \mathrm{i}Z)^2} + 4Y_\mathrm{L}^2}\right). \tag{1.15}$$

Substituting (1.14) in (1.12) for the vector

$$\boldsymbol{E}^* = \begin{pmatrix} E_\mathrm{o} \\ E_\mathrm{e} \end{pmatrix},$$

we obtain the Fösterling system [Fösterling, 1942]:

$$\frac{\mathrm{d}^2\boldsymbol{E}^*}{\mathrm{d}z^2} + k^2\mathbf{H}\boldsymbol{E}^* - 2kg\mathbf{G}\frac{\mathrm{d}\boldsymbol{E}^*}{\mathrm{d}z} = 0\,, \tag{1.16}$$

where

$$g = \frac{\mathrm{i}Y_\mathrm{T}^2Y_\mathrm{L}}{4(1+\mathrm{i}Z-X)^2Y_\mathrm{L}^2+T_\mathrm{T}^2}\,\frac{X'-\mathrm{i}Z'}{k}\,,$$

H and **G** are matrices

$$\mathbf{H} = \begin{pmatrix} n_\mathrm{o}^2+g^2 & -\dfrac{g'}{k} \\ -\dfrac{g'}{k} & n_\mathrm{e}^2+g^2 \end{pmatrix}, \quad \mathbf{G} = \begin{pmatrix} 0 & 1 \\ 1 & 0 \end{pmatrix},$$

and n_o and n_e are refractive indices for ordinary and extraordinary waves:

$$n_\mathrm{o}^2 = 1 - \frac{X}{1+\mathrm{i}Z-\dfrac{Y_\mathrm{T}^2}{2(1-X+\mathrm{i}Z)}+\sqrt{\dfrac{Y_\mathrm{T}^4}{4(1-X+\mathrm{i}Z)^2}+Y_\mathrm{L}^2}}\,, \tag{1.17a}$$

$$n_\mathrm{e}^2 = 1 - \frac{X}{1+\mathrm{i}Z-\dfrac{Y_\mathrm{T}^2}{2(1-X+\mathrm{i}Z)}-\sqrt{\dfrac{Y_\mathrm{T}^4}{4(1-X+\mathrm{i}Z)^2}+Y_\mathrm{L}^2}}\,. \tag{1.17b}$$

The parameter g is characteristic of the coupling of the two equations. In (1.16), singularity of the kind of $1/x$ (at $Z=0$) remains only in index n_e. If the wave propagates parallel or perpendicular to the geomagnetic field, $g=0$, and systems (1.12) and (1.16) divide into pairs of independent equations. In most cases, coupling between magnetoionic components is weak. However, in the presence of steep gradients or sharp jumps in electron density, such a coupling should be taken into account.

More scrupulous consideration of the plasma as a medium of propagating electromagnetic waves based on the kinetic theory of gases, allowing for velocity distributions of electrons, ions, and neutrals, leads to refined material equations; in this case, the form of (1.12) remains the same, but corrections are allowed to the coefficients dependent on the statistical distribution of particle velocity as well as on the wave frequency and hyrofrequency [Davies, 1969].

During more than a 70-year history, the methods for vertical radio probing by dynasonds yielded a wealth of data on the structure of the ionosphere. The basic principle of vertical radio sounding is rather simple [Reinish, 1996]: HF radio waves of frequency f are reflected at the cutoff ionospheric magnetoplasma which are given by $X=1$ for an ordinary wave, $X=1-Y$ for an extraordinary wave, $X=1+Y$ for a z-wave. For diagnostics for the E- and F-layer ionosphere, frequencies of the range from 1 to 20 MHz are usually

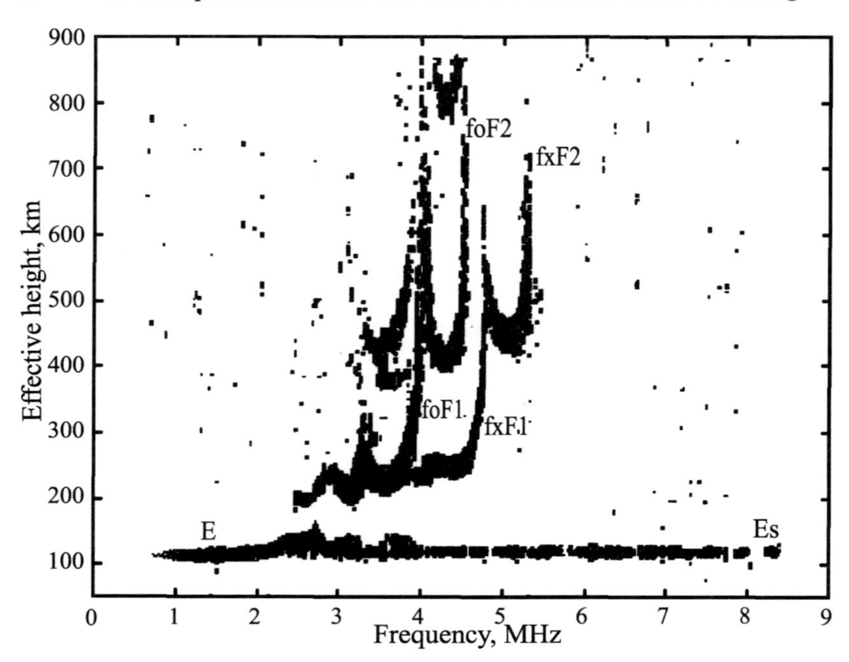

Fig. 1.5. An example of an ionogram obtained near Murmansk by the PGI ionosond on June 21, 1996 at 11:16 UT

used in dynasonds. An ionosond radiates a modulated radio signal that, after having been reflected by the ionosphere, is measured and analyzed. Different modulation types are used in ionosonds such as amplitude, frequency, and phase modulation. Basically, impulse ionosonds are employed with frequency varying from pulse to pulse. The recording device in the receiver output measures the time lag of the signal reflected from the ionosphere as a function of sounding frequency. The frequency dependency of the effective reflection height (i.e., the height obtained under the assumption that the wave propagates at the speed of light) is called the height–frequency characteristic, or ionogram. Figure 1.5 shows an example of an ionogram obtained near Murmansk by an ionosond of the Polar Geophysical Institute. In the ionogram, reflections are seen of two magnetoionic components and the z-wave from the E, E_s, F1, F2 layers as well as multiple reflections. In the figure, the positions of critical frequencies of different ionospheric layers are indicated for both magnetoionic components foF1, foF2, fxF1, fxF2. Reflection traces practically merge in the E region, whereas in the F region, separation of the traces is seen. The example shown displays the complexity of analyzing ionograms containing various magnetoionic components and multiple reflections. The situation becomes particularly difficult in the presence of diffuse reflections when the traces of the components broaden and overlap significantly.

Some difficulties are also encountered in reconstruction of electron density profiles from ionograms, for example, the problem of taking into account the interaction of magnetoionic components, bending of their ray trajectories even in a plane-stratified ionosphere, and the "valley problem" (valley here means the region between the maxima of neighboring layers). But the principal limitation of the standard method of vertical sounding is the assumption of a stratified ionosphere. If the ionosphere is not stratified and contains noticeable irregularities within the coverage of the radiation pattern (which is usually rather wide, up to 40–60° and wider), the problem of data interpretation and density profile reconstruction loses its uniqueness.

Nevertheless, within the framework of the limitations stated, modern dynasonds are capable of solving the problem of the analysis of reflected radio signals and reconstruction of electron density profiles; therefore, dynasonds are an extremely versatile tool for the study of the bottomside ionosphere [Hunsucker, 1991; Reinish, 1996].

There are also lots of other methods for probing the ionosphere: the partial reflection method, Faraday rotation, and so forth [Rishbeth and Garriott, 1969; Ratcliffe, 1972; Hunsucker, 1991]. However, generally, these methods are only episodically used in ionospheric research and are less suitable for continuous monitoring of the ionosphere. At present, the ionosphere is monitored by dynasonds and incoherent scatter radars.

In this monograph, we shall consider the use of satellite radio probing and tomographic methods for diagnostics and monitoring of the ionosphere. Modern techniques make it possible to perform VHF/UHF satellite radio probing with different geometries of transmitting/receiving systems and to apply tomographic methods. Such systems based on radio probing that "illuminates" the ionosphere and subsequent measurement of forward-scattered signals make it possible to use low-power transmitters. On the other hand, the use of VHF/UHF signals allows us to avoid the difficulties caused by magnetoactivity and to solve scalar inverse problems. At high probing frequencies, the nondiagonal terms of $\hat{\varepsilon}$ in (1.11) are negligible since, as seen from (1.13), they do not exceed the squared ratio of plasma frequency $f_{\rm N}$ to the probing frequency, that is, $(f_{\rm N}/f)^2\,(f_{\rm H}/f)$ ($f_{\rm H}$ here is hyrofrequency) [Ginzburg, 1961]. For example, a typical value of the maximum electron density in the natural ionosphere is $N \sim 10^6$ cm^{-3}. For this at $f > 50\,$MHz, $(f_{\rm N}/f)^2(f_{\rm H}/f) \lesssim 10^{-3}$. Similarly, at high frequencies, the last "depolarization" term in (1.11) can also be neglected; its order is defined by the ratio of the wavelength of the emitted signal to the typical scale of changes in electron density. Thus, at high sounding frequencies, the vector equation (1.11) splits into three equations, and the problem reduces to a scalar problem for each component of the field.

2. Ray Radio Tomography

2.1 Statement of the Problems and Methods of Ray Radio Tomography

The problems of ray radio tomography of large-scale structures are usually formulated as follows: to recover the structure of some ionospheric region from linear integrals measured along a series of rays intersecting this region. Since the sizes of large-scale irregularities, both natural (such as, e.g., the ionospheric trough) and artificial (spacecraft traces, technological emissions), are of the order of dozens to thousands of kilometers, diffraction effects can be neglected in VLF/UHF probing.

Similar tomographic problems, when the structure of the medium is reconstructed from linear integrals measured along rays intersecting the given region, arise as well in a number of different fields of science and techniques. At present, there is likely no scientific domain, kind of radiation, or wave type that tomographic methods have not yet been tried on, at least theoretically. Here, it makes no sense to enumerate all modifications of the tomographic approach. It may be worthwhile to give only a brief description of tomography applications closest to the tomography of geophysical structures. By now, quite well developed are the methods and equipment for seismic tomography, where the measurements of propagation times of seismic waves are used for reconstructing the seismic "slowness" – quantity reciprocal of the wave velocity. The bibliography on seismic tomography is rather large, and here we refer the reader only to several reviews [Dines and Lytle, 1979; Ivansson, 1986; Nolet, 1987]. In seismic tomography, a lot of solution methods and algorithms are valid also for other types of waves. At present, tomographic methods are applied to investigation as well of other Earth "spheres." Acoustic tomography of the ocean from measurements of propagation times of acoustic waves is being actively developed now [Spindel, 1982; Munk and Wunsch, 1983]. Results are reported on radio and optical tomography of the atmosphere [Phinney and Anderson, 1968].

Various implementations have been proposed of ionospheric radio tomography. In [Krasnushkin, 1981] the tomographic method is discussed for determination of the local attenuation factor as a function of geographical coordinates from data on integral attenuation factors along paths covering the investigated area densely enough. Emitters and receivers far from the region to be reconstructed provide straight propagation paths. It was suggested to use the Radon transform, which is infeasible because of the small number of rays available; besides, the question of the height dependence of the attenuation factor remains to be settled. Radio tomography methods based on measurement of satellite signals were first suggested in [Austen et al., 1988] where the distribution of electron density was supposed to be reconstructed from the total electron content along a set of rays. The data acquisition system is sketched in Fig. 2.1. In this case, three (or more) receivers are located in a plane of the satellite path, and the rays are assumed linear. The schemes of tomography from linear integrals for different wave types are practically undistinguishable. The main requirement is merely that as many rays as possible should intersect the given region over as wide range of angles as possible. This means that as many receivers and transmitters as possible should be used, or they should be moved with respect to each other within a wide range of possible locations. The approach described in [Austen et al., 1988; Afraimovich et al., 1989] that makes use of ray integral electron content is identical, for instance, to seismic interstitial tomography [Gustavsson et al., 1986], where the sources of seismic waves are placed in boreholes and the receiver is moved along the surface. In the further investigations, the

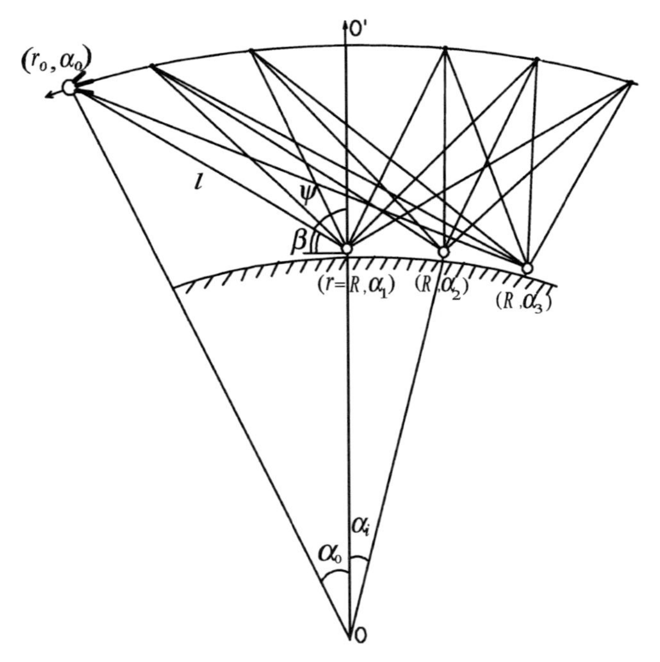

Fig. 2.1. A sketch of the ray RT experiment

methods of ionospheric radio tomography using total electron content (TEC) have been developed by many authors.

The suggested method for radio tomography based on TEC data is a particular case of tomography from linear integrals, since it was known for a long time that the reduced phase difference of two coherent frequencies is proportional to the TEC. The main question is how to find this absolute phase or linear integral, since only relative phases can be directly measured. In other words, in determination of the absolute phase proportional to the TEC, a problem of determining the "initial phase" arises, and a considerable mistake by a constant may occur, which would lead to an inconsistency in tomographic data and give a poor reconstruction. The authors of the first publications [Austen et al., 1988; Afraimovich et al., 1989] on ray radio tomography expounded the general idea of the tomographic reconstruction of the electron density distribution from the measurements of linear integrals proportional to the absolute phase of a radio signal at a reduced frequency. However, in the mentioned works, the possibility has not been demonstrated of such a reconstruction with typical errors in absolute phase determination, i.e., no modeling of the influence of errors in the initial phase determination on the reconstruction results has been carried out. In these works, only the possibility of the functional reconstruction from linear integrals has been modeled. However, such a simulation, up to the replacement of words ("group delay" by TEC, "distribution of seismic slowness" by "electron density distribution" and so on) had been carried out earlier in many works on seismic and other kinds of tomography.

In contrast to seismics, however, where a linear integral (a group delay) is measured directly, in ionospheric investigations, only methods for the approximate determination of a linear integral are available. Here, a specific error appears – an error by a constant, which is essential for tomography. In direct measurements of linear integrals, this error is usually noiselike which does not affect the results significantly. As shown earlier [Kunitsyn and Tereshchenko, 1991; Kunitsyn et al., 1994b] and will be seen further on, phase tomography leads to poor results in the case of typical errors in determining the absolute phase or TEC. Therefore, some years ago [Kunitsyn et al., 1990; Kunitsyn and Tereshchenko, 1991; Andreeva et al., 1992], we proposed phase-difference radio tomography of the ionosphere.

Theoretically, ray RT is based on the well-known relations [Kunitsyn and Tereshchenko, 1991] for phases and amplitudes of radio waves in geometric optics approximation. The following pair of equalities determines linear integrals from the electron concentration distribution N and effective collision frequency ν:

$$\begin{aligned} \lambda r_{\mathrm{e}} \int N \,\mathrm{d}\sigma &= \phi\,, \\ -\frac{\lambda r_{\mathrm{e}}}{\omega} \int N\nu \,\mathrm{d}\sigma &= \chi\,, \end{aligned} \tag{2.1}$$

where λ is the length of the probing wave, $\omega = kc$; k is the wave number in free space; c is light velocity; and $\int \mathrm{d}\sigma$ is the symbol of integrating along the ray. Linear integrals here are phase difference $\phi = \Phi_0 - \Phi$ and amplitude level χ – the logarithmic ratio of amplitudes $\chi = \ln(A/A_0)$ of the measured field $[E = A\exp(\mathrm{i}\Phi)]$ to the probing wave $[E_0 = A_0\exp(\mathrm{i}\Phi_0)]$. Measurements of ϕ and χ are realized by receiving signals at two coherent frequencies from navigation satellites. As will be shown in Chap. 4, the above formulas can be also obtained as well as asymptotics of the relations for diffraction tomography.

Before estimating the errors in phase measurements, let us introduce a series of parameters in polar coordinates (r, α) characteristic of the geometry of the recording scheme. In Fig. 2.1, (r_0, α_0) are coordinates of the satellite, (R, α_i) are coordinates of one of the ground-based $(r = R)$ receivers, β is the elevation angle of the satellite, $\psi = (\beta - \pi/2)$ is the explement of the elevation angle to the vertical, O is the center of Earth, $O - O'$ is the polar axis, and positive angles are counted clockwise. Of course, the real proportions of Earth's radius to the satellite height are not obeyed in the figure. Based on simple geometric relationships for any point in the ionosphere with coordinates (r, α) at distance l from the receiver, the following equalities are satisfied:

$$\frac{l}{\sin(\alpha_i - \alpha)} = \frac{r}{\sin\left(\frac{\pi}{2} + \beta\right)} = \frac{R}{\sin\left[\pi - (\frac{\pi}{2} + \beta) - (\alpha_i - \alpha)\right]} .$$

From this, an equation $r(\alpha)$ is obtained for the straight (β=const) ray

$$r(\alpha) = \frac{R\cos\beta}{\cos(\beta + \alpha_i - \alpha)} \tag{2.2}$$

and also inverse relation:

$$\alpha(r) = \alpha_i + \beta - \arccos\left(\frac{R}{r}\cos\beta\right) . \tag{2.3}$$

The relationship between β and α, r stems from (2.2):

$$\tan\beta = \frac{\cos(\alpha_i - \alpha) - \frac{R}{r}}{\sin(\alpha_i - \alpha)} . \tag{2.4}$$

Making use of (2.3), we arrive at a formula for elementary ray length $\mathrm{d}\sigma$:

$$\mathrm{d}\sigma^2 = \left[1 + r^2\left(\frac{\mathrm{d}\alpha}{\mathrm{d}r}\right)^2\right]\mathrm{d}r^2 = \frac{r^2}{r^2 - R^2\cos^2\beta}\,\mathrm{d}r^2 . \tag{2.5}$$

Then, the relation for the measured linear integral (2.1) with respect to the electron concentration will have the form,

$$\phi = \lambda r_e \int N(r,\alpha)\ \mathrm{d}\sigma = \lambda r_e \int \frac{N(r,\alpha)\, r\ \mathrm{d}r}{\sqrt{r^2 - R^2 \sin^2 \psi}} \,. \tag{2.6}$$

Instead of polar coordinates (r, α), in further computations it is convenient to use orthogonal coordinates (h, τ): $h = (r - R)$ is the height above ground level, and $\tau = \alpha R$ is the "transverse" (horizontal) distance along Earth's surface in the plane containing the satellite path. In this reference system, the ray equation (2.2) is no longer a straight line:

$$h(\tau) = R \left[\cos\beta + \frac{\tau_i + \tau}{R} - 1\right] . \tag{2.7}$$

Here, $(\tau_i, h = 0)$ are coordinates of the receiver. The relation inverse to (2.7) is similar to (2.3):

$$\tau - \tau_i = R \left[\beta - \arccos\left(\frac{R}{R+h} \cos\beta\right)\right] . \tag{2.8}$$

In this case, with (2.5) taken into account, linear integrals of type (2.1) have the form

$$\int_0^{h_0} \frac{F(h,\tau)\,(R+h)\ \mathrm{d}h}{\sqrt{R^2 \sin^2\beta + 2Rh + h^2}} = I(\beta, \tau_i) \,. \tag{2.9}$$

Integration along the ray joining the ith receiver ($\tau_i = \alpha_i R$) with the satellite is replaced, in accordance with (2.5), by integration along the height from the ground level to satellite altitude h_0. Elevation angle β in (2.4) is defined by the satellite position (h_0, τ_0). The satellite orbit should not necessarily be circular; then h_0 is a function of τ_0. Linear integral $I(\beta, \tau_i)$ depends on the ith receiver and satellite elevation β. As well, the level χ of the measured signal or its total phase ϕ (2.1) can also be linear integrals; then the reconstructed function F will be proportional either to $N\nu$ or N, respectively. As the number of receivers cannot be too big and the range of angles β is limited, it is inadvisable to examine methods for analytical inversion of such linear integrals and methods for integral transformations. In the given case of aspect-limited tomography, it is pertinent from the very beginning to solve the problem in discrete form and to use algebraic reconstruction algorithms or methods for expansion into finite series.

Before we proceed with the solution of (2.9), let us consider the possibility of replacing the ray (2.7) by a straight line. The ray becomes curved after passing to the new coordinates (h, τ) convenient for solution of the discrete problem. The straight ray between the receiver ($h_i = 0$, $\tau_i = 0$) and the satellite (h_0, τ_0) is defined by the function $h'(\tau) = \tau \cot \psi_0$, which differs from the dependence

$$h(\tau) = R\left[\frac{\sin\psi}{\sin\left(\psi - \frac{\tau}{R}\right)} - 1\right],$$

where $h(\tau_0) = h'(\tau_0) = h_0$. Expanding in powers of minor component $\tau/R \ll \psi$, we find that the height difference Δh between the two trajectories is expressed by the formula,

$$\begin{aligned}\Delta h &= h'(\tau) - h(\tau) \\ &\simeq \frac{\tau_0}{R}\left(\frac{1}{2} + \cot\psi_0^2\right)\tau - \frac{\tau^2}{R}\left(\frac{1}{2} + \cot^2\psi_0\right) + O\left(\frac{\tau^3}{R^2}\right).\end{aligned}$$

In the middle $\tau = \tau_0/2$ of the trajectory at $\psi_0 = \pi/4$, $h_0 = 1000\,\mathrm{km}$, the difference Δh is about 60 km. Therefore, when dividing the ionosphere into vertical increments, the latter should significantly exceed Δh if not taking account of the curvature of the ray in coordinates (h, τ) or, otherwise, if the "curvature" of polar coordinates in the region of reconstruction with a straight ray used is not allowed for. The total phase will be even more sensitive to the curvature of the ray not taken into account, if we attempt to reconstruct N from the total phase data, because the lengths of the curved and straight rays will differ significantly. In brief, consideration of the curvature of the polar coordinates in the region of reconstruction or of the curvature of the ray in coordinates (h, τ) in ray ionospheric RT of global structures is necessary, which is unfortunately not taken into account in a number of works [Afraimovich et al., 1989; Saenko et al., 1991].

The phase method involves measuring a linear integral of the form (2.1) multiplied by a constant of the order of unity [Kunitsyn and Tereshchenko, 1991]. This constant appears when recalculating the phase from one frequency to another and is insignificant here. The basic difficulty in the determination of linear integral (2.6) is that the phase value is very high. For typical values $N \sim 10^{-12}\,\mathrm{m}^{-3}$, $\lambda = 2\,\mathrm{m}$ and ray length in the ionosphere of the order of 1000 km, ϕ is about 1000 radians. Thus, the problem arises of separating the "initial phase" $\phi_0 = 2\pi n$ to be added to the measured (within 2π) $\Delta\phi$ to obtain the absolute (total) phase $\phi = \phi_0 + \Delta\phi$ or the linear integral (2.6).

To explain the difficulties arising, let us examine the possibility of separating the initial phase in the presence of minor horizontal gradients. The concentration can be represented as an expansion $N(r, \alpha) = N_0(r) + N'(\alpha - \alpha_m)$, where the regular spherically symmetrical background $N_0(r)$ is picked out; $N'(r) = \left.\frac{\partial N}{\partial \alpha}\right|_{\alpha=0}$; $\alpha_m(\psi)$ is the angle of intersection of the ionospheric maximum with the ray; its vicinity mainly contributes to integral (2.6). In this case, introducing a new variable $h = r - R$ – that is the height above ground level and retaining the first terms in the power series of h/R in (2.6), we obtain

$$\begin{aligned}\phi \simeq \frac{\lambda r_e}{\cos\psi}\int N_0(h)\ \mathrm{d}h - \lambda r_e \frac{\tan^2\psi}{\cos\psi}\int \frac{h}{r}\ N_0(h)\ \mathrm{d}h \\ + \frac{\lambda r_e}{\cos\psi}\int N'(h)\,(\alpha(h) - \alpha_m)\ \mathrm{d}h + \ldots\,.\end{aligned} \tag{2.10}$$

If there were no horizontal gradient, the $\Delta\phi(\psi)$ could be measured directly for different angles ψ over the range $\Delta\phi$, and a linear system could be obtained yielding $\int N_0\ \mathrm{d}h$, $\int hN_0/R\ \mathrm{d}h$. In other words, the known functional dependence of ψ would then make it possible to determine the TEC and other moments of the function $N_0(h)$. However, the presence of the term containing N' greatly complicates the situation when its value becomes comparable to 2π. Let us estimate this term for nearly vertical sounding, $|\alpha|$, $|\alpha_i| \ll 1$; for this purpose, let us substitute $\frac{\psi(h-h_m)}{R+h} \simeq \frac{\psi(h-h_m)}{R}$ for $\alpha(h) - \alpha_m$ in the integrand, where h_m is the top height; this asymptotic equation (at low angles) stems from (2.4): $\psi \simeq \frac{\alpha-\alpha_m}{1-R/r} = \frac{(\alpha-\alpha_i)(R-h)}{h}$. Therefore, such methods for determining the constant component will work if the following condition is satisfied:

$$\left|\lambda r_e \Delta\psi \int \frac{N'(h)\,(h-h_m)}{R}\ \mathrm{d}h\right| \ll 2\pi\,. \tag{2.11}$$

Typical values of $\partial N/\partial\alpha$ in the presence of a "trough" in the ionosphere are $\partial N/\partial\alpha \sim 10^{13}\,\mathrm{m}^{-3}\,\mathrm{rad}^{-1}$; then (2.11) is valid only for $\Delta\psi \ll 10^{-2}$. But at angles of fractions of a degree, it is practically impossible to determine the functional dependence of ψ in the presence of noises. Inequality (2.11) relates to the determination of the TEC along the vertical. The limitation on the horizontal gradient becomes even more strict in oblique sounding. A detailed analysis of a number of traditional techniques for determing the constant phase is made in [Solodovnikov et al., 1988]. These methods include combined techniques making use of simultaneous Doppler and Faraday measurements. The results of numerical modeling [Solodovnikov et al., 1988] showed that in the presence of a trough and steep typical gradients $\partial N/\partial\alpha \sim 10^{13}\,\mathrm{m}^{-3}\,\mathrm{rad}^{-1}$, the error in determining the constant varies within 100–1000%! This agrees with estimate (2.11) and indicates that practical determination of the linear integral of electron density is unreal in the presence of typical horizontal gradients in the ionosphere, which is exactly the case of interest in RT.

There is one more method for determining the TEC in the presence of horizontal gradients [Leitinger et al., 1975, 1984; Solodovnikov et al., 1988] which is based on recording satellite signals by a pair of spaced receivers. With data on the phases measured at the given base, it is possible to construct a pair of linear equations with nearly identical last terms (2.10), that is, the rays from the satellite to different receivers should intersect the top of the ionosphere at the same point. Similarly, a pair of equations for another instant leads to a system for the initial phase, where the influence of the gradient term will be weakened. Otherwise, condition (2.11) is replaced by a less strict one. However, this method is incapable of complete elimination of the effect of the horizontal gradient and higher order derivatives on the results of the initial phase determination. Of course, it is always possible to propose a recording scheme allowing determination of the initial phase, but this involves multifrequency measurements by several receivers.

Thus, due to the nature of phase measurements, it is inadvisable to reduce the problem of ionospheric RT to a problem in linear integrals. Determination of the initial phase by the simplest recording systems leads to major errors, whereas the use of complex multiposition and multifrequency systems is not worthwhile here since another solution of the problem is possible. Here, a method is proposed for solution of the RT problem without determination of the initial phase, based on phase-difference or Doppler measurements only.

Before presenting the method of phase-difference RT let us consider the scheme of phase RT allowing determination of linear integral (2.1). It is possible to find the absolute phase incursion at an inhomogeneity in reconstructing sufficiently large localized artificial (releases, heating, etc.) or natural inhomogeneities arising in the ionosphere during the time interval between satellite passes. Such a formation being localized in space provides the possibility of solving the problem without involving additional a priori assumptions on the irregularity. By measuring before and after the disturbance appeared, one can find the contribution of the localized inhomogeneity being reconstructed by subtraction of the data, if the ionosphere changes little between flights.

Let us begin consideration of the problems of phase and phase-difference RT with a discretization procedure for equalities (2.1). Perform digitization of the linear integrals $I(\beta, \tau_i)$ (2.9) with respect to the ith position of the satellite dependent on the coordinate τ_{0j} or the angle $\alpha_{0j} = \tau_{0j}/R$. The set of satellite coordinates τ_{0j} is recalculated, according to (2.4), into a series of satellite elevation angles β_{ij} with respect to the ith receiver:

$$\tan \beta_{ij} = \frac{(R + h_0)\cos(\alpha_i - \alpha_{0j}) - R}{(R + h_0)\sin(\alpha_i - \alpha_{0j})} .$$

The sets of elevations at all receivers define a series of discrete values of linear integrals $I_{ij} \equiv I(\beta_{ij}, \tau_i)$. The simplest method for digitization of the sought function $F(h, \tau)$ in a fixed rectangular $(m_0 \times n_0)$ grid is to replace it by a piecewise-constant approximation or, otherwise, to represent F as a system of $(m_0 \times n_0)$ basis functions equal to unity within a certain rectangle and zero within all others. The rectangular reconstruction region is divided into m_0 heights $(m \leq m_0)$ and n_0 horizontal samples $(1 \leq n \leq n_0)$. Let the value of $F(h, \tau)$ within a fixed $(m \times n)$ rectangle be F_{mn}. The point within the $(m \times n)$ rectangles where the function $F(h, \tau)$ is sampled is not too important; this may be either the middle of the rectangles or grid nodes.

The problem of tomographic reconstruction from linear integrals is to determine the set of discrete values $\{F_{mn}\}$ on the known grid from the set $\{I_{ij}\}$. Designating the length of the (i,j) ray within the (m,n) cell as $L_{i,j}^{m,n}$, we obtain a linear system:

$$L_{i,j}^{m,n} F_{m,n} = I_{i,j} \quad \text{or} \quad L_J^M F_M = I_J \,. \tag{2.12}$$

Here, "renumbering" of the rays $(i,j) \rightarrow J$ and ionospheric cells $(m,n) \rightarrow M$ is performed in the second equation. The repeating indexes are to be understood as a summation. The number of rays is defined by the parameters of the recording system. Coefficients L_{JM} are calculated according to the given rays and cells into which the ionosphere is divided. System (2.12) may be either overdetermined or subdefinite.

Therefore, if it is possible to determine linear integrals (2.1), then by the discretization procedure, the problem of ray RT reduces to the solution of systems of linear equations. Consideration of methods for solving linear systems as applied to ray RT problems is a very extensive topic that will be somewhat touched on in Sect. 2.3. Since the modeling results and conclusions described below are practically independent of specific methods for solving linear equations, we shall use the well-known algorithms of the algebraic reconstruction technique (ART) only.

The principal statements of the present section will be illustrated below by the results of numerical modeling. Here and further on, it is pertinent to describe the errors of numerical simulation in terms of numbers δ that characterize a relative deviation of the reconstructed function $\tilde{F}$ from a model (true) function F: $\delta = \dfrac{\|F - \tilde{F}\|}{\|F\|}$. The norms in l^2 and l^∞ spaces [$\delta(l^2)$ and $\delta(l^\infty)$] are most often used:

$$\delta(l^2) = \frac{\sqrt{\sum_i \left(F_i - \tilde{F}_i\right)^2}}{\sqrt{\sum_i F_i^2}} \,, \quad \delta(l^\infty) = \frac{\max_i |F_i - \tilde{F}_i|}{\max_i |F_i|} \,. \tag{2.13}$$

Phase radio tomography makes it possible to successfully reconstruct localized objects. By subtracting the phase data corresponding to two close in time satellite passes, one can obtain a finite object for reconstruction if the difference between the two electron density distributions is spatially localized. In this case, a problem of initial phase determination no longer exists. In particular, in [Kunitsyn et al., 1994b], a reconstruction of such a finite object as a heated ionospheric lens has been simulated. A zero initial guess was used. The pixel size was 10 km × 10 km. The reconstruction only faintly differs from the model structure; relative errors of reconstruction are $\delta(l^2) = 0.07$, $\delta(l^\infty) = 0.09$.

Quite a different situation arises in the reconstruction of nonlocalized structures – cross sections of the ionosphere. Here, as has already been dis-

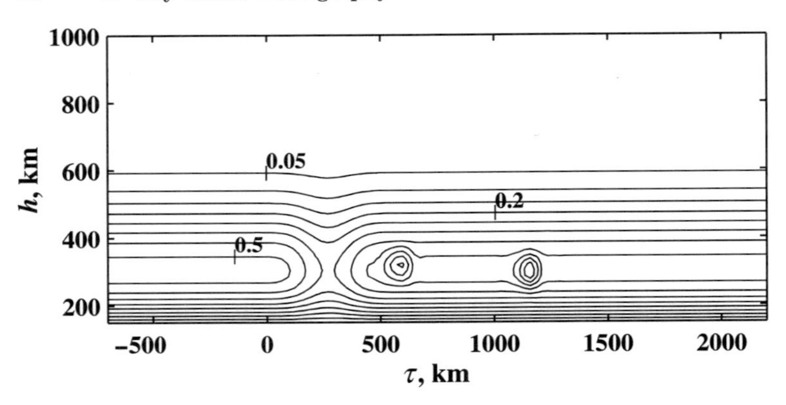

Fig. 2.2. Model ionosphere (2-D section)

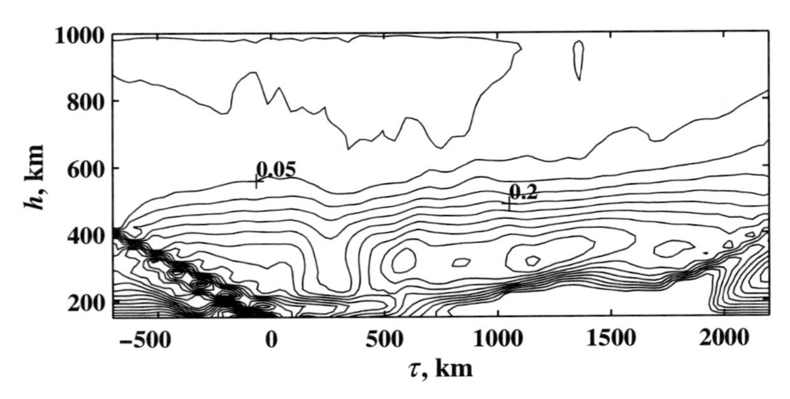

Fig. 2.3. Phase RT reconstruction of the model in Fig. 2.2 with 3% accuracy in the TEC determination

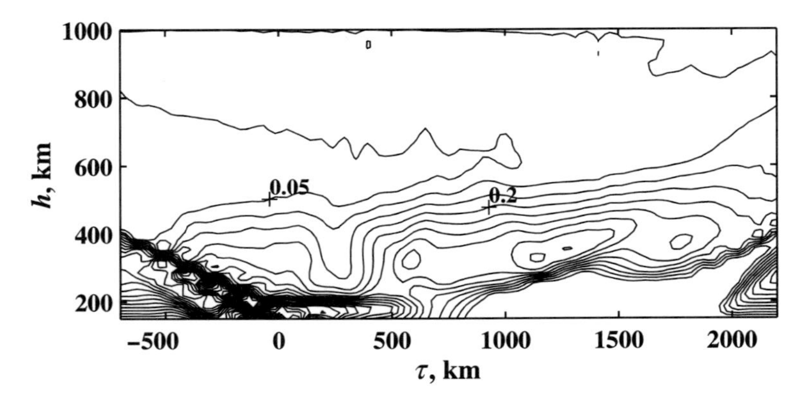

Fig. 2.4. Same as Fig. 2.3, with 5% accuracy in the TEC determination

cussed above, it is rather difficult to measure directly the absolute phase (TEC), and the methods used [Leitinger et al., 1975, 1984; Solodovnikov et al., 1988] have a constant error of about 10% or at least a few percent. Here are the results of numerical simulation of the reconstruction of the ionospheric section for typical errors in finding the absolute phase (TEC). A simple model of the ionosphere with a trough and two positive irregularities to the left of the trough is displayed in Fig. 2.2 in contours of $10^6\,\mathrm{cm}^{-3}$ units. It was assumed that satellite radio probing (with the satellite flying at $h_0 = 1000\,\mathrm{km}$) is performed at the frequencies of 400 and 150 MHz ($\lambda = 2\,\mathrm{m}$) and the receivers are located at sites with the coordinates $\tau_1 = 0\,\mathrm{km}$, $\tau_2 = 423\,\mathrm{km}$, and $\tau_3 = 1435\,\mathrm{km}$. A homogeneous ionosphere having no trough was used as an initial guess. Figures 2.3 and 2.4 show the results of the reconstruction by phase RT with ±3% and ±5% accuracy in determining the TEC (which corresponds to errors of $18\,\pi$ and $30\,\pi$ in absolute phase); the signs of the errors alternate at different receivers. These figures illustrate an extremely poor quality of reconstruction using the phase RT method with typical errors in determining the TEC: even the principal features of this simple model structure are not recovered, and at the same time, some heavy artifacts are present. Figure 2.5 portrays the dependences of reconstruction errors $\delta(l^2)$ and $\delta(l^\infty)$ on the error in determining the absolute phase ($\pm 2\pi m$), on the number m. It can be seen that even errors of units to a few dozens of 2π lead to low quality of reconstruction. If the signs of the errors in determining the absolute phase are the same for all receivers, the reconstruction qual-

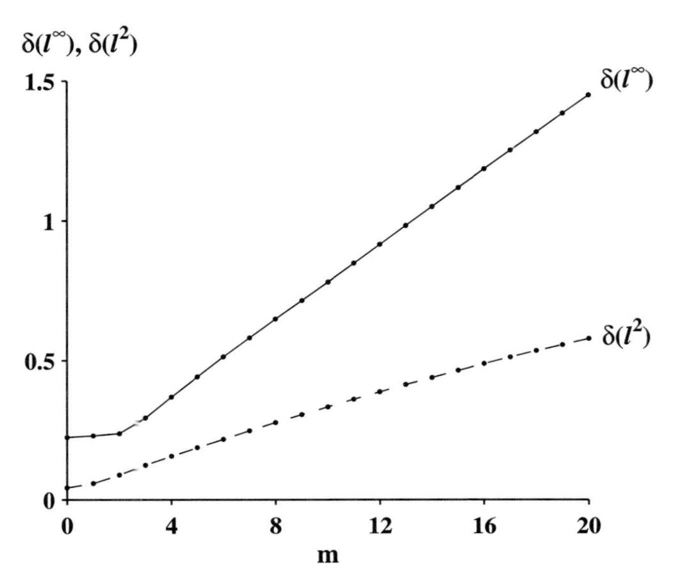

Fig. 2.5. Reconstruction errors $\delta(l^2)$ and $\delta(l^\infty)$ as functions of errors in the absolute phase

ity is somewhat better. Nevertheless, at an error level higher than 10%, the reconstruction quality is still poor. Here, the scheme of the first radio tomography experiment with receivers installed between Murmansk and Moscow was used in modeling (see Sect. 2.7). In further discussion, we shall also report the results of modeling obtained for geometries of different real experiments. Simulation results as well as the results of subsequent real experiments depend on the number of receivers and the spacing between them. However, the key qualitative results to be obtained in this chapter are weakly dependent on the geometry of the experiment. In particular, errors in the initial phase determination will also ruin the reconstruction, even if a larger number of receivers is involved.

Next we proceed to phase-difference tomography and, at first, we will show that the problem of ionospheric RT based on phase difference or Doppler measurements cannot be solved by the scheme discussed using a piecewise-constant approximation. The point is that the data here will be derivatives of linear integrals of type (2.9): $d = \mathrm{d}I/\mathrm{d}\alpha_0$, or finite-difference ratios of the increment ΔI of the linear integrals to the increment $\Delta\alpha_0$ of the satellite coordinate. The Doppler frequency $\Omega = \mathrm{d}\phi/\mathrm{d}t$ measured in the experiment is defined by the phase derivative (2.1). The relation between the angle α_0 of a satellite moving along a circular orbit with constant speed v_0 and time $\alpha_0 = \dfrac{v_0 t}{R+h_0}$ makes it possible to express the Doppler frequency Ω in terms of the derivative with respect to the angle of the satellite:

$$\Omega = \frac{v_0}{R+h_0}\frac{\mathrm{d}\phi}{\mathrm{d}\alpha_0};$$

hence, the input data for phase-difference tomography are proportional to $\Delta I/\Delta\alpha_0$. The derivatives of the linear integrals within a piecewise-constant approximation of the sought function F will be discontinuous. This results from the fact that each linear integral is the sum of integrals over the set of cells. As the elevation of the satellite changes, the ray encounters a new cell; the integral of this cell with respect to unity is a continuous function of the angle of the satellite α_0, but the derivative of the linear integral with respect to α_0 will contain a discontinuity when the ray contacts the corner of any cell. On the other hand, the Doppler frequency measured in the experiment is a continuous fuinction of satellite elevation. Therefore, the piecewise-constant representation of the function to be reconstructed does not allow analyzing the phase-difference problem.

Doppler or phase-difference measurements require higher order interpolation than the piecewise-constant representation of the recorded function. Correspondingly, the matrix L_{JM} of the conversion from reconstructed function to linear integrals should be calculated in a different way so as to ensure continuity of linear integrals with respect to the satellite coordinate – α_0 (or elevation β). If the matrix of the forward problem L_{JM}: $F_M \leftarrow I_J$ is continuous with respect to the angle of the satellite α_0, then instead of system (2.12), it is possible to obtain a system for phase-difference or Doppler data by differentiation of (2.12) with respect to angle α_0:

$$D_{JM}F_M = d_J \,. \tag{2.14}$$

Here, $d_J \equiv \Delta I/\Delta\alpha_0$ are Doppler data and $D_{JM} = \Delta L_{JM}/\Delta\alpha_0$ is the finite-difference ratio (or derivative) of matrix L_{JM} to the angle increment. Contributing to the Doppler data are not only the changes in total phase related to the integral electron concentration along the ray, but also the local electron concentration N_s at the satellite location. The correction for the Doppler data is N_s times the along-ray component of the satellite velocity: $\lambda r_\mathrm{e} N_\mathrm{s}\cos(\alpha_i + \beta - \alpha_0)$. This correction can be inserted into the iteration algorithm, and the resulting values of N_s at the boundary $h = h_0$ of the ionosphere will be constantly "retouching" the measured Doppler frequency values [Kunitsyn and Tereshchenko, 1991; Kravtsov et al., 2000]. Note that the Doppler frequency shift caused by local electron density may reach fractions of a Hertz.

Examples of constructing the conversion matrices for passing from the reconstructed function to linear integrals (matrices of projection operators) suitable in phase-difference RT will be given in the next section. Here (in Fig. 2.6), we represent the result of reconstruction of the model shown in Fig. 2.2 by the phase-difference RT method (the matrix was built using bilinear approximation). One can easily see that the main features of the ionospheric section are reconstructed quite well. Numerical simulation of reconstructing various ionospheric structures that we also carried out proved the noticeable advantage of the phase-difference RT method over phase RT with typical errors in determining the absolute phase. Phase-difference RT provides quite satisfactory results even if experimental data contain considerable errors. Figure 2.7, similarly to Fig. 2.6, shows the reconstruction results obtained by phase-difference RT, but in this case, data with a noise level of 20% maximum amplitude of Doppler data were used. One can see that random measurement errors slightly affect the reconstruction results. This

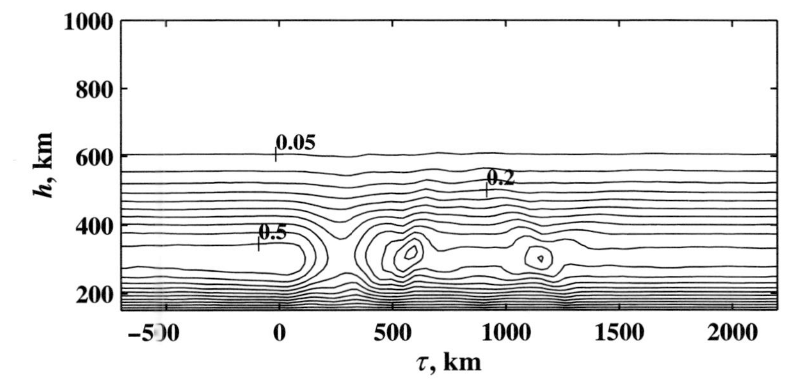

Fig. 2.6. Phase-difference RT reconstruction of the model in Fig. 2.2

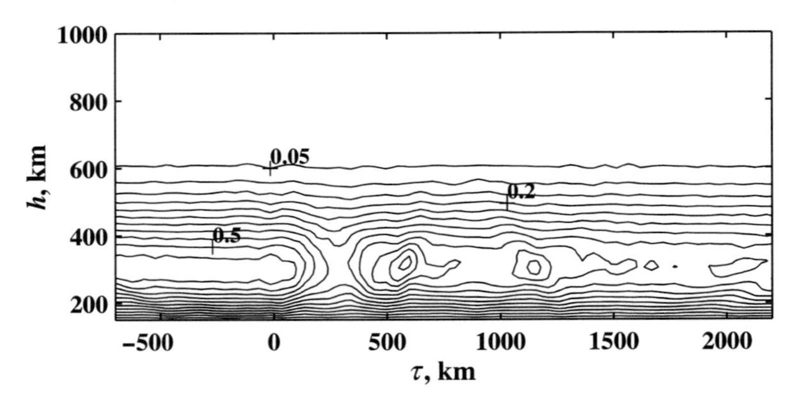

Fig. 2.7. Same as Fig. 2.6 (20% noise in the input data)

conclusion is confirmed by Fig. 2.8 showing the dependence of reconstruction errors upon the noise level. Even 50% noise level makes it possible to obtain reconstructions within the Doppler error level which can be explained by mutual compensation and effective "averaging" of noise in the process of tomographic reconstruction. In our experiments, the noise level did not exceed a few percent.

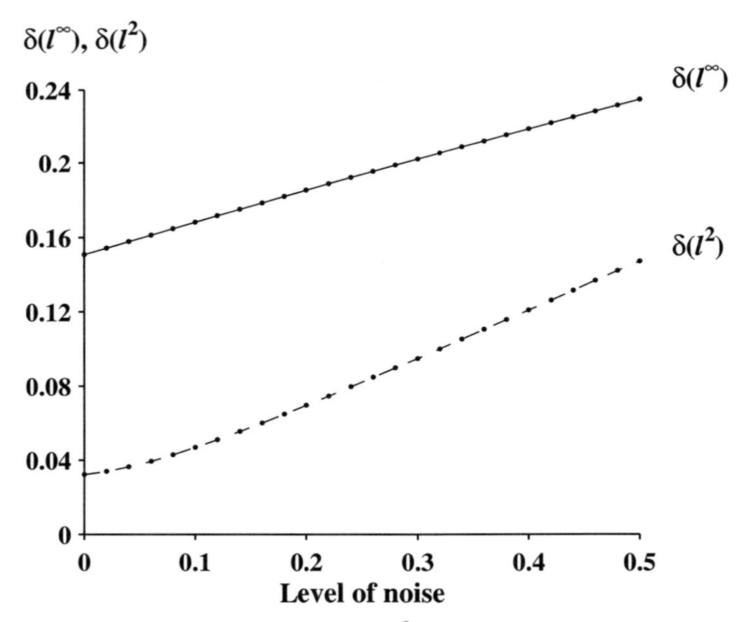

Fig. 2.8. Reconstruction errors $\delta(l^2)$ and $\delta(l^\infty)$ as functions of noise level

Here, we shall not consider in detail the effects of the initial guess on the reconstruction results and make only some brief remarks. In simulations, the uniform (in τ) ionosphere with different levels of concentration maxima were used as an initial guess. Changes in initial guesses produce small changes in the background level, but the spatial structure of the local extrema remains the same. The spatial structure recovered by phase-difference RT may be supposed to depend weakly on the initial guess. Generally speaking, the aim is to generate an assembly of solutions satisfying (2.14) with a given accuracy defined by experimental errors. By now, we have developed methods for generating such a solution assembly (by varying the algorithms, initial guess, etc.) which makes it possible to get an "assembly-averaged" solution and to estimate the reconstruction error distribution. These questions will be considered in Sect. 2.4.

In the earliest works on ray RT, it was proposed that the TEC should be used as the initial data. We have already criticized that approach [Kunitsyn and Tereshchenko, 1992; Kunitsyn et al., 1994b] because of typical errors in determing the absolute phase (as high as a few percent) that, in practice, does not allow the realization of RT reconstruction. Here we shall not dwell on that criticism since there is a rather simple way to avoid the necessity of determining the absolute phase. For this purpose, it is sufficient, instead of (2.1), (2.12), to use a different equation where by subtracting the data along two neighboring rays in the right-hand part, it is possible to obtain the phase difference (or Doppler) as reported in [Andreeva et al., 1992; Kunitsyn and Tereshchenko, 1991]. The alternative of subtracting for rather distant rays is also possible (for example, a ray with the minimal phase can be subtracted from all of the rays of the same receiver), as was shown by [Fremouw et al., 1994b; Raymund et al., 1994b]. Such phase-difference RT methods make it possible to avoid the need to determine the absolute phase and to obtain an acceptable quality of RT reconstructions. Here, it is perhaps worthwhile to compare different modifications of this phase-difference approach. Which kind of subtraction is preferable: on neighboring rays, which corresponds to the phase derivative (the Doppler frequency), or on relatively distant, which corresponds to the relative phase?

It is obvious that a priori neither of those techniques can be deemed preferable in all cases because the methods for constructing the operator matrix for the forward problem vary. However, these methods will undoubtedly differ in sensitivity, i.e., in ability to resolve relatively small irregularities faintly contributing to the phase. Doppler data are more sensitive to small inhomogeneities which have little effect on the phase [Kunitsyn et al., 1995a]. For example, when the ray scans the inhomogeneity ΔN of size a, the phase changes (2.1) by $\Delta\phi \sim \lambda r_e \Delta N\, a$; here, the relative change of the phase $\Delta\phi/\phi \sim \Delta N\, a/N_t \sim \Delta N\, a/N_m L \sim (\Delta N/N_m)(a/L)$, where $N_t \sim N_m L$ is the TEC along the ray, N_m is the value at the maximum electron concentration, and L is the ray length parameter. As a result, relative variations in the phase are proportional to the ratio of the TEC of the inhomogeneity to the TEC of the whole ionosphere. One should not expect the methods of solution of (2.12), (2.14) to be more sensitive to changes in the right-hand part of less than a few percent. Therefore, phase methods would not distinguish even sufficiently strong $\Delta N/N_{\mathrm{m}} \sim 0.1$ inhomogeneities of the size $a \leq L/10$ ($\leq 100\,\mathrm{km}$), since they produce only 1% of phase variations. It is not accidental, in our opinion, that in the reported reconstructions using phase methods [Pryse and Kersley, 1992; Pryse et al., 1993; Raymund et al., 1993], details with dimensions of less than a few hundred kilometers are not revealed. This is not the case for phase-difference measurements. Here, as long as $\Delta\phi \sim \lambda r_{\mathrm{e}} \int N\,\mathrm{d}l$ and $\Delta t = L/v$, total Doppler variations are proportional to $\mathrm{d}\phi/\mathrm{d}t \sim \Delta\phi/\Delta t \sim \lambda r_{\mathrm{e}} N_{\mathrm{m}} L/(L/v)$ (v – velocity of satellite), and Doppler variations at an inhomogeneity are $\sim \lambda r_{\mathrm{e}} \Delta N\, a/(a/v)$. Then, relative Doppler variations are proportional to the ratio of electron concentrations $\sim \Delta N/N_{\mathrm{m}}$. Thus, phase-difference methods allow the reconstruction of inhomogeneities of a few percent against the background, regardless of the size of an inhomogeneity. This is fully supported by our experimental results [Andreeva et al., 1990, 1992; Kunitsyn and Tereshchenko, 1991, 1992; Foster et al., 1994].

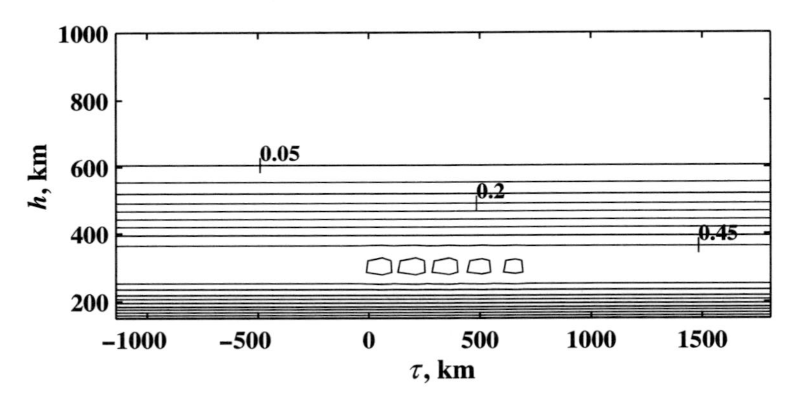

Fig. 2.9. A model of the ionosphere containing a chain of irregularities

To illustrate the estimation of the sensitivity of the methods, we shall consider the reconstruction results of the model shown in Fig. 2.9 with the chain of irregularities (5 "pimples" with smooth profile $\sim \cos^2$) of the size $a = 100\,\mathrm{km}$ and variations $\Delta N = 0.025 \cdot 10^{12}\,\mathrm{m}^{-3}$ (the first three irregularities on the left) and $\Delta N = 0.015 \cdot 10^{12}\,\mathrm{m}^{-3}$ (the next two irregularities), which amount to 3–5% of the maximum $N_{\mathrm{m}} = 0.5 \cdot 10^{12}\,\mathrm{m}^{-3}$. Due to irregularities, variations of the total phase are a few fractions of a percent and are practically invisible in Fig. 2.10. In accordance with the above estimations, Doppler variations are equal to a few percent and are distinctly seen in Fig. 2.11. The

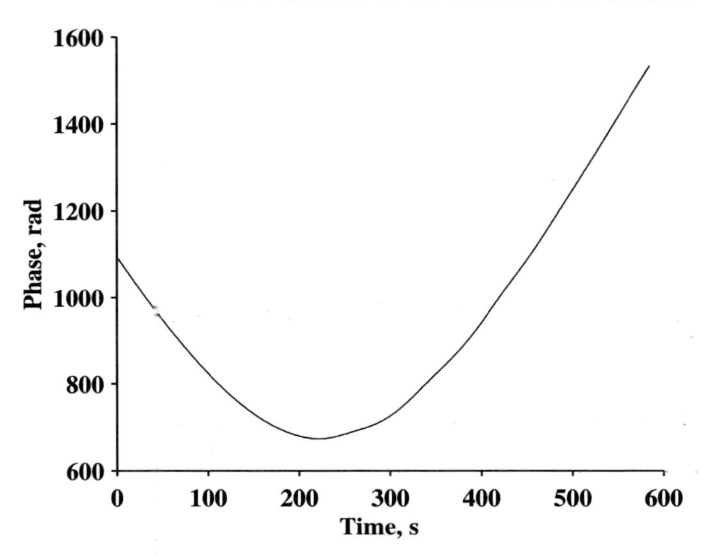

Fig. 2.10. Total phase variations calculated for the ionospheric model shown in Fig. 2.9

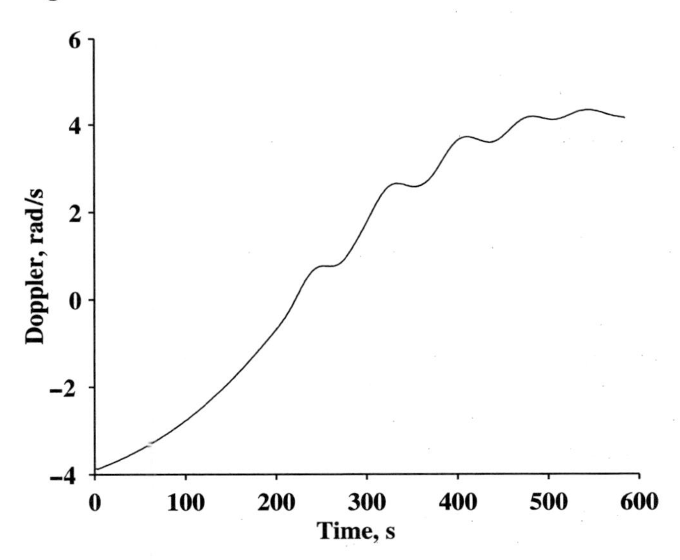

Fig. 2.11. Doppler variations calculated for the ionospheric model shown in Fig. 2.9

RT reconstruction results for this model are represented in Fig. 2.12 (phase RT) and Fig. 2.13 (phase-difference RT). The geometry of the experiment was the same as that of RATE-93 (Russian American tomography experiment-93) [Foster et al., 1994] [altitude of satellite flight $h_0 = 1000\,\text{km}$, the coordinates of receivers (distance along Earth's surface in kilometers) in a plane of satel-

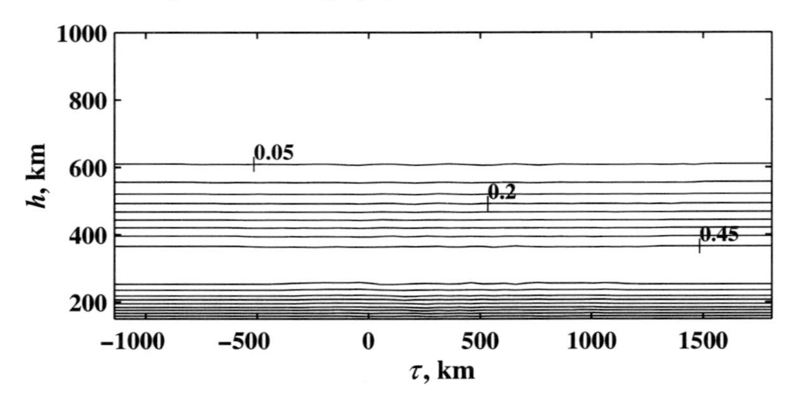

Fig. 2.12. Phase RT reconstruction of the ionospheric model in Fig. 2.9

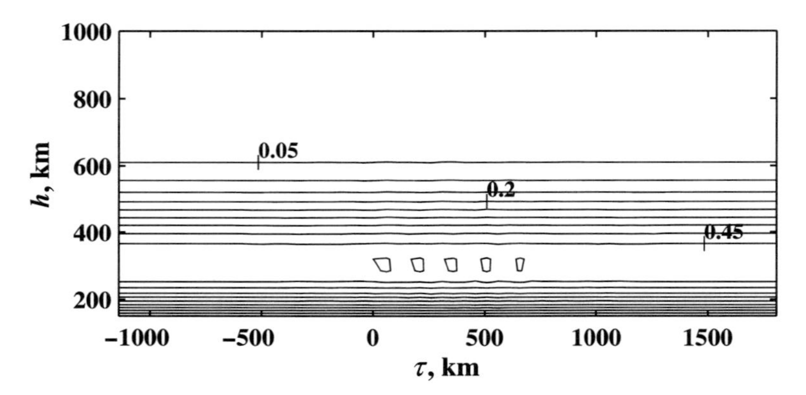

Fig. 2.13. Phase-difference RT reconstruction of the ionospheric model in Fig. 2.9

lite flight $\tau_1 = 0\,\text{km}$, $\tau_2 = 380.4\,\text{km}$, $\tau_3 = 624.9\,\text{km}$, $\tau_4 = 809.3\,\text{km}$]. As an initial guess, we used here a very good approximation which coincides with the background ionosphere without five irregularities. In spite of this good initial guess, the phase method does not reveal the irregularity chain, whereas the phase-difference method does identify the given structure quite satisfactorily.

In both cases, the ART algorithm and 15 iterations were used in modeling. If the initial guess is exact with high accuracy (of the order of a few percent in the right-hand part), hundreds of iterations can be carried out until the solution starts to deteriorate. In this case, details such as these pimples can be revealed. However, in real experimental situations and an inhomogeneous ionosphere, it is impossible to hit an initial guess at better than 20–30%; then, the large number of iterations will make the result even worse.

Quite often experimental data exhibit small variations in the phase corresponding to ionospheric irregularities of relatively small scale. However, the contribution of such disturbances is quite distinct in Doppler data. For ex-

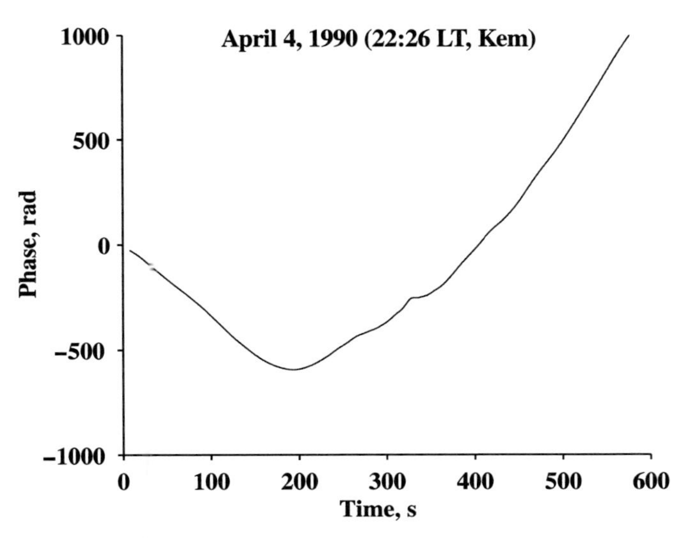

Fig. 2.14. An example of experimental data on relative phase variations produced in an ionosphere with irregularities

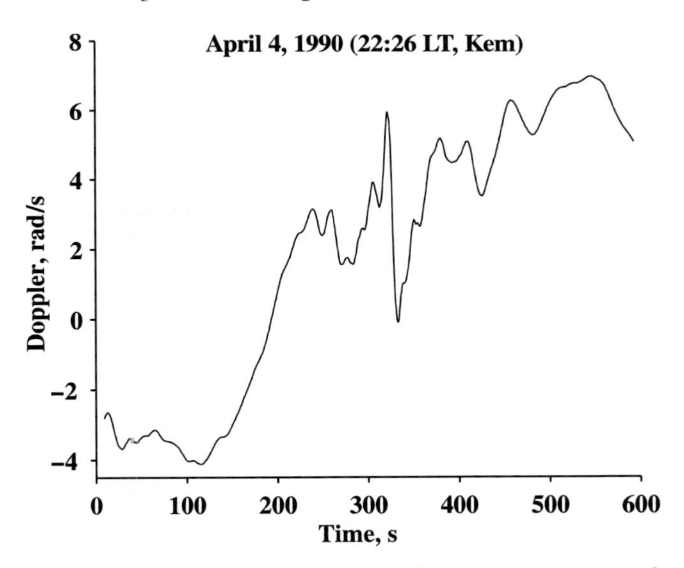

Fig. 2.15. Doppler data record for the same experiment shown in Fig. 2.14

ample, such experimental data are shown in Fig. 2.14 (relative phase) and Fig. 2.15 (Doppler). The RT reconstruction made using these data will be discussed in Sect. 2.7.

In conclusion of this section devoted to the statement of ray RT problems, let us briefly consider the question of the uniqueness of RT reconstructions.

It is well known that if the number of projections is finite, invisible phantoms appear in tomographic problems. The topics of basic RT limitations related to incompleteness of the data, limited observation angles, and the small number of projections available were considered in a series of works [Yeh and Raymund, 1991; Raymund et al., 1994b; Saksman et al., 1997; Andreeva et al., 2001a]. In particular, in the two latter papers, the phantoms invisible over all projections were constructed. These phantoms have the form of structures with alternating signs, with a special symmetry associated with the location of receivers. The sizes of phantoms increase with a growing number of receivers, and in this case the phantom elements will be found also at great altitudes within the region of small values of the reconstructed function. Taking into account the necessary condition of positiveness of the reconstructed function, one can arrive at quite a strong restriction on the oscillation range of the phantom alternating structures. In other words, such phantoms little change the reconstructed functions within regions of large values. Note that the condition of positiveness of the reconstructed function is exact. As established earlier [Andreeva et al., 2001a], such phantoms are overwhelmed in RT reconstructions.

Mathematical statement (2.1) assumes that ray R is a straight line connecting the transmitter and the receiver. In practice, radio rays follow curves that are determined by unknown distribution N. Therefore, the problem of ray RT is generally nonlinear because of the dependence $R[N]$. Analytical estimates and computer simulation have shown that refractive effects are determined by variations in N and are insignificant for peak values of electron density as high as $1.5\text{–}2 \cdot 10^{12}\,\mathrm{m}^{-3}$. Refractive effects restrict both the pixel size and the resolution of the linear problem to 30–40 km. Another limitation is the effect of radio wave diffraction that provides a lower resolution limit of about 10 km (the Fresnel radius is about 2 km). Therefore, the solution of the nonlinear problem of ray RT of the ionosphere allowing for refraction makes it possible to improve the resolution slightly, up to 10–20 km [Andreeva et al., 1999].

2.2 Construction of Projection Operator Approximations

Here we shall consider some methods for constructing projection operators of the RT forward problem and compare them. All possible operators can hardly be analyzed here; therefore, we shall consider only those based on discretization using a rectangular grid. As for other methods of constructing operators, only a few general remarks will be made. One of the possible approaches is to construct operators based on finite expansions in known continuous functions. For example, one-dimensional empirical orthonormal functions may be used to reconstruct vertical ionospheric profiles [Fremouw et al., 1992]. For rather simple typical ionospheric cross sections, this method has an evident advantage because it makes it possible to use a comparatively small number of expansion terms and to reduce the dimensionality of the system of linear equations (SLE). However, when a complicated cross section with various irregularities is reconstructed, such an approach becomes less preferable compared to ordinary local discretization. The fact is that obtaining high resolution involves a sufficiently large number of expansion terms for a given system of functions (the number of functions used is of the order of the number of resolution discretes). The required coefficients of the expansion will be obtained after solving the corresponding SLE. The sensitivity of the reconstructed function to disturbances in the coefficients will grow with their number. Therefore, the results of RT reconstruction will be highly unstable to disturbances in coefficients with large numbers: small variations in these coefficients will cause a great number of irregularities or quasi-wave structures, etc. appearing and disappearing in the reconstruction.

Let us consider some ways of constructing the forward problem projection operators based on different kinds of interpolation of the function between the nodes of a rectangular grid. We shall consider reconstruction of function F in the coordinates h – the altitude above Earth's surface (in kilometers) and τ – the distance across Earth's surface (in kilometers) which is proportional, for example, to the latitude of a given point. In the previous section an example was shown of constructing the projection operator on the basis of a piecewise-constant approximation, that is, with the function F represented in terms of a set of $(m_0 \times n_0)$ local basis functions equal to unity within the given rectangle and zero in all others. The value of function $F(h, \tau)$ within a fixed rectangle is F_{mn}. The problem of tomographic reconstruction from linear integrals I (or D) based on (2.12) or (2.14) is to find a set of discrete values $\{F_{mn}\}$ on the known grid in accordance with the set of linear integrals $\{I_{ij}\}$ from a system of equations like (2.12):

$$L_{i,j}^{m,n} F_{m,n} = I_{i,j} \quad \text{or} \quad L_J^M F_M = I_J \,.$$

Here, in the second equation, rays $(i, j) \to J$ and ionospheric cells $m, n \to M$ are reindexed. Repeating indexes are understood as summations. Coefficient L_{JM} is calculated in accordance with the given rays and the cells intersected by these rays.

If it is possible to determine the linear integrals, the problem of ray RT (2.12) reduces, after data discretization, to the solution of a system of linear equations (SLE). But, as stated above, the problem of ionospheric RT based on phase-difference or Doppler measurements, cannot be solved by using a scheme of piecewise-constant approximation because the derivatives of linear integrals will be discontinuous. Phase-difference experiments require a higher order interpolation than piecewise-constant representation of the function. Correspondingly, the transfer matrix L_{JM} for passage from the reconstructed function to linear integrals should be calculated in another way to provide smoothness of linear integrals with respect to satellite angle α_0. If a forward problem matrix $L_{JM} : F_M \to I_J$ is a smooth function with a first derivative continuous in satellite angle α_0, then after differentiating (2.12) with respect to α_0, one can obtain, instead of (2.12), a system (2.14) for phase-difference or Doppler data.

Let us consider examples of constructing continuous projection operators of the forward problem [Kunitsyn et al., 1994b]. A L_{JM} matrix should be constructed so as to provide continuity of the linear integral with respect to the satellite angle. In the beginning of the section, we consider contribution on the basis of triangular elements; at the end of the section, other possible schemes will be outlined. Proceed to calculation of the matrix D_{JM} of the difference problem, which, as already noted, should be determined from the increment of the matrix L_{JM} that is continuous with respect to the angle of the satellite. The continuity of matrix L_{JM} can be assured by introducing finite triangular elements for representation of the function $F(h, \tau)$, that is, when the sought function is replaced by a piecewise-planar approximation. The smooth function $F(h, \tau)$ is replaced by a continuous polyhedral approximation surface, which makes the derivatives of the linear integrals with respect to the satellite angle already continuous functions.

Triangular elements are obtained naturally from a grid of rectangles by dividing each of them in half diagonally. The function $F(h, \tau)$ within each triangular element is replaced by linear approximation

$$F(h, \tau) = a + b\tau + ch\,. \tag{2.15}$$

The values of the coefficients (a, b, and c) in each finite element are determined from a system of three equations for three boundary points. One can easily write straight off the expressions for the coefficients in a given finite element. These expressions slightly differ for triangular elements of two types: those occupying cells "below" or "above." We stipulate that the cell (m, n)

$$(\tau_m, \tau_{m+1}) \times (h_n, h_{n+1}) = \Delta\tau \times \Delta h$$

is divided by a diagonal $(\tau_m, h_{n+1}) - (\tau_{m+1}, h_n)$, running downward and left to right, into two triangular elements, the "lower" and "upper" elements. Then, in the lower (m, n) element,

$$F(h, \tau) = F_{m,n} + \frac{F_{m+1,n} - F_{m,n}}{\Delta\tau}(\tau - \tau_m) + \frac{F_{m,n+1} - F_{m,n}}{\Delta h}(h - h_n) \tag{2.16}$$

and in the upper (m, n) element,

$$\begin{aligned} F(h, \tau) = F_{m+1,n+1} &+ \frac{F_{m+1,n+1} - F_{m,n+1}}{\Delta\tau}(\tau - \tau_{m+1}) \\ &+ \frac{F_{m+1,n+1} - F_{m+1,n}}{\Delta h}(h - h_{n+1})\,. \end{aligned} \tag{2.17}$$

As before, to simplify the notation, we will renumber the values of the following samples: $F_{m,n} \to F_M$, $(m+1,n) \to (M+1)$, $(m,n+1) \to (M+\Delta M)$, $(m+1,n+1) \to (M+\Delta M+1)$, where ΔM is the number of cells horizontally in one row.

The linear integral I_j (2.9) is the sum of the integrals with respect to all finite elements that intersect the ray J:

$$I_J = \sum_M \int \gamma(h)\, F(h,\tau)\; dh\,,$$

where $\gamma(h) = (R+h)\left[R^2 \sin^2\beta + 2Rh + h^2\right]^{-1/2}$ $F(h,\tau)$ and $F(h,\tau)$ is represented in the form of piecewise-planar approximations (2.16), (2.17) in each finite element. The result of integrating such an approximation in lower element M is

$$\int \gamma(h)\, F\; dh = J_0 F_M + J_\tau(F_{M+1} - F_M) + J_h(F_{M+\Delta M} - F_M)\,, \tag{2.18}$$

and in upper triangular element M,

$$\begin{aligned}\int \gamma(h)\, F\; dh = {} & J_0{}' F_{M+\Delta M+1} + J_\tau{}'(F_{M+\Delta M+1} - F_{M+\Delta M}) \\ & + J_h{}'(F_{M+\Delta M+1} - F_{M+1})\,.\end{aligned} \tag{2.19}$$

Here, J_0, J_τ, J_h, $J_0{}'$, $J_\tau{}'$, $J_h{}'$ are the following integrals:

$$\begin{aligned}
J_\tau &= \frac{1}{\Delta\tau}\int_{h_n}^{h} \gamma(h)\,[\tau(h) - \tau_m]\; dh\,,\\
J_\tau{}' &= \frac{1}{\Delta\tau}\int_{h}^{h_{n+1}} \gamma(h)\,[\tau(h) - \tau_{m+1}]\; dh\,,\\
J_h &= \frac{1}{\Delta h}\int_{h_n}^{h} \gamma(h)\,[h - h_n]\; dh\,,\\
J_h{}' &= \frac{1}{\Delta h}\int_{h}^{h_{n+1}} \gamma(h)\,[h - h_{n+1}]\; dh\,.
\end{aligned} \tag{2.20}$$

Integration with respect to a lower finite element starts at the lower boundary of the cell $h = h_0$ and ends at the height h, where the nth ray leaves the lower element. Integration with respect to the upper finite elements begins at this height and ends at the height of the upper boundary of the cell. Remember that $\gamma(h)$ and all intervals with respect to the cell M are functions of the elevation β, α_0, or the number of the ray J.

After integrating (2.18) with respect to ray J in the lower element M, the quantity $(J_0 - J_\tau - J_h)$ enters into the coefficient L_{JM}, as it is a coefficient for F_M. Correspondingly, J_τ appears in $L_{J,M+1}$ and J_h in $L_{J,M+\Delta M}$. However, in integration with respect to only the lower finite element M, these coefficients L are still not completely determined. It can be easily understood that each sample F_M falls into three lower and three upper finite elements adjacent to M. Only after integration with respect to ray J over all finite elements is it possible to construct coefficient L_{JM} completely from six neighboring F_M where the ray fell. Integration with respect to the upper finite element M, with respect to ray J (2.19), makes the contribution $({J_0}' + {J_\tau}' + {J_h}')$ to coefficient $L_{J,M+\Delta M+1}$, the contribution $(-{J_0}' + {J_\tau}' + {J_h}')$ to coefficient $L_{J,M+\Delta M+1}$, the contribution $(-{J_\tau}')$ to coefficient $L_{J,M+\Delta M}$, and the contribution $(-{J_h}')$ to coefficient $L_{J,M+1}$. The integrals with respect to all rays of type (2.20) can be calculated by various numerical methods; in view of the smoothness of $\gamma(h)$ and the piecewise-planar approximation of F, it is sufficient to use the trapezoid or Simpson method. Here, at each integration step Δh, it is necessary to make sure that the ray does not exceed the limits of the finite element.

Performing numerical integration with respect to all rays, we obtain matrix L_{JM}. The latter is related to the set $\{\alpha_0\}$ of satellite positions and the corresponding series of rays. It is also possible to calculate matrix L' for another set of close positions of the satellite with a fixed increment $\{\alpha_0 + \Delta\alpha_0\}$. After this, we determine the matrix for the phase-difference tomography problem $D_{JM} = \frac{L'_{JM} - L_{JM}}{\Delta\alpha_0}$.

The projection operator or L_{JM} matrix can also be built on the basis of higher order approximations than that in (2.15). For example, one can use a two-dimensional approximation in the form of the product of linear functions or the product of cubic splines. Then, the function $F(h, \tau)$ assumes the form,

$$F(h, \tau) = \sum_{m,n=0}^{3} a_{mn} \tau^m h^n. \tag{2.21}$$

Inside an arbitrary (m, n) rectangular, expressing the function in terms of normalized coordinates $x = \frac{\tau - \tau_m}{\Delta\tau}$, $y = \frac{h - h_0}{\Delta h}$, we can obtain the following representation in terms of the functional values at the four corner points $(x, y) = \{(0,0);\ (0,1);\ (1,0);\ (1,1)\}$:

$$F(x, y) = F_{00} P_{00} + F^x_{00} P^x_{00} + F^y_{00} P^y_{00} + F_{01} P_{01} + \dots .$$

Here, the subscripts refer to the coordinates (x, y) of the corner points; F is the functional value at the corresponding point; F^x, F^y are the values of the partial derivatives of the function with respect to x, y; and F^{xy}_{00} is the value of the second-order partial derivative of the function with respect to x, y. The total sum, see (2.21), will contain 16 summands; $P^y_{\alpha\beta}(x, y)$ are corresponding polynomials of a power not higher than 3. We shall not write here the mentioned polynomials completely; four examples are enough for illustration:

$$\begin{aligned}
P_{00} &= 4x^3y^3 - 6x^2y^3 - 6x^3y^2 + 9x^2y^2 + 2y^3 - 3y^2 + 2x^3 - 3x^2 + 1\,,\\
P_{00}^{x} &= 2x^3y^3 - 4x^2y^3 - 3x^3y^2 + 6x^2y^2 + 2xy^3 - 3xy^2 + x^3 - 2x^2 + x\,,\\
P_{00}^{y} &= 2x^3y^3 - 3x^2y^3 - 4x^3y^2 + 6x^2y^2 + 2x^3y - 3x^2y + y^3 - 2y^2 + y\,,\\
P_{00}^{xy} &= x^3y^3 - 2x^2y^3 - 2x^3y^2 + 4x^2y^2 + x^3y - 2x^2y + xy^3 - 2xy^2 + xy\,,\\
&\quad \ldots\,.
\end{aligned}$$

Then, by integrating the given polynomials over each cell, we can produce the corresponding elements of the matrix, as in (2.16)–(2.20). Note that now not only the values of the function F but also the values of the mentioned derivatives are unknowns, that is, such representation makes it possible to find the function and its first derivative. The matrix for the product of linear approximations can be constructed in a similar and even simpler way, since it is a particular case of that described above, where the summation in the formula goes up to 1 rather than up to 3.

Let us denote the different approximations of projection operators (matrices) of the tomographic problem as follows:

a – is built with the piecewise-constant approximation,
b – is built with the piecewise-planar approximation,
c – is built with the linear product bilinear approximation,
d – is built with the cubic spline product approximation.

As an illustration to compare the results between different operators, we will use a simple model of irregularity comprising three Gaussians (Fig. 2.16). An 8×8-pixel grid was used in the reconstruction; however, a denser grid (32×32-pixel) was used in displaying the results to show the essential difference in

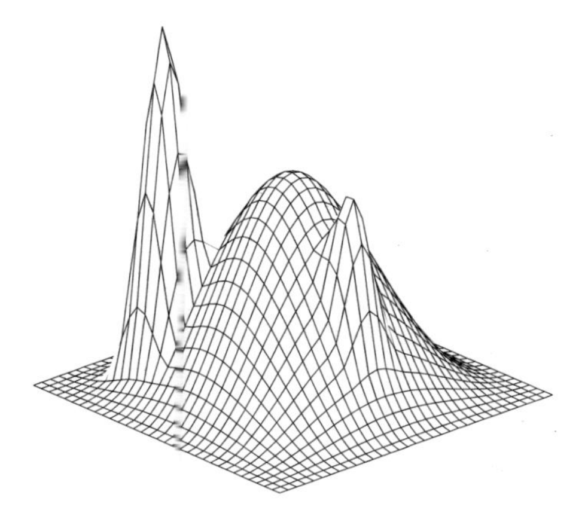

Fig. 2.16. A model of an irregularity composed of three Gaussians

the representations corresponding to the different approximations applied [Kunitsyn et al., 1995a].

Projection operators with higher order approximations make it possible to improve approximations of the operator in the forward problem, that is, we obtain operator values closer to exact ones in the forward problem. Table 2.1 presents several examples of the forward problem calculation based on the model considered above, using the different operators, a, b, c, and d. Errors of calculation, as earlier, are described by numbers δ (2.13). It can be seen from the results shown in the table that the calculation accuracy increases with the growth in the order of approximation.

Table 2.1. Forward problem errors

Operator type	δ_∞	δ_2
a	0.371	0.258
b	0.124	0.053
c	0.062	0.035
d	0.034	0.025

One can see that passing to higher orders makes it possible to improve significantly the exactness of the forward problem solution. However, as the approximation order increases, the matrix becomes more complicated and less sparse, which can impair the solution of the inverse problem. One cannot

Table 2.2. Inverse problem errors (phase RT)

Operator type	δ_∞	δ_2	Δ_∞	Δ_2
a	0.465	0.499	0.510	0.464
b	0.109	0.181	0.207	0.146
c	0.109	0.129	0.128	0.098
d	0.087	0.130	0.103	0.105

Table 2.3. Inverse problem errors (phase-difference RT)

Operator type	δ_∞	δ_2	Δ_∞	Δ_2
a	5.802	6.161	5.896	6.291
b	0.179	0.270	0.265	0.204
c	0.099	0.148	0.103	0.099
d	0.094	0.121	0.102	0.103

say in advance which operator will suit best in solving the inverse problem, because at the beginning of the increase in the approximation order, the function is approached better, but the matrix properties for solving the inverse problem become worse. Specific operators should be chosen by computer simulation (to be illustrated further). Tables 2.2 and 2.3 display examples of reconstructions of the model shown in Fig. 2.16 obtained by using different approximations of the projection operators, a, b, c, and d. The errors of numerical simulation are described in the terms of δ_2 and δ_∞ (2.13). The results of reconstruction should be compared in the norms approaching the integral ones, i.e., the norms L^2 instead of l^2 and L^∞ should be used instead of l^∞ (and corresponding numbers Δ_2 and Δ_∞ instead of numbers δ_2 and δ^∞).

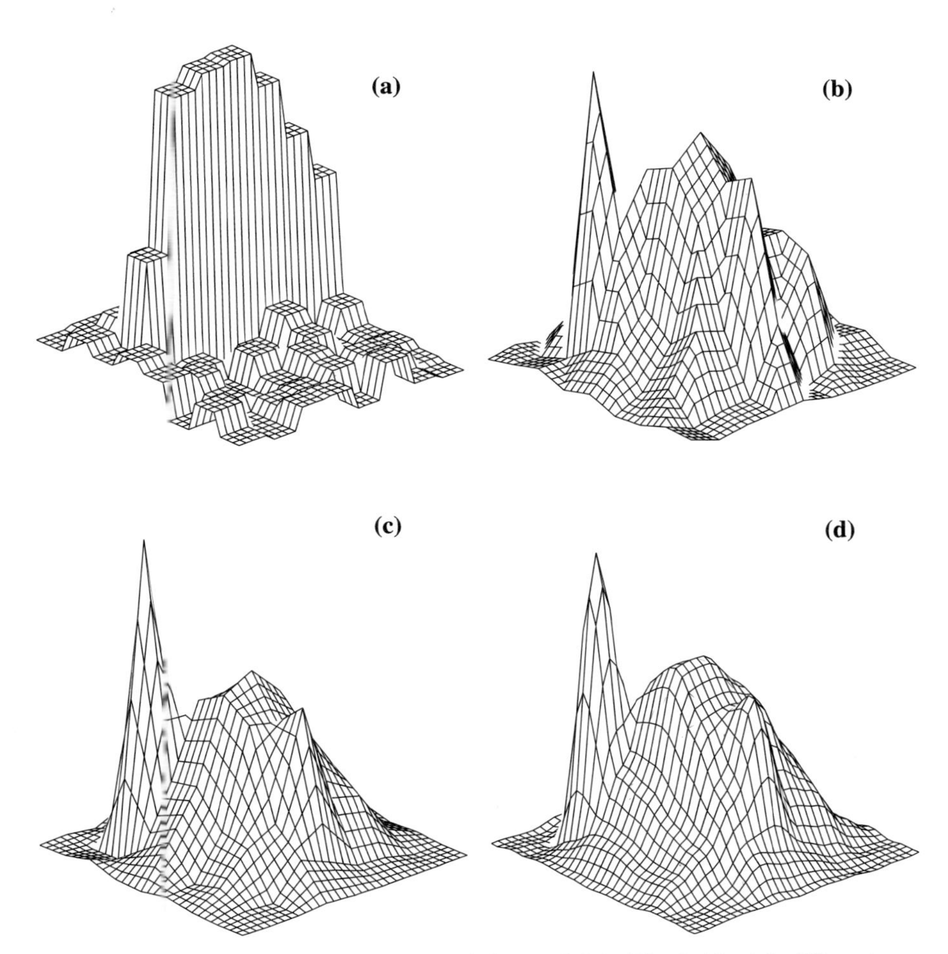

Fig. 2.17. Phase RT reconstructions of the model in Fig. 2.16 with different approximations of the projection operator

In other words, the functions should be compared using a grid other than that used for reconstruction. Otherwise, if the results are compared in reconstruction points only, we may come to erroneous conclusions. Figure 2.17 shows the RT reconstruction results for different approximations (a, b, c, d, respectively) by the phase method. Figure 2.18 (a, b, c, d) shows the RT reconstruction results for different approximations obtained using the phase-difference technique. From the reconstruction results and Tables 2.2, 2.3, one can see the increase in the accuracy of solving the problem of RT reconstruction for operators with higher orders of approximation. Noticeable improvement is observed when passing from approximation (a) to (b) and certain amelioration takes place when passing from (b) to (c). However, the results of solving the tomographic problem with approximations (c) and (d)

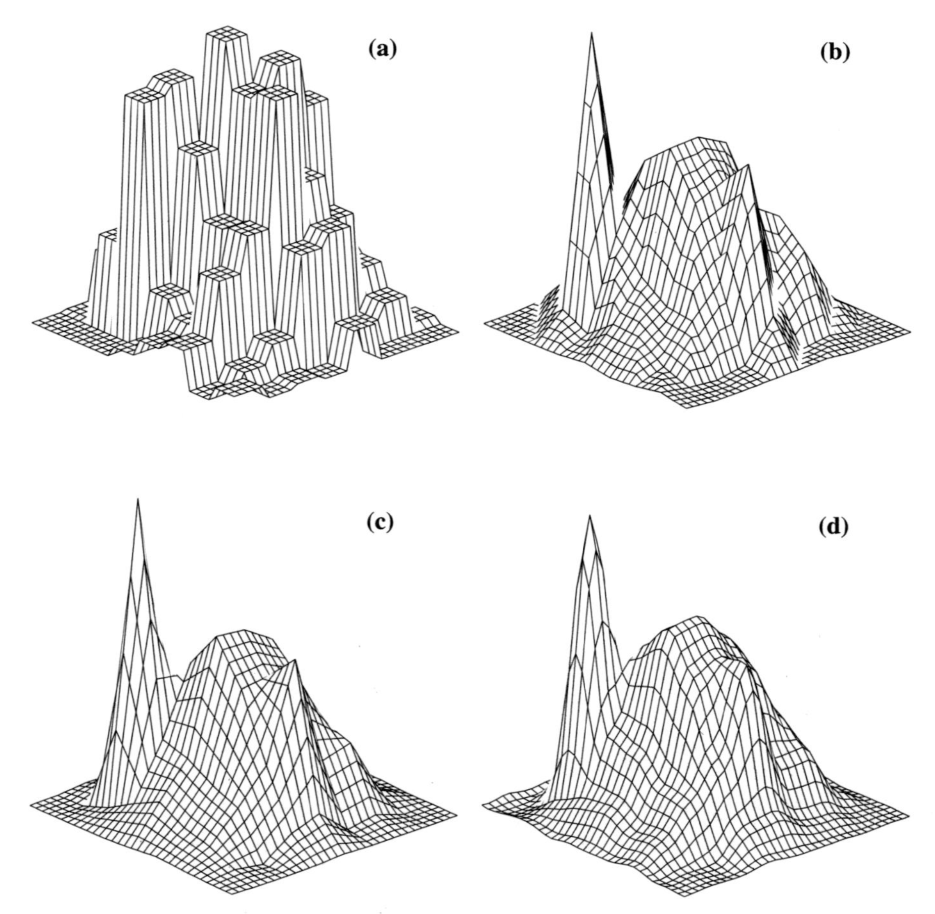

Fig. 2.18. Phase-difference RT reconstructions of the model in Fig. 2.16 with different approximations of the projection operator

are comparable now, which is due to a deterioration in the characteristics of the projection operator matrix. As our numerical experiments showed, usually it is sufficient to employ a bilinear approximation (of type "c") of the projection operator.

Here, it should be emphasized that the RT reconstruction results should be represented with the same approximation used when constructing the forward problem operator, as shown in Figs. 2.17–2.18. Many authors use the (a) type operator with a piecewise-constant approximation, find the values of the function at the grid and then, using these discrete data, draw a smooth function with an approximation of a higher order. This is an incorrect approach. It means that if the expansion coefficients are obtained using one system of functions and applied to another system of functions, then dif-

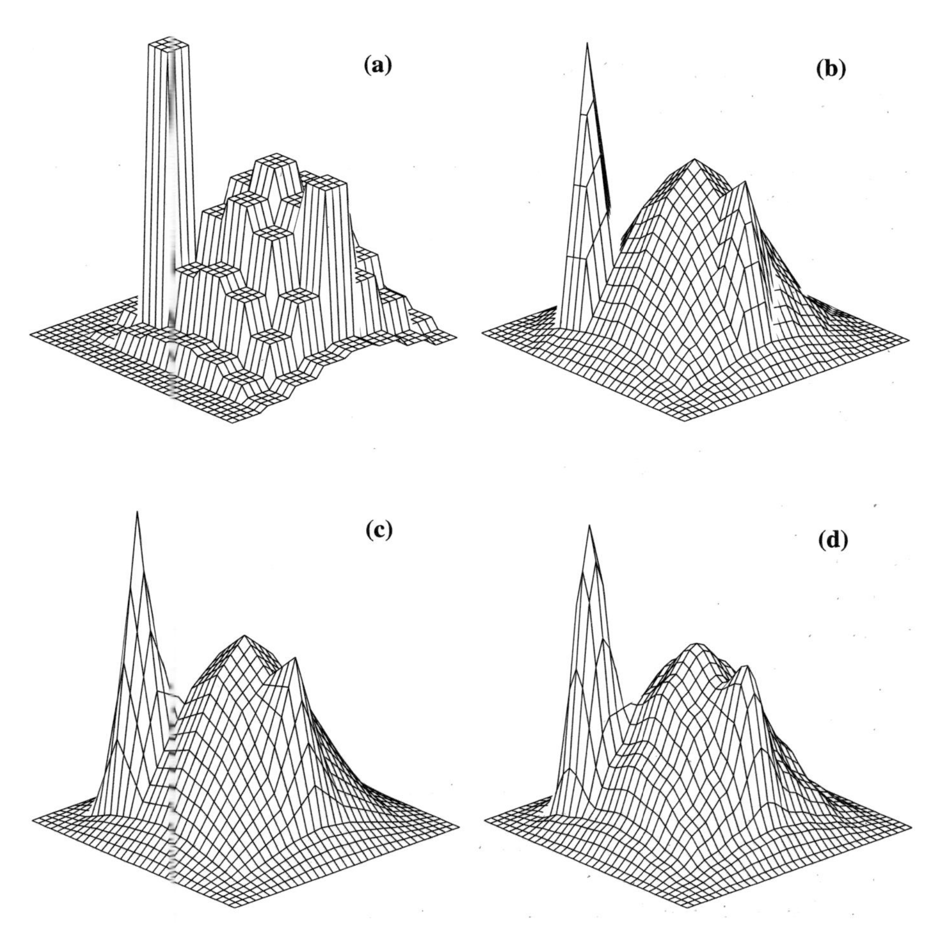

Fig. 2.19. Images of the function in Fig. 2.16 obtained with different approximations on an 8×8-pixel grid

ferent systems of functions (interpolations of the results represented) will produce different results. The form of representation of results must exactly correspond to the form in which the coefficients of the functional expansion were obtained. Therefore, for the piecewise-constant approximation, it is necessary to use representations of the type shown in Figs. 2.17a and 2.18a. To illustrate these differences, representations are portrayed in Fig. 2.19 of the function from Fig. 2.16 with different approximations using exact values of this function in the grating pixels (8×8). Errors in δ_2 and δ_∞ for all of these representations are equal to 0; however, it can be easily seen how large the difference is between these representations for the applications where the behavior of the function between the pixels is important (errors in Δ_2 and Δ_∞ for different representations differ greatly).

2.3 Algorithms for Solving Systems of Linear Equations

Most tomographic problems ultimately reduce to solving systems of linear equations (SLE). Hence it becomes clear how important is to choose proper methods (algorithms) for solving SLE that allow for the specificity of tomographic problems and yield the best results in solving such problems. In this section, we will briefly describe and analyze various algorithms for solving SLE in RT problems. Solving SLE in ray RT of type (2.12), (2.14) is a complex computational problem. If ionospheric sections with sizes of a few thousand kilometers are reconstructed with dozens of kilometer digitization steps, the rank of the SLE matrix is of the order of a few thousand and contains about 10^6–10^7 elements but is sparse enough, with a large number of zero elements. A lot of both direct and iteration algorithms for solving such SLE are known. However, even now the theory and practical algorithms for such SLE solution are being actively developed. Many of them have been tested in ray RT problems. Here, we can mention the algebraic reconstruction technique (ART), including those with relaxation and for systems of inequalities (ART2); the simultaneous iterations reconstruction technique (SIRT); the multiplicative algebraic reconstruction technique (MART); block-iteration algorithms; algorithms for regularization of rms error; image entropy optimization algorithms, etc. [Censor, 1983]. In seismic tomography, good practical results have been obtained by the method of minimizing iteration correction in various matrices, including modifications based on weight correction, ray weighting, and interiteration smoothing [Dines and Lytle, 1979]. Studies have been carried out of matrix spectra characteristics, the resolving power of the systems, and uniqueness of reconstruction approachable "in the limit", i.e., with an unbounded increase in the number of measurements [Nolet, 1987].

A series of reviews is devoted to the analysis of SLE solution algorithms as applied to the ray RT of the ionosphere [Raymund, 1994, 1995; Raymund et al., 1994b Pryse et al., 1998b; Leitinger, 1999]. To obtain RT reconstructions, earlier we used both the various iteration algorithms for SLE solution [Kunitsyn et al., 1994b; Kunitsyn et al., 1995a; Andreeva et al., 2001a] and the Bayesian approach, see Sect. 2.5) and simulated RT problems under different conditions. Prior to proceeding to the analysis of the iteration algorithms used, we will present formulas for the basic ones.

Let us put the tomographic SLE in terms of the column vector $\boldsymbol{x}$ of the unknowns and the column vector $\boldsymbol{m}$ of the measured data:

$$\hat{A}\boldsymbol{x} = \boldsymbol{m} \quad \text{or} \quad \sum_{j=1}^{N} A_{ij} x_j = m_i \,, \quad i = 1, \ldots, L \,. \tag{2.22}$$

Here, the matrix $\hat{A}$ can be either the phase RT matrix (2.12) with various approximations, or the phase-difference RT matrix (2.14). Denote as a vector $\boldsymbol{A}^i = (A_{ij})_{j=1}^{N}$. Here, the iteration formulas are written for a series of basic algorithms used in modeling. In all of the iteration algorithms below, rays are exhausted cyclically.

ART The initial guess is chosen arbitrarily:

$$\boldsymbol{x}^{k+1} = \boldsymbol{x}^k + \frac{m_i - \langle \boldsymbol{A}^i, \boldsymbol{x}^k \rangle}{\langle \boldsymbol{A}^i, \boldsymbol{A}^i \rangle} \boldsymbol{A}^i \,,$$

where $\langle \boldsymbol{U}, \boldsymbol{V} \rangle = \sum_{j=1}^{N} U_j V_j$ is a scalar product for any vectors $\boldsymbol{U}, \boldsymbol{V} \in \mathbb{R}^N$. This algorithm can be also interpreted geometrically in the space of image vectors. The manifold

$$H_i = \{\boldsymbol{x} \in \mathbb{R}^N\} \,, \quad \langle \boldsymbol{A}^i, \boldsymbol{x} \rangle = m_i \,,$$

that contains all solutions of the ith equation of the original system, is a hyperplane in the space $\mathbb{R}^N$, the vector $\boldsymbol{A}^i$ being its normal. The point $\boldsymbol{x}^{k+1}$ (in the space $\mathbb{R}^N$) is nothing else but an orthogonal projection of point $\boldsymbol{x}^k$ onto hyperplane H_i. Therefore, in the ART algorithm, a sequential projecting of approximate solutions onto hyperplanes H_i corresponds to the equations of the system, this process is executed cyclically.

ART with relaxation The initial guess is arbitrary:

$$\boldsymbol{x}^{k+1} = \boldsymbol{x}^k + \lambda_k \frac{m_i - \langle \boldsymbol{A}^i, \boldsymbol{x}^k \rangle}{\langle \boldsymbol{A}^i, \boldsymbol{A}^i \rangle} \boldsymbol{A}^i \,.$$

Relaxation parameters are a sequence of real numbers, the elements of which usually are within the interval

$$\varepsilon_1 \le \lambda_k \le 2 - \varepsilon_2 \,,$$
$$\varepsilon_1, \varepsilon_2 > 0 \,.$$

The use of relaxation parameters makes it possible either to reduce or to enlarge the orthogonal projection by ART, which is exceedingly important in practical realization of the algorithm.

ART2 (ART for inequalities)
The initial guess is arbitrary.
Let us replace the system

$$\langle \boldsymbol{A}^i, \boldsymbol{x} \rangle = m_i\,, \quad i = 1, \ldots, L$$

by a system of inequalities:

$$m_i - \varepsilon_i \leq \langle \boldsymbol{A}^i, \boldsymbol{x} \rangle \leq m_i + \varepsilon_i\,, \quad i = 1, \ldots, L\,.$$

By multiplying the left-hand part of the inequality by (-1), we obtain a system of unilateral linear inequalities twice as big (in the number of inequalities):

$$\begin{aligned}
&\langle \boldsymbol{B}^i, \boldsymbol{x} \rangle \leq C_i\,, \quad i = 1, \cdots, 2L\,,\\
&\boldsymbol{x}^{k+1} = \boldsymbol{x}^k + s_k \boldsymbol{B}^i\,, \text{where}\\
&s_k = \min\left(0, \lambda_k \frac{C_i - \langle \boldsymbol{B}^i, \boldsymbol{x}^k \rangle}{\langle \boldsymbol{B}^i, \boldsymbol{B}^i \rangle}\right)\,,\\
&\varepsilon_1 \leq \lambda_k \leq 2 - \varepsilon_2\,,\\
&\varepsilon_1, \varepsilon_2 > 0\,.
\end{aligned}$$

ART3
The initial guess is arbitrary:

$$\boldsymbol{x}^{k+1} = \boldsymbol{x}^k + s_k * \frac{\boldsymbol{A}^i}{\langle \boldsymbol{A}^i, \boldsymbol{A}^i \rangle}\,,$$

$$s_k = \begin{cases} 0, & \text{if } \left| m_i - \langle \boldsymbol{A}^i, \boldsymbol{x}^k \rangle \right| \leq \varepsilon_i \\ m_i - \langle \boldsymbol{A}^i, \boldsymbol{x}^k \rangle, & \text{if } \left| m_i - \langle \boldsymbol{A}^i, \boldsymbol{x}^k \rangle \right| \geq 2\varepsilon_i \\ 2\left(m_i + \varepsilon_i - \langle \boldsymbol{A}^i, \boldsymbol{x}^k \rangle\right)\,, & \text{if } m_i + \varepsilon_i, \langle \boldsymbol{A}^i, \boldsymbol{x}^k \rangle < m_i + \varepsilon_i \\ 2\left(-m_i + \varepsilon_i + \langle \boldsymbol{A}^i, \boldsymbol{x}^k \rangle\right)\,, & \text{if } m_i - 2\varepsilon_i, \langle \boldsymbol{A}^i, \boldsymbol{x}^k \rangle < m_i - \varepsilon_i\,. \end{cases}$$

In acccordance with ART3, each hyperstrip is "enveloped" by a larger one. Next, the following rules are used:

- If $\boldsymbol{x}^k$ lies within the ith hyperstrip, then $\boldsymbol{x}^{k+1} = \boldsymbol{x}^k$.
- If $\boldsymbol{x}^k$ lies outside the ith hyperstrip but inside the corresponding "enveloping" hyperstrip, then, to obtain $\boldsymbol{x}^{k+1}$, orthogonal mapping of $\boldsymbol{x}^k$ is carried out with respect to the nearest hyperplane of the ith hyperstrip.
- If $\boldsymbol{x}^k$ lies outside the "enveloping" hyperstrip, then $\boldsymbol{x}^{k+1}$ is obtained by orthogonal projecting of $\boldsymbol{x}^k$ onto the initial ith hyperplane.

SIRT The initial guess is arbitrary:

$$x_j^{k+1} = x_j^k + \frac{1}{P_j} \sum_i \lambda_k \frac{m_i - \langle \boldsymbol{A}^i, \boldsymbol{x}^k \rangle}{\langle \boldsymbol{A}^i, \boldsymbol{A}^i \rangle} \boldsymbol{A}^i , \quad j = 1, \ldots, N ,$$

where P_j denotes the number of nonzero elements in the jth column of matrix $\hat{A}$, or, physically, the number of rays intersecting the jth cell. In this method, the approximate solution is modified only after all equations have been processed. There are many methods of this kind (averaging forms).

MART As an initial guess, a homogeneous, smooth ionosphere $\boldsymbol{x}^0$ is used:

$$x_j^{k+1} = x_j^k * \left(\frac{m_i}{\langle \boldsymbol{A}^i, \boldsymbol{x}^k \rangle} \right)^{\frac{\lambda_k A_j^i}{\sqrt{\langle \boldsymbol{A}^i, \boldsymbol{A}^i \rangle}}} , \quad j = 1, \ldots, N .$$

Another representation of the MART algorithm [Pryse et al., 1998b] is

$$x_j^{k+1} = x_j^k * \left(\frac{m_i}{\langle \boldsymbol{A}^i, \boldsymbol{x}^k \rangle} \right)^{\frac{\lambda_k A_j^i}{A_{\max}}} , \quad j = 1, \ldots, N .$$

$A_{\max}$ is the maximum path-pixel intersection length in the grid.

DART The initial guess is arbitrary:

$$\boldsymbol{x}^{k+1} = \boldsymbol{x}^k * \left(1 + \lambda_k \frac{m_i - \langle \boldsymbol{A}^i, \boldsymbol{x}^k \rangle}{\langle \boldsymbol{A}^i, \boldsymbol{A}^i \rangle} \boldsymbol{A}^i \right) ,$$

$$\varepsilon_1 \le \lambda_k \le 2 - \varepsilon_2 ,$$

$$\varepsilon_1, \varepsilon_2 > 0 .$$

MART2 (DART – decomposition of MART2) A homogeneous, smooth ionosphere $\boldsymbol{x}^0$ is used as an initial guess:

$$x_j^{k+1} = x_j^k * \left(\frac{m_i}{\langle \boldsymbol{A}^i, \boldsymbol{x}^k \rangle} \right)^{\lambda_k A_j^i \frac{\langle \boldsymbol{A}^i, \boldsymbol{x}^k \rangle}{\langle \boldsymbol{A}^i, \boldsymbol{A}^i \rangle}} , \quad j = 1, \ldots, N ,$$

$$0 < \varepsilon \le \lambda_k \le 1 .$$

The DART algorithm (decomposed algebraic reconstruction technique) was proposed not very long ago [Kunitsyn et al., 1995a] and proved to give good results in RT problems. The algorithm can be interpreted as the first two terms of the Taylor expansion of the MART2 algorithm. DART gives approximately the same results as MART, but it works two to three times faster. However, its principal advantage is that, in contrast to the MART algorithm, here a positiveness of the data and matrix elements is not required. This makes the DART algorithm applicable in phase-difference RT. To our knowledge, the MART2 algorithm has not been presented earlier; however, it is a version of the MART class algorithms and works comparably with other similar algorithms.

Note that the factor of principal importance in successfully solving the RT problem is the use of the positiveness condition of the solution of a given SLE. Iteration algorithms make it possible to allow easily for this positiveness.

To compare different modifications of RT, simulation was carried out with a large number of quasi-real models of the ionosphere. In our opinion, it is pointless to compare separately the algorithms of an SLE solution for RT problems. The result of reconstruction depends on the choice of the RT method (phase or phase-difference RT), the approximation of the projection operator (the discrete projection operator construction of which was described in the previous Sect. 2.2), the algorithm chosen for solving the SLE obtained, and some additional conditions (the grid choice, initial guess, and others). When comparing the different schemes of RT problem solutions, the whole triad "method-operator-algorithm" along with additional conditions should certainly be taken into account. If we, for example, choose the best algorithm for a single type operator without taking into account the distortions caused by approximation, this will be the best algorithm for solving only that given SLE but not the best one for solving the RT problem on the whole (and vice versa). This topic will be discussed in Sect. 2.4.

However, if the method (either phase or phase-difference RT) and the approximation of the projection operator are fixed, then only different algorithms of a SLE solution remain to be compared. In the overwhelming majority of works on ionospheric RT, the approach of phase RT with piecewise-constant approximation of the projection operator was used. Therefore, in the reviews of the methods for solving RT problems [Raymund et al., 1994b; Raymund, 1995; Leitinger, 1999; Pryse et al., 1998b] only SLE solution algorithms are analyzed and compared.

A great number of experiments on numerical modeling carried out by our teams showed that there are cases when certain RT modifications have unquestionable advantages over others (for instance, phase-difference tomography in sensitivity compared to phase RT) [Kunitsyn et al., 1994b, 1995a; Andreeva et al., 2001a]. However, as a rule, it is impossible to give a preference in advance for any one combination of methods and algorithms. Some combinations of methods and algorithms happen to prove preferable for certain kinds of structures but others turn out more suitable for other types of structures. However, one could still advance several general and valid statements on iteration algorithms for SLE solution in most typical ionospheric situations.

- MART and DART algorithms are more sensitive; they are capable of reconstructing smaller and weaker irregularities than ART. However, MART and DART do not always surpass ART in reconstruction quality. These algorithms work better in an interval of large values of the reconstructed function. If, for example, strong enough irregularities are located at high altitude within a region of low electron density, then outside the domain of high values of reconstructed function, the MART algorithm starts distorting the results as if trying to move these irregularities into the region of high concentrations, or, in other words, to attach the irregularities to the maximum of the layer. An example of such distortion is shown in [Kunitsyn et al., 1995a]. This distortion of the results is undoubtedly a serious defect of MART and DART algorithms.
- ART and ART2 algorithms provide larger artifacts than MART, DART, and SIRT, particularly in the region of small values of reconstructed functions.
- The SIRT algorithm is much less sensitive to data errors owing to intermediate averaging of the results during the iteration process. However, the SIRT algorithm in contrast to MART badly reconstructs small irregularities. Numerical experiments carried out for a series of models showed that in most cases SIRT is not suitable in practice because of high reconstruction errors. This algorithm is worth using if data errors are big ($\geq$ 10–20%). However, in this case, SIRT reconstructs only strongly "averaged" objects and even such prominent structures as troughs are often not reconstructed.
- On the whole, reconstruction results obtained using ART, MART, and DART are comparable. In some cases, ART is advantageous, and in others MART is more suitable (only in phase RT), or DART is preferable.
- In most cases, phase-difference RT with a combination of ART and DART algorithms has an advantage over the other combinations of methods and algorithms

All of the above mentioned iteration algorithms minimize the residual (right-hand part of SLE) $\|\hat{A}\boldsymbol{x} - \boldsymbol{m}\|$ tending to zero with iterations. But the behavior of reconstruction error δ is more complicated. Since the reconstruction errors are defined both by the number of iterations and measurement errors, each algorithm can be described in terms of a 2-D "portrait" of the algorithm representing the dependence of the relative error of reconstruction $\delta(l^2)$ (in percent) on the number of iterations N_{iter} and on the relative error in the right-hand part of equation γ (in percent). Such a portrait of the error is characteristic of the convergence rate, the accessible minimum of reconstruction error for different error levels γ. Although the portraits of algorithms depend somewhat on the ionospheric model used, this dependence is quite weak, and the portraits are mostly descriptive of just algorithms. Typical portraits of ART and SIRT algorithms are shown in Fig. 2.20 (for the model in Fig. 2.2). It can be seen that for big errors, the ART algorithm (like MART and DART) converges only at the first few iterations and then

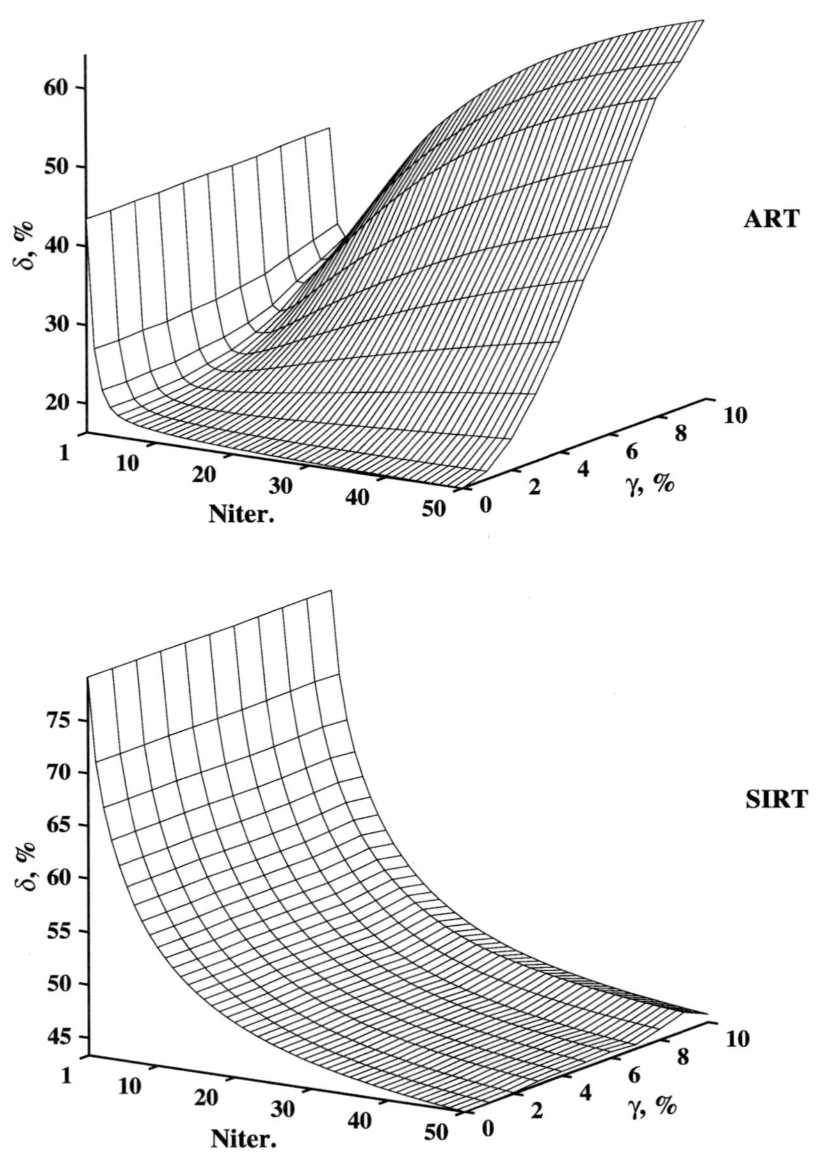

Fig. 2.20. Examples of "portraits" of ART and SIRT algorithms

diverges rather quickly. The iteration process, thus, should be confined within a region of minimum reconstruction error, the location of which is defined by the errors γ in the right-hand part of the SLE (2.22). In the case of small errors in the right-hand part of SLE (less than 1–2%), the number of iterations should be large (dozens to hundreds), and the reconstruction error can achieve up to 10% and less. With errors of an average level (3–6%), it is possible to reach quite fast (by 5–10 iterations) the minimum of reconstruction error that can be 10–20%. If errors in the right-hand part are big (higher than 10%), three to five iterations are enough to minimize the reconstruction error. The SIRT algorithm, to judge from the portrait of the algorithm in Fig. 2.20, is faintly sensitive to the errors in γ; however, a very high level ($\sim$40%) of reconstruction errors is inherent in this algorithm. Note that the errors γ in the right-hand part are defined not only by the noise and influence of scattering from small-scale irregularities but mainly by errors of projection operator approximation (see Sect. 2.4). The algorithms manifest similar behavior with respect to the initial guess error. In the case of a complex irregular ionosphere, it is difficult to hit the initial guess more accurately than 20–30% because without a priori data we use a smooth regular ionosphere as an initial guess. Therefore, in iteration algorithms, 3–10 iterations are mostly used for each algorithm of a combination of algorithms.

Concluding our consideration of algorithms for SLE solution applied to RT problems, let us make several remarks on the bibliography. At present, iteration algorithms are often used in practical RT reconstructions. Perhaps, today, MART and derivatives are still the most widely used iterative algorithms [Raymund et al., 1990; Bust et al., 1994; Mitchell et al., 1995; Cook and Close, 1995; Pryse et al., 1995; Vasicek and Kronschnabl, 1995]. In [Mitchell et al., 1997a,b,c; Pryse et al., 1998a], it is DART that was actually used. However, the authors referred to it as ART type algorithms. In our view, this algorithm obviously belongs under multiplicative algorithms and is closer to MART since, as mentioned above, it constitutes the first two terms of a MART2 expansion. However, as DART has a highly valuable property in RT problems, it is applicable also in the case of negative data and matrix elements, and therefore, it deserves to be singled out into a separate group. Also, block iterative algorithms [Sutton and Na, 1996a] and, simultaneous multiplicative algebraic reconstruction technique (SMART) [Kuklinski, 1997] were employed.

Among noniteration algorithms, the singular value decomposition (SVD) and modified truncated SVD [Raymund et al., 1994a; Kunitake et al., 1995; Zhou et al., 1999] should be mentioned first of all. North West Research Associates [Fremouw et al., 1992, 1994a, 1997] use the weighted damped least squares solution which they obtained from [Menke, 1990]. They use nonlocalized basis functions: sines and cosines in the latitudinal direction and empirical orthogonal functions (EOFs) in the vertical direction. The EOFs are based on model ionospheres. In this algorithm, the minimum norm principle is good enough because of the clever use of basis functions. In [Fehmers, 1996], this method was shown to be equivalent to the standard Tikhonov regularization [Tikhonov and Arsenin, 1977], where the minimum norm principle constitutes the prior information. The orthogonal decomposition algorithm [Sutton and Na, 1994; Na and Lee, 1991] incorporated the a priori information, derived from model ionospheres, in the spatial frequency domain. As in the NWRA algorithm mentioned above, the unknown electron density is assumed to be described by a model. In [Fougere, 1995], they used a maximum entropy algorithm, inserted the background information into the maximum entropy method by calculating TEC values along fictitious horizontal paths through modeled ionospheres, and used these as input together with the measured TEC. Quadratic Programming involves finding the feasible solution of a series of linear equations or inequalities for which a quadratic objective function relating the unknowns is either maximized or minimized. In addition, prior constraints or boundary values may be applied to the unknowns [Spenser et al., 1998]. A model-less, constrained algorithm [Fehmers, 1994, 1996] is noniterative, in that it searches for a weighting factor that appropriately balances the solution between features reconstructed from the data using minimum norm least squares and several constraints, including smoothness in the vertical, flatness in the horizontal, and zero along the top and bottom of the image. The algorithm is based on constrained optimization. The Ionospheric Data Assimilation 3-D algorithm was used for analyzing miscellaneous experimental data [Bust et al., 2000, 2001].

Estimation of the accuracy and limitations inherent in RT is of considerable interest for practice. The continuous inverse theory has been used by [Yeh and Raymund, 1991] to illustrate the effect of missing horizontals. In several works, a useful analysis of geometry, limitations, and resolution of RT is made [Na and Lee, 1991; Raymund et al., 1994b; Na and Sutton, 1994; Mitchell et al., 1997b; Sutton and Na, 1995, 1996b; Biswas and Na, 2000; Mitchell et al., 1997a,c; Na et al., 1995; Leitinger, 1996].

2.4 Ray Radio Tomography Problems and Assembly of Solutions

In its original formulation, the ray RT problem reduces to a system of linear equations like (2.1) which can be denoted in terms of a projection operator $\hat{P}$ mapping continuous 2-D distribution $\boldsymbol{x}$ (electron density N or product $N\nu$) into a set (along different rays) of phase differences or logarithmic relative amplitude – the measured quantities $\boldsymbol{m}$

$$\hat{P}\boldsymbol{x} = \boldsymbol{m}\,. \tag{2.23}$$

On the other hand, after discretization of the system of linear integral equations (2.23), we solve SLE of type (2.22) with a discrete operator $\hat{A}$. Such a replacement, of course, is not equivalent. If the system of linear integral equations (2.23) quite exactly represents the real experimental situation, then the use of (2.22) instead of (2.23) is associated with a noticeable error. Estimations to be given below show that relative accuracy of representation of the RT problem by a system of linear integral equations (2.23) is of the order of (10^{-3}–10^{-4}). The representation error here is related to the experimental noise, the noises caused by scattering from small-scale irregularities, the diffraction effects, more exactly – the remnants of these effects remaining after appropriate filtering. All of these quantities constitute a quasi-noise component ξ that can be additionally augmented to the right-hand part of (2.23). With this component taken into account, system (2.23) can be represented as

$$\hat{P}\boldsymbol{x} = \boldsymbol{m} + \boldsymbol{\xi} \quad \text{or} \quad \hat{A}\boldsymbol{x} = \boldsymbol{m} + \boldsymbol{\xi} + (\hat{A}\boldsymbol{x} - \hat{P}\boldsymbol{x})\,. \tag{2.24}$$

From this it follows that, when solving system (2.22) instead of SLE (2.23), we neglect the additional term $\boldsymbol{\zeta} = (\hat{A}\boldsymbol{x} - \hat{P}\boldsymbol{x})$ in the right-hand part of (2.24) that is an approximation error and depends on the very solution $\boldsymbol{x}$.

In other words, in SLE (2.22), a summary error $\varepsilon = \xi + \zeta$ is always present that contains both the quasi-noise component ξ and the correlated (in time and rays) error of approximation. Note that the error ζ usually exceeds the level of quasi-noise component ξ.

To estimate the value of ζ, an appropriate simulation was carried out. Figure 2.21a portrays the model ionosphere containing a trough, two irregularities on both sides of the trough, and finger-like structures. Structures like this are often observed in real experiments; they represent a difficult test for RT simulation due to steep gradients present at the edges of the trough and finger-like structures at high altitudes. Figure 2.21b shows the relative error of approximation $\psi = \zeta/m$ as a function of satellite position τ, the signal being measured by a third receiver. In all, four receivers with coordinates $\tau = 0$, 300, 800, 1200 km have been employed here. The dash-dotted line depicts the relative error ψ in the piecewise-constant approximation of the projection operator, and the solid line represents the error corresponding to a piecewise-planar approximation. Figure 2.22a portrays another model of the ionosphere with two closely spaced oblique troughs and several irregularities. The error $\psi = \zeta/m$ as a function of satellite position τ (the signal being received by the third receiver) is shown in Fig. 2.22b. The error corresponding to a piecewise-constant approximation of the projection operator is depicted as a dash-dotted line, and the error in the piecewise-planar approximation is drawn as a solid line. Simulation results show that relative error of approximation ψ depends strongly on the order of approximation. In the piecewise-planar approximation, it varies within 1–2%, but with the piecewise-constant approximation used, it reaches 10%. Errors for bilinear approximation (not shown in the figures) are close to those for piecewise-planar approximation (usually a bit smaller). Errors of approximation by a product of cubic splines are less, but the matrix properties get worse, as it has been already mentioned in Sect. 2.2, and the results of solving the RT problem are generally not better than those obtained with bilinear or piecewise-planar approximation.

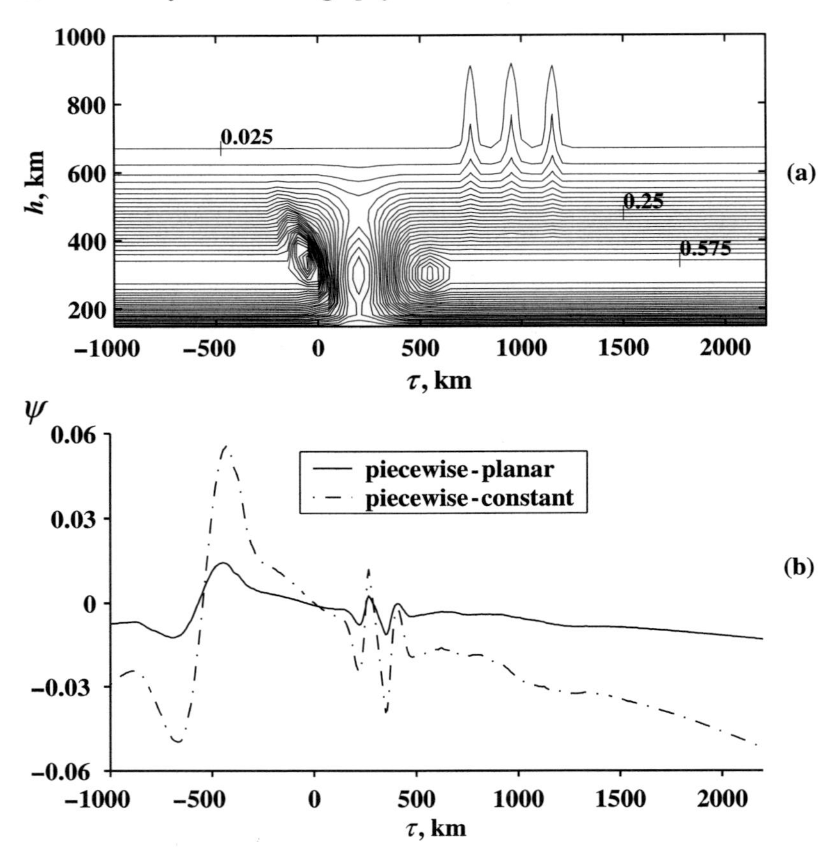

Fig. 2.21. A model of the ionosphere (**a**) and approximation relative error ψ (**b**)

A great many of algorithms for solving SLE are known; some of them we used in the modeling and enumerated in the previous section. The use of different SLE solution algorithms will yield different results of reconstruction. This is illustrated in Figs. 2.23 (a–c), 2.23 (d–f) portraying the results of reconstruction of the model shown in Fig. 2.20a. In the simulation, phase-difference RT with bilinear approximation was used for constructing the projection operator approximation. Next, the resulting SLE was solved by various algorithms. Figure 2.23 shows the results of reconstruction obtained with the following algorithms: (a) ART without relaxation and ART2, (b) SIRT, (c) ART with relaxation, (d) ART with relaxation and SIRT, (e) DART, and (f) ART with relaxation and ART2. It can be seen that all of the reconstructions are similar in general but differ in details. The trough and the two irregularities on both sides of it are reconstructed in all reconstructions (except for Fig. 2.23b where SIRT was used). The three finger-like structures are also reconstructed in all cases with Fig. 2.23e excluded where

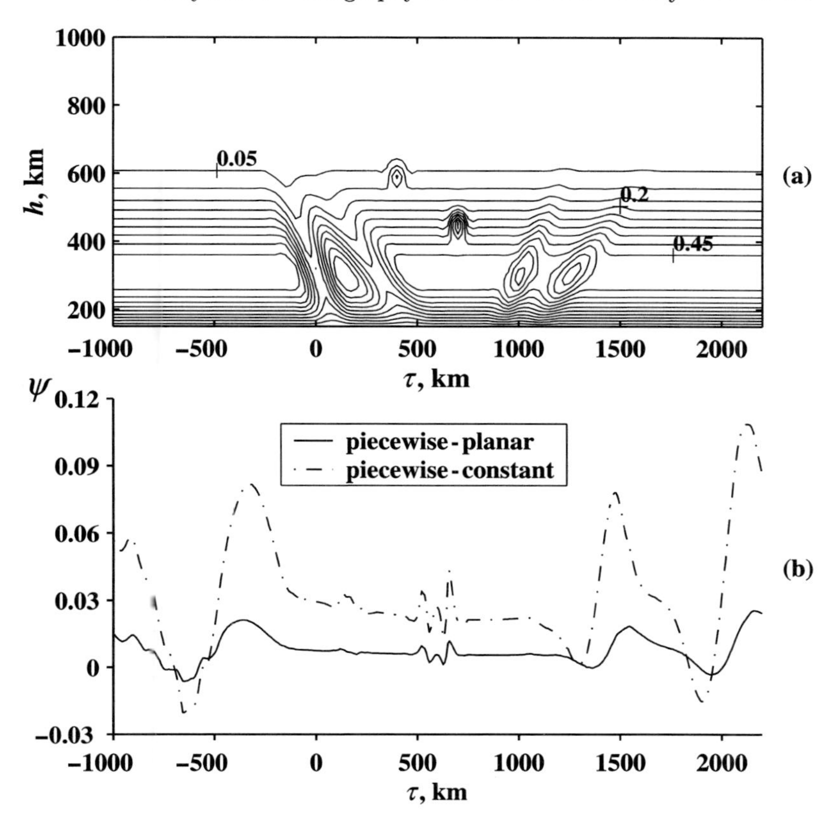

Fig. 2.22. Another model of the ionosphere (**a**) and the corresponding relative error of approximation (**b**)

DART was employed. However, the distortions and artifacts arising differ significantly. Modeling results [Kunitsyn et al., 1994b, 1995a; Andreeva et al., 2001a] showed that the choice of RT method (phase RT, phase-difference RT, or RT based on relative phase) and the choice of projection operator approximation affect the reconstruction results more strongly than the choice of the SLE solution algorithm.

It becomes clear from the above discussion that a lot of options are available to obtain RT reconstructions. Three methods exist for using phase data. To those mentioned, also a method can be added of RT based on Faraday rotation [Ganguly et al., 2001] which is close to phase methods. There are 5 to 10 various possibilities for choosing an approximation of the projection operator. From 10–20 methods for SLE solution applicable to RT problems are reported in the literature. More promising and useful is to employ combinations of these methods with various regularization procedures; in this case, the total number of combinations will reach 10^2 and even more. The

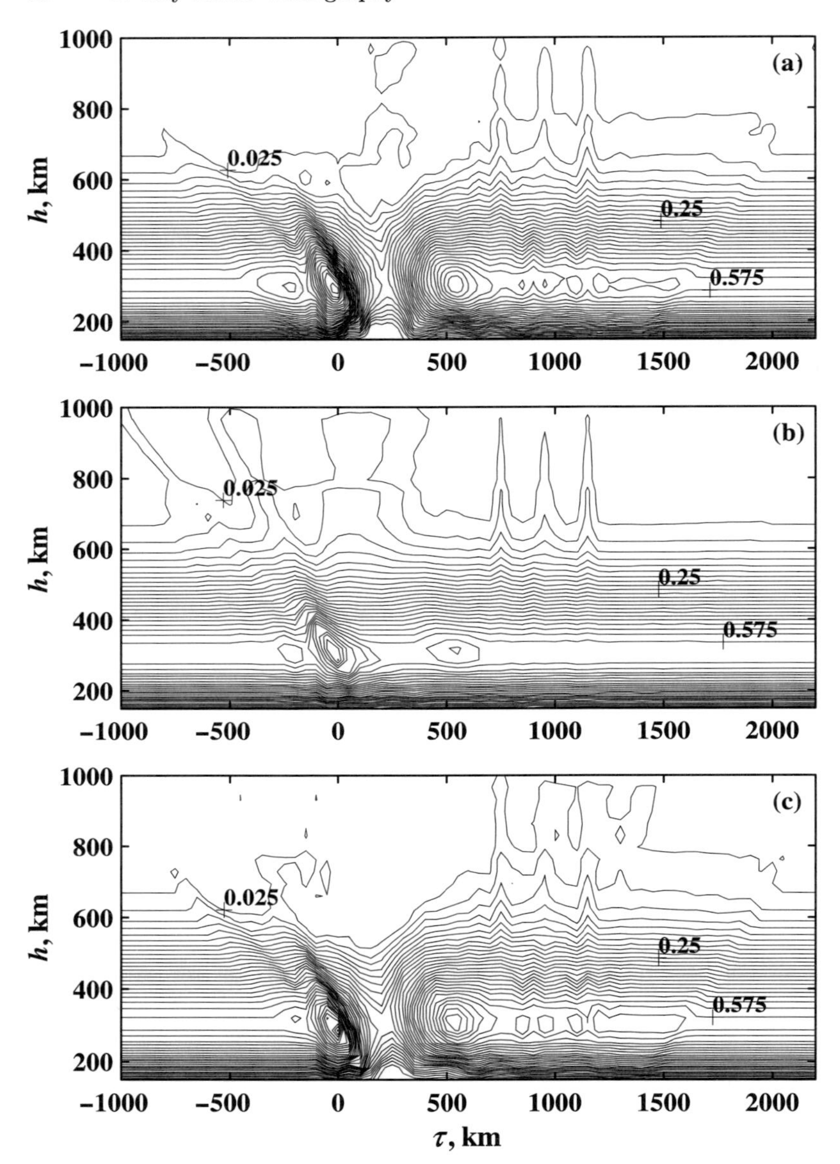

Fig. 2.23. Phase-difference RT reconstruction of the model in Fig. 2.21 obtained with different combinations (**a**, **b**, **c**) of SLE solution algorithms

choice of additional conditions (discretization grid, initial guess, and others) may amount to 10^1–10^2 scenarios. Thus, the total number of possibilities for obtaining solutions with various combinations in the triad "method-operator-algorithm" together with additional conditions amounts to a few thousand

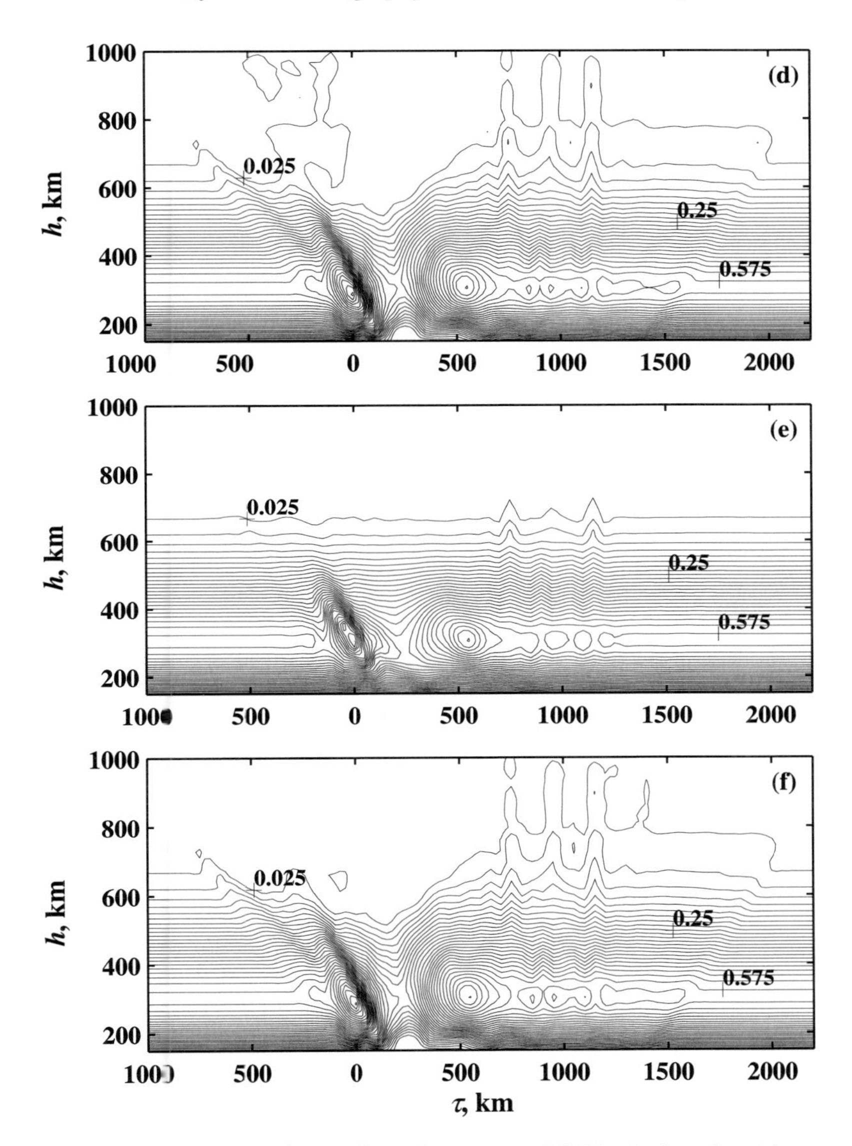

Fig. 2.23. Continued. (**d**, **e**, **f**) combinations of SLE solution algorithms

up to several hundred thousand. Any particular choice of the method, the operator, and the algorithm realizes one of several thousands of possible solutions of an RT problem. In the sense of their correspondence to the experimental data, all of these solutions are equivalent. Which one is preferable, and is it possible to point to a certain single "best" solution? The possible way to solve this question is to construct an ensemble averaged solution of the given RT problem by using various combinations in the space "method-operator-algorithm" together with additional conditions. In this case, we will obtain an ensemble-averaged solution which will be practically independent on an addition of a few new terms to the assembly. The assembly built also makes it possible to get an estimate of the solution error, the "variance," or root-mean-square deviation (rms) of the solutions of the given problem. Figure 2.24a illustrates the ensemble averaged solution of the RT problem for the model in Fig. 2.20, and the distribution of rms deviation for an assembly of more than a hundred solutions is shown in Fig. 2.24b [Andreeva

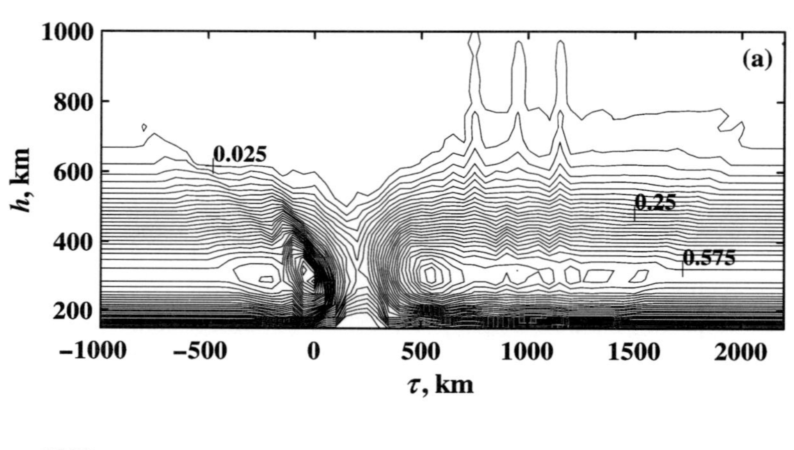

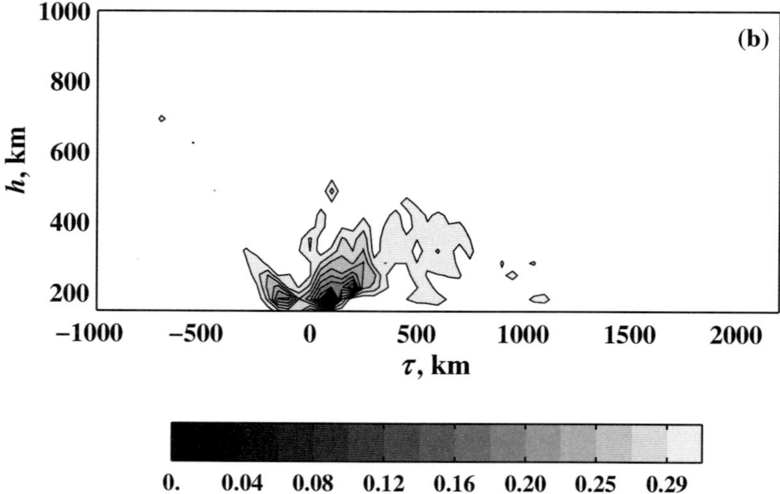

Fig. 2.24. Ensemble-averaged solution of the ray RT problem for the model in Fig. 2.20 (**a**) and rms of the assembly of solutions (**b**)

et al., 2001a]. Averaging was carried out by a summation with the weights determined in accordance with errors in the right-hand part of (2.22). A low rms points at a disappearance of artifacts that takes place when averaging the solutions, because the artifacts associated with diverse methods, operators, and algorithms differ. The rms error is high only within regions of steep gradients in the vicinity of the trough where it reaches 29%.

The ensemble-averaged solution, as we will show below, is the best one in the given assembly of reconstructions. In the modeling, it is possible to calculate errors of all reconstructions obtained, the assembly members, and the ensemble-averaged solution. The horizontal axis σ_2 in Fig. 2.25 represents the rms deviation of the reconstructed phase difference from the model phase difference, normalized by the maximum phase-difference value and calculated in percent. In an experimental situation, the model phase difference is replaced by the measured one, that is, σ_2 can be determined both by modeling and in the experiment. The vertical axis σ_1 is the rms deviation of the reconstructed electron density distribution and can be found only in simulation but not in the experiment. In the modeling, each reconstruction of the assembly is mapped as a single point on a given plane. In all, more than a hundred points corresponding to all assembly members are displayed in the figure [Andreeva et al., 2001a]. In the calculation of each reconstruction, the iteration procedure continued until σ_2 stopped decreasing or reached some certain value of the order of 12%, where, according to the results of our sim-

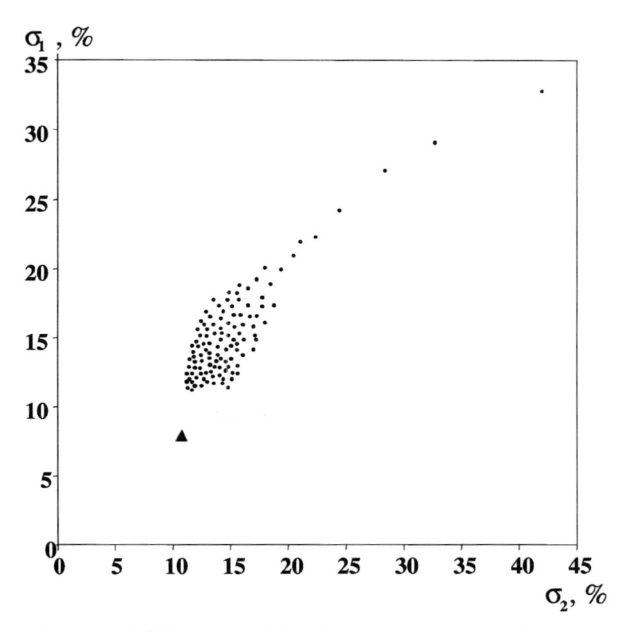

Fig. 2.25. The assembly of solutions in coordinates σ_2 (rms of the input data) – σ_1 (rms of the solutions)

ulation of such ionospheric structures, further iterations no longer lessen the rms deviation of the reconstruction σ_1, that is, the solutions start to diverge.

For those assembly elements having σ_2 less than 20%, the points start to form a group like a cloud. The better choice of the projection operator or SLE solution algorithm, etc., will yield better reconstruction; however, there were no cases encountered with an error less than 12% that defines the lower error limit of all solutions for this ionospheric model. The error is caused mainly by an approximation error ("approximation noises") and depends partly on artifacts associated with solving the SLE. When all assembly elements are summarized in obtaining the ensemble-averaged solution shown in Fig. 2.24a and represented as a triangle in Fig. 2.25, the error of this ensemble-averaged solution noticeably decreases to approximately 7.5% which is less than the error of any of the assembly elements. The fact that σ_1 for the ensemble-averaged solution is the least one makes it possible to conclude that the ensemble-averaged solution is the best one in this assembly.

It is of interest to find a reason explaining why the ensemble-averaged solution is most exact. All tomographic reconstructions contain the following main components: the true distribution, distortions and artifacts due to incompleteness of the data and the influence of summary error ε. The true distribution free of distortions can be obtained only with complete projection data available. If the data are incomplete, the reconstruction will differ from the true distribution. By virtue of incompleteness in the projection data, the artifacts are generated, and they depend on the method, the operator, and the algorithms chosen. Therefore, if one takes the ensemble-averaged solution, the true distribution is emphasized, because each assembly element contains the single uniform true distribution. On the other hand, the distortions, noises, and artifacts that depend on the methods and algorithms significantly differ in the assembly. When averaging, these method-dependent components will be neutralized and compensate for each other thus producing a more exact solution. Therefore, in our view, constructing the assembly of solutions and finding the ensemble-averaged solution makes it possible to obtain a stable solution of the RT problem and to estimate the error of this solution. It is clear that if the number of assembly elements tends to infinity, we will obtain a unique solution of the RT problem. Ensemble averaging will neutralize all of the "invisible" distributions with alternating signs; therefore also the problem of uniqueness is settled when building an assembly. From the practical point of view, it is sufficient to restrict the number of assembly elements to a hundred or a few hundred terms; in this case, the ensemble-averaged solution will practically no longer depend on an addition of some new elements to the assembly. When constructing the assembly, one should choose different combinations of "method-operator-algorithm" together with additional conditions which will cause limited errors σ_2 over the right-hand part of SLE (2.22), for example, to include in an assembly all of the solutions having errors σ_2 less than 20%. Here, the errors in all elements of the assembly should be limited in accordance with the results of reconstruction of models from a representative sampling of ionospheric structures with various irregularities. In other words, one should select a series of ionospheric models containing typical ionospheric structures (troughs, gravity waves, localized inhomogeneities, finger-like structures, equatorial anomaly crest, and so on). Then, over the chosen set of models, a hundred or more various combinations of "method-operator-algorithm" with additional conditions should be selected that give solutions with an error σ_2 of 20% or less. Note that in several models, it is quite enough to use 10 to 30 typical models since the number of types of various structures in the ionosphere is rather small. The quality of reconstruction of this or that model does not depend on the disposition of the structures in this model but depends only on the shape of structures, typical dimensions, and gradients. Therefore, it seems sufficient to include in the model only several types of the trough or crest of an equatorial anomaly, and all elements of the constructed assembly will successfully recover the structures like these, regardless of their sizes and positions. In averaging and building the ensemble-averaged solution, different elements of the assembly may have different "weights" that can be determined from the results of reconstruction of the models of the mentioned representative sampling of ionospheric structures with various irregularities.

2.5 Bayesian Approach in Ionospheric Tomography

When numerically solving integral equation (2.1), it is always reduced to a system of linear algebraic equations (2.22). One can arrange the measured results into a column-vector of measurements $\boldsymbol{m}$; the unknown values of electron density at grid points over the region of interest can be put into another column-vector $\boldsymbol{x}$ as well as unknown phase constants. After this, the problem of ionospheric tomography may be written in the form of the following equation:

$$\boldsymbol{m} = \hat{A}\boldsymbol{x} + \boldsymbol{\varepsilon}\,, \tag{2.25}$$

where $\hat{A}$ is a $m \times n$ matrix of linear coefficients and $\boldsymbol{\varepsilon}$ is a column-vector of measurement errors of $\boldsymbol{m}$. The latter could be caused by various reasons such as instrumental noise in recording the signal, the effect of small-scale irregularities on phase records, and errors due to replacing the integral equation that relates the electron density distribution to phase variations by its discrete analogue and so on.

In the Bayesian approach, the errors $\boldsymbol{\varepsilon}$ are considered random variables. Let us denote the probability density of $\boldsymbol{\varepsilon}$ by $P_{\boldsymbol{\varepsilon}}(\boldsymbol{\varepsilon})$. Obviously, $\boldsymbol{m}$ are also random variables depending on both true values of $\boldsymbol{x}$ and errors of $\boldsymbol{\varepsilon}$.

The conditional probability density of $\boldsymbol{m}$ for a given $\boldsymbol{x}$ is

$$P(\boldsymbol{m}/\boldsymbol{x}) = P_{\boldsymbol{\varepsilon}}\left(\boldsymbol{m} - \hat{A}\boldsymbol{x}\right) . \tag{2.26}$$

The unknowns $\boldsymbol{x}$ describe the state of the medium. Some random process of measurements gives us the values of $\boldsymbol{m}$. Let us assume that the conditional probability density $P(\boldsymbol{m}/\boldsymbol{x})$ is known. The aim of the statistical approach is to answer the question: what can we conclude about the $\boldsymbol{x}$ values from the measured $\boldsymbol{m}$ data?

Based on the formula for total probability [Fischer, 1976], the following expression can be written

$$P(\boldsymbol{x})\, P(\boldsymbol{m}/\boldsymbol{x}) = P(\boldsymbol{m})\, P(\boldsymbol{x}/\boldsymbol{m}) , \tag{2.27}$$

where $P(\boldsymbol{x})$ is the a priori probability density of measured $\boldsymbol{x}$, and $P(\boldsymbol{m})$ is the a priori probability density that $\boldsymbol{m}$ appears in the measurement. Therefore, using (2.27), one can write for the a posteriori probability

$$P(\boldsymbol{x}/\boldsymbol{m}) = \frac{P(\boldsymbol{x})\, P(\boldsymbol{m}/\boldsymbol{x})}{P(\boldsymbol{m})} \sim P(\boldsymbol{x})\, P(\boldsymbol{m}/\boldsymbol{x}) . \tag{2.28}$$

The denominator in (2.28) does not depend on $\boldsymbol{x}$ and is a normalizing constant that will be omitted in the further consideration. As follows from the Bayes formula (2.28), the a posteriori probability is a product of the a priori probability $P(\boldsymbol{x})$ and conditional probability $P(\boldsymbol{m}/\boldsymbol{x})$. Hence, if one of the factors in (2.28) is more informative than another, then a posteriori information is determined by this factor and is almost independent of another one. Naturally, of most interest for us is the case when the most informative factor is conditional probability $P(\boldsymbol{m}/\boldsymbol{x})$; otherwise it would make no sense to carry out experimental investigations.

Assume that $P(\boldsymbol{m}/\boldsymbol{x})$ as a function of $\boldsymbol{x}$ has a distinct sharp extremum in some space region and decreases rapidly to zero outside this area, whereas $P(\boldsymbol{x})$ only faintly varies within this volume. In this case, a posteriori probability $P(\boldsymbol{x}/\boldsymbol{m})$ will be practically independent of specific $P(\boldsymbol{x})$. Particularly, $P(\boldsymbol{x})$ may be assumed constant.

However, before we proceed to consider various methods for inserting a priori information into the solution, we should specify the shape of the function of the error probability density. Suppose that $\boldsymbol{\varepsilon}$ is not a mere random vector but a normal random vector. Note that a random normal vector is a set of random variables $\boldsymbol{\varepsilon}$ whose multidimensional density distribution is

$$P(\boldsymbol{\varepsilon}) = \frac{1}{(2\pi)^{m/2}\left|\hat{\Sigma}\right|^{1/2}} P\left(-\frac{1}{2}\,\boldsymbol{\varepsilon}^{\mathrm{T}}\hat{\Sigma}^{-1}\boldsymbol{\varepsilon}\right) , \tag{2.29}$$

where matrix $\hat{\Sigma}$ is determined by the second product central momentum of random variable $\hat{\varepsilon}$ [Mudrov and Kushko, 1976]. Supposing the zero mean value of $\boldsymbol{\varepsilon}$, i.e. $\langle \boldsymbol{\varepsilon} \rangle = 0$, one can write for Σ_{ij}

$$\Sigma_{ij} = \langle \boldsymbol{\varepsilon}\boldsymbol{\varepsilon}^{\mathrm{T}} \rangle_{ij} = \int \mathrm{d}\boldsymbol{\varepsilon}\ \varepsilon_i \varepsilon_j P(\boldsymbol{\varepsilon}) \ .$$

$\hat{\Sigma}^{-1}$ is an matrix inverse to $\hat{\Sigma}$. In the particular case when errors of different measurements are independent,

$$\Sigma_{ij} = \delta_{ij}\, \sigma_i^2 \ ,$$

where δ_{ij} is the Cronecker symbol and σ_i is the variance, one can write the two following relations:

$$\Sigma_{ij}^{-1} = \delta_{ij} \frac{1}{\sigma_i^2} \ , \qquad \left| \hat{\Sigma} \right| = \prod_{i=1}^{m} \sigma_i^2 \ ,$$

and the distribution may be represented as a product of normal distributions

$$P(\boldsymbol{\varepsilon}) = \prod_{i=1}^{m} \left(2\pi\sigma_i^2 \right)^{-1/2} P\left(-\frac{\varepsilon_i^2}{2\sigma_i^2} \right) .$$

With (2.29) taken into account, the a posteriori probability may be written as

$$P(\boldsymbol{x}/\boldsymbol{m}) \sim P(\boldsymbol{x})\, P\left[-\frac{1}{2} \left(\boldsymbol{m} - \hat{A}\boldsymbol{x} \right)^{\mathrm{T}} \hat{\Sigma}^{-1} \left(\boldsymbol{m} - \hat{A}\boldsymbol{x} \right) \right] . \tag{2.30}$$

Next, some reasonable assumptions about a priori information $P(\boldsymbol{x})$ should be made. If no information on $P(\boldsymbol{x})$ is available, one could accept the assumption that appears most natural in that specific case. The simplest assumption is that $P(\boldsymbol{x})$ does not depend on $\boldsymbol{x}$, i.e., $P(\boldsymbol{x}) = \text{const}$.

Let us see where $P(\boldsymbol{x}/\boldsymbol{m})$ in this case attains its extremum. Introduce a function

$$S(\boldsymbol{x}) = \left(\hat{A}\boldsymbol{x} - \boldsymbol{m} \right) \hat{\Sigma}^{-1} \left(\hat{A}\boldsymbol{x} - \boldsymbol{m} \right) . \tag{2.31}$$

Hence, the extremum condition for $P(\boldsymbol{x}/\boldsymbol{m})$ is $\dfrac{\partial S(\boldsymbol{x})}{\partial \boldsymbol{x}} = 0$. Taking into account that

$$\frac{\partial S(\boldsymbol{x})}{\partial \boldsymbol{x}} = 2 \left(\hat{A}^{\mathrm{T}} \hat{\Sigma}^{-1} \hat{A} \right) \boldsymbol{x} - 2\hat{A}^{\mathrm{T}} \hat{\Sigma}^{-1} \boldsymbol{m} \, ,$$

the a posteriori probability will reach the extremum at

$$\boldsymbol{x}_0 = \hat{Q}^{-1} \left(\hat{A}^{\mathrm{T}} \hat{\Sigma}^{-1} \boldsymbol{m} \right) , \tag{2.32}$$

where the matrix $\hat{Q} = \hat{A}^{\mathrm{T}} \hat{\Sigma}^{-1} \hat{A}$ is a Fisher information matrix and $\hat{Q}^{-1}$ is a matrix inverse to $\hat{Q}$. The solution of (2.32) corresponds to that obtained by the least-squares method [Mudrov and Kushko, 1976].

Let us rewrite the equation for the a posteriori probability using (2.32):

$$P(\boldsymbol{x}/\boldsymbol{m}) \sim P\left[-\frac{1}{2}\left(\boldsymbol{x}-\boldsymbol{x}_0\right)^{\mathrm{T}} \hat{Q}\left(\boldsymbol{x}-\boldsymbol{x}_0\right)\right]. \tag{2.33}$$

It follows from (2.33) that

$$\left\langle \left(\boldsymbol{x}-\boldsymbol{x}_0\right)\left(\boldsymbol{x}-\boldsymbol{x}_0\right)^{\mathrm{T}} \right\rangle = Q^{-1};$$

therefore the Fisher information matrix gives an estimate for the errors of $\boldsymbol{x}$ or the variance of the $\boldsymbol{x}_i$ component of $\boldsymbol{x}$:

$$\left\langle \left(\boldsymbol{x}_i-\boldsymbol{x}_{0i}\right)^2 \right\rangle = Q_{ii}^2 .$$

The next step is to set a less trivial a priori distribution of $P(\boldsymbol{x})$. In particular, let us suppose that the a priori probability is a zero-mean Gaussian with its variance equal to $1/\lambda$ where λ is some not large positive number. In this case, the a posteriori probability $P(\boldsymbol{x}/\boldsymbol{m})$ takes the shape

$$P(\boldsymbol{x}/\boldsymbol{m}) \sim P\left[-\lambda \boldsymbol{x}^2 - S\left(\boldsymbol{x}\right)\right]. \tag{2.34}$$

The derivative of the power index is proportional to

$$-2\left(\lambda \hat{I}+\hat{Q}\right)\boldsymbol{x}+2\hat{A}^{\mathrm{T}}\hat{\Sigma}^{-1}\boldsymbol{m}$$

and accordingly becomes zero at

$$\boldsymbol{x}_0 = -\left(\hat{Q}+\lambda\hat{I}\right)^{-1}\hat{A}^{\mathrm{T}}\hat{\Sigma}^{-1}\boldsymbol{m}. \tag{2.35}$$

Comparing (2.35) with (2.32), one can see that the matrix $\hat{Q}$ contained in solution (2.32) is replaced by $(\hat{Q}+\lambda\hat{I}\,)$ which is just a Tikhonov regularization of the solution [Tikhonov, 1963]. Therefore, it seems not quite correct to divide methods for solving algebraic equations into those using regularization and those without regularization. The case is merely that the methods using regularization use nontrivial a priori information. The necessity to use regularization in solving inverse problems is caused by the fact that usually an algebraic system, to which the initial problem is reduced, is ill-conditioned and their solutions in the form of (2.32) depend strongly on both variations of errors in the coefficients and on measurement errors.

To illustrate how Tikhonov regularization affects the solution of a system, let us consider a specific case when all of the variances of errors are equal and their covariance matrix is diagonal. Then $\hat{\Sigma}=\sigma^2\hat{I}$, $\hat{\Sigma}^{-1}=\hat{I}/\sigma^2$, $\hat{Q}=\hat{A}^{\mathrm{T}}\hat{\Sigma}^{-1}\hat{A}=\hat{A}^{\mathrm{T}}\hat{A}/\sigma^2$, and from (2.32), one can obtain

$$\boldsymbol{x}_0=\sigma^2\left(\hat{A}^{\mathrm{T}}\hat{A}\right)^{-1}\left(\frac{\hat{A}^{\mathrm{T}}\hat{I}\boldsymbol{m}}{\sigma^2}\right)=\hat{A}^{-1}\left(\hat{A}^{\mathrm{T}}\right)^{-1}\hat{A}^{\mathrm{T}}\boldsymbol{m}=\hat{A}^{-1}\boldsymbol{m},$$

which is the result of inverting of $\hat{A}\boldsymbol{x} = \boldsymbol{m}$. The solution of equations of this type is usually constructed as a linear transform

$$\boldsymbol{x}_0 = \hat{H}\boldsymbol{m}\,. \tag{2.36}$$

Substitution of $\boldsymbol{m} = \hat{A}\boldsymbol{x}$ in (2.36) gives

$$\boldsymbol{x}_0 = \hat{R}\boldsymbol{x}\,,$$

where $\hat{R} = \hat{H}\hat{A}$ is a resolution matrix. If $\hat{R}$ is an identity matrix, the solution is unambiguous, and all incomplete parameters are solvable. The row of $\hat{R}$ determines the flatness of the parameter corresponding to this row. Therefore, calculation of $\hat{R}$ allows us to estimate the flatness of the solution.

Let us now show that an increase in the regularization parameter results in a more smoothed solution. One can write for the solution in the form (2.35)

$$\hat{R} = \left(\hat{Q} + \lambda\hat{I}\right)^{-1} \hat{A}^{\mathrm{T}}\hat{\Sigma}^{-1}\hat{A} = \left(\frac{\hat{A}^{\mathrm{T}}\hat{A}}{\sigma^2} + \lambda\hat{I}\right)^{-1} \frac{\hat{A}^{\mathrm{T}}\hat{A}}{\sigma^2}$$

$$= \hat{I} - \lambda\sigma^2\left(\lambda\sigma^2\hat{I} + \hat{A}^{\mathrm{T}}\hat{A}\right)^{-1}.$$

Hence, an increase in λ leads to a higher deviation of the resolution matrix from an identity matrix.

We described above the Tikhonov regularization based on the substitution of $\hat{Q} + \lambda\hat{I}$ instead of $\hat{Q}$. This method can be extended to the generalized case when an appropriate nondiagonal matrix is substituted instead of diagonal matrix $\hat{I}$ [Nygrén et al., 1997]. Regularization is necessary to avoid sharp jumps of inversion results from one point to another and thus to make the solution steady. Therefore, one should develop a method for reducing such jumps.

Let x_i and x_j be values of the electron density in some neighboring elementary cells of the grid over the region under study. If the difference between the values of the electron density is supposed to be close to zero, one can write, with an accuracy of some error $\varepsilon^{(ij)}$,

$$0 = x_i - x_j + \varepsilon^{(ij)}\,. \tag{2.37}$$

Extending this equation over all horizontal and vertical differences between neighboring grid points, the equation can be written in matrix form:

$$\boldsymbol{0} = \hat{A}_{\mathrm{r}}\boldsymbol{x} + \boldsymbol{\varepsilon}_{\mathrm{r}}\,. \tag{2.38}$$

In this equation, $\hat{A}_{\mathrm{r}}$ is a matrix with elements $+1$, -1, and 0 set at appropriate positions so that each component of (2.38) is of the form (2.37). Combining (2.38) and (2.25), we obtain the system

$$\begin{pmatrix} \boldsymbol{m} \\ \boldsymbol{0} \end{pmatrix} = \begin{pmatrix} \hat{A} \\ \hat{A}_{\mathrm{r}} \end{pmatrix} \boldsymbol{x} + \begin{pmatrix} \boldsymbol{\varepsilon} \\ \boldsymbol{\varepsilon}_{\mathrm{r}} \end{pmatrix}. \tag{2.39}$$

If $\boldsymbol{\varepsilon}_{\mathrm{r}}$ is assumed to be a random zero-mean Gaussian variable, the further solution procedure is identical to that described at the beginning of this section, given $\hat{A}_{\mathrm{r}} = \hat{0}$ and $\boldsymbol{\varepsilon}_{\mathrm{r}} = 0$.

Using the notation $\hat{\Sigma}_{\mathrm{r}} = \langle \boldsymbol{\varepsilon}_{\mathrm{r}}\, \boldsymbol{\varepsilon}_{\mathrm{r}}^{\mathrm{T}} \rangle$, one can write an expression for the a posteriori probability density in a way similar to (2.30):

$$P\left[\boldsymbol{x}/\begin{pmatrix}\boldsymbol{m}\\ \boldsymbol{0}\end{pmatrix}\right] \sim P\left[-\frac{1}{2}\left(\hat{A}_{\mathrm{r}}\boldsymbol{x}\right)^{\mathrm{T}} \hat{\Sigma}_{\mathrm{r}}^{-1}\left(\hat{A}_{\mathrm{r}}\boldsymbol{x}\right)\right]$$
$$\times P\left[-\frac{1}{2}\left(\boldsymbol{m}-\hat{A}\boldsymbol{x}\right)^{\mathrm{T}} \hat{\Sigma}^{-1}\left(\boldsymbol{m}-\hat{A}\boldsymbol{x}\right)\right]. \qquad (2.40)$$

The most probable values of the unknowns obtained from this function will be of the following form:

$$\boldsymbol{x}_0 = \left(\hat{Q} + \hat{A}_{\mathrm{r}}^{\mathrm{T}} \hat{\Sigma}^{-1} \hat{A}_{\mathrm{r}}\right)^{-1} \hat{A}^{\mathrm{T}} \hat{\Sigma}^{-1} \boldsymbol{m}. \qquad (2.41)$$

Comparing (2.41) with (2.35), one can see that the regularization matrix is no longer diagonal. Comparison of (2.30) and (2.40) shows that the first Gaussian factor in (2.40) may also be thought of as insertion of the a priori information into calculations.

The mathematical formalism described implies that the most probable value of variations of electron density from one point to another is zero. Notice that positive and negative changes in electron density are equiprobable but the magnitudes of these are controlled by a priori variances composing the $\hat{\Sigma}_{\mathrm{r}}$ matrix. If the variance of electron density corresponding to some specific step within the grid is not high, it is quite likely that the changes in electron density will be not large, too. In the case of high variances, large changes in electron density are possible. Hence, by choosing an appropriate (this or that) regularization profile for the variances, regularization can be used as an advantageous tool to control the shape of an ionospheric layer.

The regularization method presented can be modified if use is made of the fact that the electron density is zero at very low and very high altitudes. Thus, two extra lines can be added in (2.38) for each grid point: one for the ground level and another for the height of the satellite, where the electron density is zero with a very small error. Adding the two rows will change $\hat{A}_{\mathrm{r}}$ and $\hat{\Sigma}_{\mathrm{r}}$ and, correspondingly, the result (2.41).

The described inversion method does not contain any positive limitation on the value of the electron concentration; therefore, negative values of electron density could appear in the reconstruction. However, these artifacts can arise in regions where the electron density is low or if too incorrect regularization densities are used. Naturally, there is no negative electron density in the ionosphere, but its appearance resulting from the inversion procedure indicates that the absolute error is higher than the density itself.

2.6 Experimental Tools in Ionospheric Radio Tomography

In the previous sections, it was shown that solving the problem of ionospheric tomography including diffraction and statistical tomography, requires data on phase and amplitude distribution of the measured field on the ground surface. However, practical implementation of phase and amplitude measurements over an area of a few square kilometers is still difficult even now, in spite of the development of modern methods for processing the information. Hence, some simplifying assumptions should be made about the features of the ionospheric volume studied.

For example, in the statistical description of the medium, it is quite natural to assume statistical homogeneity of electron density ionospheric irregularities of some specified scale sizes. With such a simplification, measurement of the field in only two perpendicular directions is quite enough for the reconstruction of the statistical parameters of the scattering medium (in particular, the correlation function of electron density fluctuations), which otherwise would require square measurements of the field. Another possible assumption is constant ionospheric structure during some time interval. With this assumption, it is possible to use the motion of the receiver (or transmitter) during an interval of stationary state of the medium and to replace the required measurements in some specified direction by those at a single point moving in this direction (i.e., to synthesize the aperture). Let us now consider one of the possible implementations of this approach based on measurement of coherent radio signals from the satellite.

The ground reception of signals from satellite radio beacons is the most commonly used method for investigating ionospheric irregularities [Davies, 1980, 1990]. This method has a lot of advantages over other methods for studying irregularities. During a comparatively short time interval, the satellite signal comes through a rather big ionospheric region, which makes it possible to measure signal scintillations caused by ionospheric irregularities of various sizes and to synthesize the aperture in the direction of the satellite movement.

Technically simplest are experiments on ray tomography. For their implementation, it is quite enough to measure the changes in the integral content of the electron density as the satellite moves over a chain of ground receivers arranged along the ground projection of the satellite path [Andreeva et al., 1990; Raymund et al., 1990]. The total electron content can be determined from the data on phase or group delay of the signal or from measurements of the changes in the wave polarization [Davies, 1990]. However, Faraday rotation [Ganguly et al., 2001] has not become widely practiced in ionospheric radio tomography. Most often used are data on ionospheric Doppler frequency shift obtained by measuring radio signals at two coherent frequencies. The advisability of receiving two coherent frequencies can be explained as follows. When a single frequency is used, besides phase changes due to passing through the ionosphere, the signal also contains a phase shift caused by the motion of the satellite. When two frequencies are received simultaneously, the higher frequency signal can be used as a reference since it is only faintly affected by ionospheric irregularities. This helps to remove the Doppler frequency of the phase changes caused by the motion and rotation of the satellite and various instabilities of the reference oscillator as well. Measurements of radio signals from a series of satellites make it possible to pass from case studies to regular daily observations capable of investigating the evolution of irregularities and the dynamics of ionospheric processes. In high-latitude experiments, it is convenient to employ polar and nearly polar orbiting satellites.

The foregoing requirements are well met by the "Transit" satellites or the Russian navigational satellites [Wood and Perry, 1980] transmitting coherently at approximately 150 and 400 MHz (with the frequency ratio 3/8 between the two carriers). The satellites have nearly circular polar orbits at about 1000 km above ground level. The 400 MHz signal can be assumed only faintly subject to diffraction at ionospheric irregularities and may therefore serve as a reference wave. Variations in a lower frequency signal caused by irregularities can be measured with respect to the reference 400 MHz wave, and thus the data can be obtained on the amplitude and phase structure of the probing wave. Just such a record of mutual influence of the two waves (reference and probing waves) is used to compose a radio hologram further employed for tomographic reconstruction.

The structure of the simplest receiving system capable of measuring a probing wave at a single ground-based receiver is shown in Fig. 2.26.

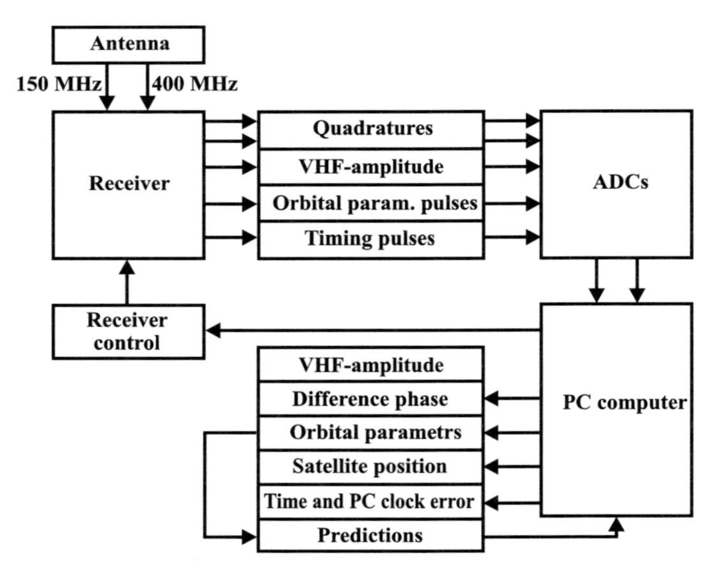

Fig. 2.26. A scheme of a RT receiver

The receiving antenna is a combined system of two simple cross-dipole pairs fixed above a ground screen which provide circular polarizations matched to those on the satellite. The receiver performs signal searching and employs analogue phase-locked loops to track the received UHF signal as its Doppler shift changes during a satellite path (typically ± 8 kHz over a pass duration of about 15 minutes). The receiver has two channels, one for each frequency. Several different channel frequency pairs at around 150 MHz and 400 MHz transmitted by different satellites, each with a 3/8 frequency ratio between the two coherent carriers, are derived by an internal frequency synthesizer which is fed by a stable oscillator. The UHF signal (employed as a phase reference) is mixed down to its 5-MHz version, and the VHF signal is converted to an intermediate frequency value of 1.875 MHz. These two RF signals come to a quadrature detector giving two final complex outputs containing both the dispersive phase and the VHF amplitude. The receiver has also a separate analogue detector of the VHF amplitude. The output band is $0 \ldots 100$ Hz. All of the analogue outputs are sampled at a rate of 50 samples per second and stored on the hard disk of an on-line PC computer.

The beacons employ 50 bit/s 150-MHz carrier modulation at 3, 5 and 7 kHz to decode Winter Moscow Time (WMT, UT+3h), the satellite's position, and its velocity in geocentric Cartesian coordinates with the corresponding rates of change and their orbital parameters (OP). The 3 and 5 kHz convey binary OP. The 7-kHz modulation frequency provides time synchronization every second. The receiver has a modulation decoding device which extracts encoded OP and positions pulses from subcarriers at 3 and 5 kHz as well as time syncronization pulses at 7 kHz which also come to the ADCs and to the PC.

A signal processing program (SPP) carrying out satellite signals recording automatically performs the following:

(1) It reads preprocessed prediction times and performs data acquisition ON/OFF.
(2) It switches the receiver synthesizer to defined frequency values corresponding to beacon carriers.
(3) It controls the ADCs and reads the outputs, the quadratures to compute the phase difference and VHF amplitude, the analogue VHF amplitude, OP, satellite positions, their changes (velocity components), and time synchronization pulses.
(4) It decodes the binary modulation words to OP values, positions, and their changes, and verifies the check sums.
(5) It computes the timing shift between satellite syncronization time and the PC's current clock value.
(6) It stores samples of the quadratures, OP, positions, their changes and PC clock error to the PC's disk as a binary raw data file.

Raw data files can be sent to the central station from tomographic sites via modem or other means. All measured parameters stored by the SPP are used later in ray tomographic analysis as well as in studying parameters of small-scale irregularities.

The prediction program creating data used by the SPP estimates the start/stop acquisition times of a satellite record as well as the azimuth and elevation of the culmination satellite position (the highest satellite elevation and the shortest distance to an observer).

Less than 3 seconds is needed for the PC (Pentium 133) to process the information received from the satellite and to switch over the waiting mode for the next record.

Despite their simplicity, measurements of satellite radio signals at a single ground-based receiver permit investigation of the irregular structure of the ionosphere. First, within the scope of a deterministic approach, single-receiver measurements can be used for location of separate scattering volumes in the ionosphere and determination of the linear sizes of irregularities in the direction of satellite motion (Chaps. 3 and 4). Second, if the medium consists of extended statistically homogeneous regions containing ionospheric irregularities, it is possible to position these regions in the ionosphere using the data on the second-order coherence functions as well as to obtain a two-dimensional cross section of the correlation function (spectrum) of electron density fluctuations (Chap. 5).

Multireceiver experiments at a chain oriented along the ground projection of the satellite trajectory are informative of the two-dimensional structure of large-scale (of hundreds km order) electron density irregularities that can be reconstructed by the tomographic method. As well, a two-dimensional cross section of the variance of electron density fluctuations can be obtained in such experiments.

Another kind of measurement consist of simultaneous reception of satellite signals at a receiving chain perpendicular to the ground projection of the satellite path, or, more generally, at a set of such chains perpendicular to the ground projection of the satellite path and parallel to each other, spaced some distance apart. In such experiments, a three-dimensional reconstruction of the irregular structure of the ionosphere becomes feasible. Naturally, implementation of these experiments is a more difficult task compared with single-receiver measurements. However, the first experiments of this kind have been already carried out in the middle and high latitudes by now [Stone, 1976a; Tereshchenko, 1987; Tereshchenko et al., 1983].

Consider the transformation of satellite radio signals in a coherent receiver. Assume that the satellite transmits two coherent radio waves at lf and mf frequencies, where f is the frequency of the radio beacon reference oscillator. For navigational satellites, l and m are 3 and 8, respectively, and f_0 is about 50 MHz. If there were no irregularities causing diffraction effects on the scattered field, then the changes in the structure of the field in the ionosphere could be described within the scope of a geometric optics approximation. In this case, the outputs of the two-channel coherent receiver are

$$\mathcal{E}_l = A_l \exp\left(-3\mathrm{i}\omega_0 t + 3\mathrm{i}\phi_0 + \mathrm{i}\frac{3\omega_0}{c}\int n_l \,\mathrm{d}\sigma_l\right) ,$$

$$\mathcal{E}_m = -A_m \exp\left(-8\mathrm{i}\omega_0 t + 8\mathrm{i}\phi_0 + \mathrm{i}\frac{8\omega_0}{c}\int n_m \,\mathrm{d}\sigma_m\right) .$$

A two-channel coherent receiver changing the frequency of the reference carrier in the ratio of 3/8 produces two quadrature outputs [Tereshchenko, 1987]:

$$I_\mathrm{c} = K_\mathrm{R}\left(E_l E^*_{m/l} + E^*_l E_{m/l}\right) ,$$

$$I_\mathrm{s} = K_\mathrm{R}\left(E_l E^*_{m/l} - E^*_l E_{m/l}\right) ,$$

where $E_{m/l}$ is the field of the reference wave mixed down to the frequency of the probing wave, that is,

$$A_m \exp\left(-3\mathrm{i}\omega_0 t + 3\mathrm{i}\phi_0 + \mathrm{i}\frac{3\omega_0}{c}\int n_m \,\mathrm{d}\sigma_m\right) ;$$

K_R is some constant defined by the parameters of the receiving system. We can assume it is unity since its value is insignificant for the further consideration. It should only remain constant during the measurement.

Thus, the quadrature components will be as follows:

$$I_\mathrm{c} = 2A_l A_m \cos\left[\frac{\omega_0}{c}\left(\int n_l \,\mathrm{d}\sigma_l - \int n_m \,\mathrm{d}\sigma_m\right)\right] = 2A_0 \cos\phi_\mathrm{d} ,$$

$$I_\mathrm{s} = 2A_l A_m \sin\left[\frac{\omega_0}{c}\left(\int n_l \,\mathrm{d}\sigma_l - \int n_m \,\mathrm{d}\sigma_m\right)\right] = 2A_0 \sin\phi_\mathrm{d} .$$

For signals transmitted from navigational satellites, $\omega_{l,m} \gg \omega_\mathrm{p} = qc^2$; hence, the refractive index can be replaced by its approximate value: $n_{l,m} \gg 1 - \omega_\mathrm{p}^2/2\omega_{l,m}^2$. Also, it can be assumed that the waves propagate along the line from the satellite to the receiver, that is, $\mathrm{d}\sigma_{l,m} \simeq \mathrm{d}L$. With these assumptions, the phase ϕ_d in the undisturbed ionosphere is completely determined by the sole ionospheric difference Doppler effect. It depends only on the total electron content in the direction of the wave propagation and can be written as follows:

$$\phi_\mathrm{d} = -\frac{55}{64}\lambda r_\mathrm{e}\int N \mathrm{d}L , \tag{2.42}$$

where λ is the wavelength of the lower frequency signal (at 150 MHz).

Note that each term $E_l E^*_{m/l}$ and $E^*_l E_{m/l}$ can be measured independently. For example,

$$E_l E^*_{m/l} = \frac{I_\mathrm{c} + \mathrm{i} I_\mathrm{s}}{2} .$$

If there are also smaller irregularities in the ionosphere with scale sizes less than or of the order of the Fresnel radius of the given experiment, then the outputs will be

$$I_\mathrm{c} = 2(A_0 + \Delta A) \cos(\phi_\mathrm{d} + \Delta\phi) ,$$
$$I_\mathrm{s} = 2(A_0 + \Delta A) \sin(\phi_\mathrm{d} + \Delta\phi) ,$$

where Δ designates the changes caused by the influence of irregularities on the probing wave. As we supposed that the reference channel at 400 MHz does not produce any distortion in the output quadrature components, all changes are attributable to the probing wave. For this case, one obtains the data required for holographic reconstruction by introducing the quantity I:

$$E_l E^*_{m/l} = (A_0 + \Delta A) \exp(\mathrm{i}\,\phi_\mathrm{d} + \mathrm{i}\,\Delta\phi) . \tag{2.43}$$

No common method exists for separating the pure contribution produced by scattering, and the trend is usually filtered out by a low-frequency filter with the bandwidth chosen in accordance with the specificity of the problem under study and the given experimental setup.

Suppose that regular variations in amplitude A_0, caused by the shift of the satellite, polarization, and other effects as well as ϕ_d, are filtered out or separated by minimizing deviations of the fitted curves from the standard ones. Then, by normalizing the amplitude by its regular value and subtracting ϕ_d from the total phase, one obtains the formula that relates measured quantities to the input data for holographic reconstruction: $E(\boldsymbol{r}_\mathrm{R}) \to E_\mathrm{H}(\boldsymbol{r})$ or $\Phi_1(\boldsymbol{\rho}_\mathrm{R}) \to \Phi_\mathrm{H}(\boldsymbol{r})$,

$$\Phi_1(\boldsymbol{\rho}_\mathrm{R}) + 1 = \frac{E(\boldsymbol{r}_\mathrm{R})}{E_0(\boldsymbol{r}_\mathrm{R})} = \frac{A_0 + \Delta A}{A_0} \exp(\mathrm{i}\,\Delta\phi) .$$

2.7 Experimental Images

To begin with, let us note that the RT reconstructions presented here are meant generally only for illustrating the possibilities of the ray RT technique. We did not strive to show and discuss in detail the geophysical aspects of the results presented, only brief relevant comments will be given.

Before we start discussing the experimental reconstructions, let us make a few remarks about the means of representing the results. Electron density distribution in a tomographic cross section is a 2-D function that can be imaged in different ways: in contours, as a surface (axonometry), in conventional colors, in particular, in gray-scale shades. Subjective perception of the reconstructed results noticeably depends on the way the results are displayed. It is difficult to give a preference to one specific method of their representation; therefore, in the general case, the analysis of RT reconstructions obtained should be based on different representations of the results. It should be pointed out only that usually conventional colors (with a large enough number of gradations ~50–200) are somewhat advantageous over the contour representation. The defect of the latter is that the image contains only comparatively few isolines (up to 10–20) and individual perception strongly depends on the choice of these gradations. Gray-scale shading also seems suitable if many shades of gray, also including those mapping weak variations of the reconstructed 2-D function, are distinguishable in the image.

In this section, we will show and discuss some of experimental reconstructions that have been obtained in different experimental campaigns since 1990. Geographical schemes of these experiments were reproduced earlier. The results will be pictured in coordinates h, km (altitude above the ground level) – geographical latitude. The date and time of the experiment (the instant when the satellite passes the middle of the section) are indicated in the present images (except for those for the experiment on heating the ionosphere where the time interval of the whole satellite pass is indicated in the reconstructions). First, RT images were obtained after experiments in March–April 1990 [Kunitsyn et al., 1990; Kunitsyn and Tereshchenko, 1991; Andreeva et al., 1990]. The receiving sites in the Russian experimental campaigns of 1990–1999 referred to here were Verkhnetulomsky (near Murmansk, 68.59°), Kem (64.95°), or Lehta (64.26°), Moscow (55.67°).

Let us start representation of the RT results with one of most interesting ionospheric structures – the main ionospheric trough. The experiments carried out showed that the trough may have various shapes; its width, slope, and depth vary within a wide range.

The appearance of the trough is portrayed in Plate 1a,b. Here, Plate 1a illustrates quite a typical case (Apr. 4, 1990, 18:59 LT) of a smooth ionosphere with a noticeable gradient in electron density from north to south. In the south, the electron concentration is about $0.93 \cdot 10^6$ m^{-3}, whereas in the north it is 2.5 times lower. Within 67–70° latitudes, the origin of the trough is observed. Near 68° latitude, 3.5 hours later (Plate 1b) a rather distinct trough arises, and the electron concentration in the north increases up to $0.5 \cdot 10^6$ m^{-3}. Next morning, a quiet smooth ionosphere is observed (Plate 1c).

The example in Plate 2 illustrates how a wide enough trough is modified within 2.5 hours. If originally (Apr. 7, 1990, 22:05 LT) the trough has a local maximum within and a steep gradient at its southern edge (Plate 2a), then in 2.5 hours (Plate 2b) the local maximum practically merges with the northern edge that is getting steeper than the southern one. The TEC appreciably decreases during these 2.5 hours to the night. Next morning, a quiet regular ionosphere with a typical gradient from north to south and no trough (Plate 2c) is encountered.

Plate 3 shows an example of the evolution of the trough observed in the experiments of 1991. Plate 3a represents an RT image of a typical trough. The trough has a rather ordinary shape with an electron density gradient at the northern edge steeper than at the southern one. Notice also a downward "bend" of the southern edge of the trough. An hour later (Plate 3b), the trough transforms – it widens. A local inhomogeneity is being detached from the "bend" of the southern edge while the northern edge moves southward. The appearance of local maxima inside the trough is not an infrequent phenomenon. At about midnight (Plate 3c), the trough becomes narrow and deep; its width is 50–100 km. Note that the revealing of the 50-km wide trough is a good illustration of the transverse resolution of the ray RT method.

By means of RT, it is also possible to observe various wave and quasi-wave structures arising in the ionosphere. Figure 2.27 depicts (in contours) the ionospheric region from Plate 3a where quasi-wave disturbances on both sides of a narrow trough are clearly seen. These disturbances are likely due to the transformation of the trough.

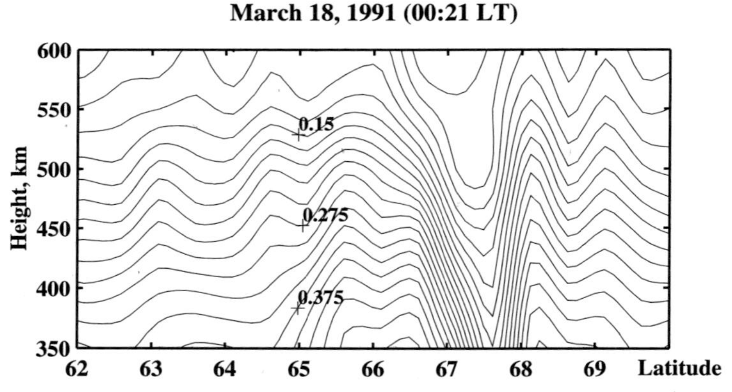

Fig. 2.27. A fragment of an RT reconstruction with wavelike disturbances in the neighborhood of a trough

An example of an almost regular ionosphere with faint perturbations arising is shown in Plate 4a. This electron density distribution was observed 25 hours 20 minutes later than that containing a narrow trough (Plate 4a) at 68.7° latitude. Here, a weak trough is scarcely noticeable at about 68.7°; that is probably a stage of a main trough transformation. This small trough and its location are quite well distinguishable in the contour plot shown in Fig. 2.28. By the evening of that day (Mar. 19, 1991, 20:01 LT), a distinct trough with a steeper northern edge appears. A steeper southern edge can be seen in the reconstruction of the trough obtained in February 1999 (Plate 4c). It is believed that gradients at the northern edge of the trough are steeper

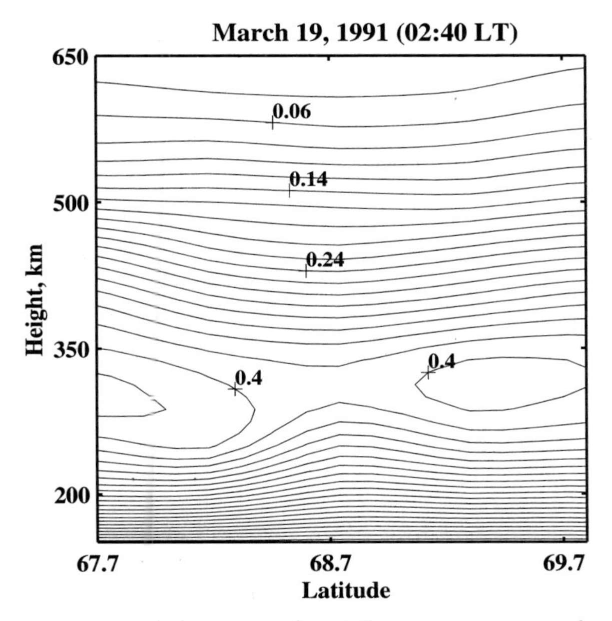

Fig. 2.28. A fragment of an RT reconstruction with a faint trough

than those at the southern edge. However, as seen in the above examples, opposite situations are also not infrequent.

The examples shown are sufficient to make one sure of the complexity and variety of the shapes of the electron density trough. An extensive series of tomographic observations is required to study the main features of this phenomenon. Note that RT makes "visible" even fine details of the trough. For example, an ionosond is not capable of detecting a narrow trough or revealing the inner structure of the trough if the radiation pattern of the ionosond transmitter is wider than the trough width.

The above images are obtained under rather quiet ionospheric conditions. During periods of disturbances, bends of the ionosphere, local extrema, and wavelike structures usually arise. Examples of RT reconstructions of the disturbed ionosphere ($K_{\mathrm{p}} > 4$) are shown in Plate 5a,b. Here, also long (200–400 km) and comparatively thin (50–100 km) fingerlike structures can be seen at altitudes from 500 to 1000 km. Similar fingerlike structures are seen in Plate 3c, too. So that one should have no doubt that these structures are not artifacts of RT reconstruction algorithms, let us note that the relative integral contribution of these patterns to the absolute phase, compared to the contribution of the main ionospheric F layer, is large enough. For instance, the relative integral contribution of the "fingers" in Plate 3c and Plate 5a,b reaches 5–20%, whereas the relative integral contribution of the structure in Plate 5b (located at 68° latitude at 600–800 km altitudes) is about 25%. The relative contribution of Doppler frequency variations is even higher, and that is why these structures are seen in the experimental data and the reconstruction algorithms are "sensitive" to these structures. Note that such fingerlike structures have been identified earlier by means of IS radar [Tsunoda, 1988]. Plate 5a,b portrays an example of a quite exotic ionosphere during a period of disturbance. Here, a region of enhanced electron density with quasi-wave perturbations is observed in the north.

In the fall of 1993, a joint Russian-American tomographic experiment (RATE) was carried out. The aim of this experiment was to compare RT reconstructions with the ionospheric cross sections obtained by IS radar in Millstone-Hill. Tomographic receivers were installed in the northwest USA and in Canada: Block Island (41.17° N), Nashua, NH (42.47° N), Jay, VT (44.93° N), and Roberval, Canada (48.42° N). The radar/tomography RATE experiment campaign window was scheduled to overlap the prediction of recurrent solar induced activity [Foster and Rich, 1998]. A severe midlatitude magnetic storm took place early on November 4, 1993 causing the planetary A index (A_{p}) to reach 77 on that day. RT proved to be useful for the observations of the early November 1993 geomagnetic storm [Knipp and others, 1998], data interpretation and integration. Radar and RT observations were made through the interval immediately preceding the storm onset and continuing through the subsequent period of rapid ionospheric trough formation and the enhancement of lower F-region ionization at midlatitudes caused by low-energy particle precipitation [Foster et al., 1994, 1998].

Plate 6a displays an example of RT reconstruction of an ionospheric cross section before the onset of the early November 1993 geomagnetic storm. In this reconstruction, a wide trough with a steep gradient at the northern edge is clearly seen. In Plate 6b (November 4, 00:45 UT), a pronounced trough is apparent near 45° latitude, and ionization enhancement is seen at altitudes between 200 km and 350 km which accompanied storm-induced particle precipitation between 46° N and 50° N geographic latitude (57°–61° N magnetic latitude). The radar field of view limits observations of the ionization enhancement to altitudes >180 km [Foster et al., 1998]. Comparison with the RT reconstruction of the ionospheric density coincident with the radar scan [Foster et al., 1994] confirmed that the radar signal enhancement in this region corresponded to a real increase in the lower F-region plasma density and was not produced by coherent backscatter accompanying increased convection of electric field strength poleward of the radar [Foster et al., 1992]. In this RT section, the uplift of the F region south of 46° latitude is also clearly seen. In [Foster and Rich, 1998], the evidence is discussed for a penetrating eastward electric field near 00:30 UT on November 4 which produced this further uplifting of the F layer and a set of ionospheric phenomena at latitudes equatorward of the trough. The RATE tomographic/radar experiments found the narrow (< 20 km) tilted trough to be at 42°–43° at 04:56 UT (Plate 6c). Note that the trough is considerably inclined to the horizon (~40–50°). In the absence of an F-region ionization source, a nighttime density depletion (such as a storm-produced midlatitude trough) will remain quite distinct even after trough formation mechanisms have ceased to operate [Foster et al., 1998].

The next three RT sections (Plate 7) illustrate rapid changes in the ionosphere that took place during the period of the early November 1993 geomagnetic storm. It can be seen that the ionospheric structure within the trough region undergoes rather fast alteration; the gradients at the equatorial and polar edges of the trough are varying quickly.

It should be pointed out that the geometry of RATE was not the best one; the extreme receivers were spaced too closely – only 800 km apart; however, in this experiment, good and interesting RT results were obtained. Comparison between RT results and the IS data demonstrated their agreement within the accuracy of both methods. In [Foster et al., 1994], a few sections are presented demonstrating the high quality of RT reconstructions. It was shown that RT has a higher horizontal resolution; in particular, in [Foster et al., 1994], it is shown that a border of about 50 km at the edge of the trough is revealed by RT but is invisible to radar. On the other hand, thin (< 50 km) extended horizontal layers are resolved by radar but not seen in the RT reconstructions. Here, inadequate (small) dimensions of the receiving system and missing quasi-tangential (to Earth) rays are telling. In general, RATE proved that RT is capable of providing high quality results dependent on the techniques applied. In our opinion, RATE demonstrated also the remarkable advantage (in sensitivity and quality) of phase-difference RT over phase RT based on linear integral determination.

A series of experimental campaigns was carried out by the Russian Polar Geophysical Institute in Scandinavia and Finland 1993–1999 [Kunitsyn et al., 1994a; Nygrén et al., 1997, 2000; Tereshchenko et al., 2000a]. RT receivers were installed in Tromsø, Norway (69.66° N, 18.95° E), Esrange; Sweden (67.89° N, 21.12° E); Kokkola, Finland (63.84° N, 23.16° E); and Nurmijarvi, Finland (60.50° N, 24.68° E). An example of an RT section in this region on January 17, 1993 is shown in Plate 8. In the first reconstruction (Plate 8a) at 11:46 UT, TIDs (travelling ionospheric disturbances) with a typical slope of about 45° are clearly seen in the southern part of the section. The evolution of a TID in the south and an electron density trough in the north can be seen about 3 hours later (Plate 8b). By 2 in the afternoon (Plate 8c), the TID relative level decreases, and disturbances almost dissipate.

Similar TIDs are not infrequent in RT reconstructions, as mentioned in [Pryse et al., 1995; Cook and Close, 1995]. Let us give a few examples. Plate 9a portrays TID observed above the same chain on January 23, 1993 (11:10 UT). The depth of ionospheric modulation by the TID here is quite large – the peak concentration in TID varies from $0.1 \cdot 10^6$ m^{-3} to $0.3 \cdot 10^6$ m^{-3}. The TID observed above the chain Moscow–Arkhangelsk on December 17, 1993 (13:40 LT) and on December 23, 1993 (13:10 LT) [Oraevsky et al., 1995] are shown in Plate 9a,b. In these cases, the depth of ionospheric modulation by TID is comparatively small – about 25–30%.

To observe equatorial ionospheric structures, a low-latitude ionospheric tomography network (LITN) was set up for a period of 3 years (1994–1996). LITN consisted of a chain of six stations from 14.6° N to 31.3° N (or 3.3° N to 19.7° N geomagnetic latitudes) at a longitude of $121 \pm 1°$ E. More detailed descriptions of the LITN and early results have been published in [Huang et al., 1997, 1999]. Applying the phase-difference method to the data from the LITN receiver network, we have been able to reconstruct, with unprecedented detail, ionospheric images spanning the northern crest of the equatorial anomaly region [Andreeva et al., 2000b].

Plate 10 represents the reconstructed ionospheric RT images from LITN data for 3 typical days. Superposed on the images are the sample geomagnetic field lines drawn in white. The analysis of RT images allowed us to reveal the main features inherent in the structure of the equatorial anomaly, which are clearly seen, for example, in Plate 10. In addition to the existence of the anomaly crest, the experimental results obtained exhibit a number of features listed in [Andreeva et al., 2000b]:

(1) The TEC crest is broader and occurs at a lower latitude than the $N_{\max}$ crest.
(2) The crest core is tilted with an approximate alignment to the geomagnetic field lines.
(3) An asymmetry exists between the equatorward edge of the anomaly region.
(4) The ionospheric thickness changes with latitude and shows a roughly two-cycle oscillation within the observed latitude range of 15°–30°.

Apparently, there exists a penetration of electrons to the lower heights about latitudinal range 26–28° N and a bite-out or constriction in the lower ionosphere at around 28–30° N. To our knowledge, the features listed above have not been reported before their publication [Andreeva et al., 2000b]. Some features, such as (1) listed above, are of considerable interest in applications related to satellite-based navigation and surveillance systems.

By means of RT, one can also investigate the dynamic aspect of the observations, especially the motion of the anomaly crest and accompanying changes in ionospheric density. The five panels of Plate 11 reveal the daytime behavior of the anomaly on October 7, 1994 [Yeh et al., 2001]. The top panel shows an anomaly crest already formed and located at 17.5° N at 11:10 LT. The crest density is $1.46 \cdot 10^{12}$ electrons/m^3. Half an hour later, at 11:40 LT shown in the second panel, the crest is moved poleward and appears at 19.0° N. The crest density increases up to $1.64 \cdot 10^{12}$ electrons/m^3, and there is also a hint of ionization tilt. At 13:00 LT shown in the middle panel, a characteristic anomaly core is completely developed. At this time, the crest density is $2.1 \cdot 10^{12}$ electrons/m^3. The tilt of the ionization contours and approximate alignment with the geomagnetic field lines now become quite distinct; the crest also appears at the highest latitude of 21.7° N. Thereafter, as shown in the last two panels, the crest density decreases, the ionization tilt gradually subsides, and the crest recedes toward the equator with time.

Using more than 350 ionospheric images reconstructed tomographically for the period October–November 1994, we studied the behavior of the anomaly crest. A general picture has emerged as follows [Yeh et al., 2001].

In response to the diurnal variations of the dynamo electric field [Fejer, 1981], the anomaly crest begins to form at around 09:00 LT. In the course of time, the anomaly crest intensifies and moves at about 1° per hour to a higher latitude. This speed is kept until shortly before noon when the poleward motion slows down and reverses at about 14:00 LT. During this time, the anomaly crest is most intense and exhibits the characteristic tilt, an approximate alignment of its core along the geomagnetic field lines, and asymmetrical behavior. Thereafter, the crest weakens and slowly recedes equatorward. On many days, the crest was observed to linger into the night with a smaller spread in latitude.

Statistically, the average crest latitude is fairly constant during the period from 12:00 to 14:00 LT, but its day-to-day variations are substantial (with a ±2° spread). An investigation of this spread shows a high correlation (correlation coefficient value of 0.70) with fountain-effect strength. Furthermore, on days when the crest forms at high latitude, a considerable increase in the total electron density in a latitudinal cross section in the anomaly region is observed. These results strongly speak for the great importance and predominance of electrodynamics in controlling the day-to-day variability in the equatorial ionosphere, compared to the causes of any other origin, be it either solar or thermospheric.

The RT method is capable not only of restoring the structure of large-scale phenomena of natural origin, and it also is a good tool for tracing artificial disturbances. Let us give some examples of RT images to illustrate ionospheric disturbances caused by artificial sources. Apparent in Plate 12 are wavelike structures that sprang up 20 minutes (Plate 12a) and 1 hour and 20 minutes (Plate 12b) after the rocket lifted off from the Plesetsk space-vehicle launching site. Plesetsk is situated 200 km aside of the plane containing the satellite path at 63° latitude. The structure of these disturbances is quite complicated; patterns with two different scale sizes are seen in the RT reconstruction where large scale ∼(200–400 km) disturbances are encountered along with smaller scale perturbations ∼(50–70 km). The "front" tilt of these wavelike structures also changes.

In RT experiments at the Murmansk–Moscow chain, long-lived local disturbances of ionospheric plasma density were detected above the site of surface industrial explosions [Andreeva et al., 2001b]. In the first case (April 4, 1990), an explosion of magnitude $M \sim 3$ was set off 100 km north-west of the northernmost receiving antenna. It is seen in Fig. 2.29 that the disturbance recorded 145 minutes after the explosion is localized in the vicinity of the explosion epicenter; the scale size of the disturbance is a few hundred

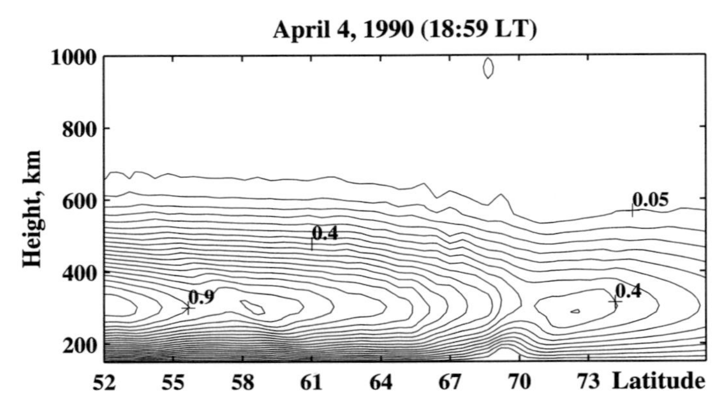

Fig. 2.29. RT reconstruction above the Murmansk–Moscow chain containing traces of ionospheric disturbances after an industrial explosion on April 4, 1990 at 16:34 LT

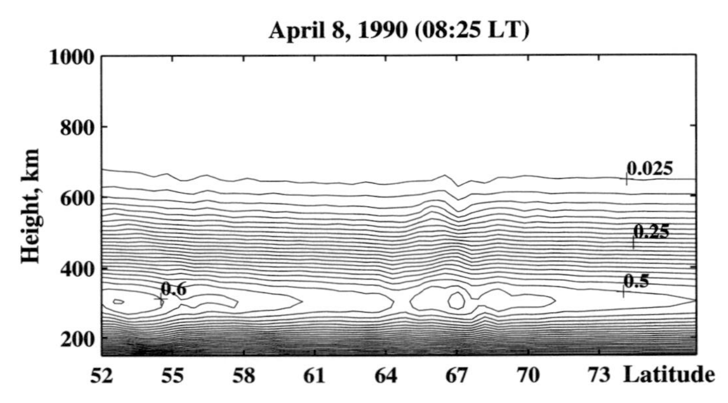

Fig. 2.30. RT reconstruction above the Murmansk–Moscow chain containing traces of ionospheric disturbances after an industrial explosion on April 8, 1990 at 07:37 LT

kilometers. In the second case (April 8, 1990), an explosion of $M \sim 2.4$ was in the immediate neighborhood of the chain approximately 100 km south of the northernmost receiving antenna. The electron density distribution at altitudes of 150–500 km observed 48 minutes after this explosion is shown in Fig. 2.30. As in the previous case, an anomaly with a characteristic scale size of a few hundred kilometers is also clearly seen in the electron density profile within the region closest to the epicenter of the explosion. If the disturbances in ionospheric plasma density observed in the two experiments described above are caused by the influence of the explosion acoustic wave on the ionosphere, we face a rather long-lasting (of the order of 10^2 min) postexplosion effect on the ionosphere. The appearance of density disturbances can be explained [Andreeva et al., 2001b] in terms of assumed turbulent (vortical) motion of a neutral component excited after passage of an acoustic impulse.

Plates 12c and 13 portray a series of successive tomographic reconstructions of 2-D electron density sections obtained in the experiment on modification of the ionosphere by powerful HF radiation transmitted by the EISCAT heating facility in November 1997 [Tereshchenko et al., 2000b]. The EISCAT Heater located in Tromsø, Norway (69.59° N, 19.22° E), was operating in the CW mode. Radiated power was up to 1.2 MW; operating frequency was mainly 4.04 MHz. The heating wave was emitted vertically upward for a 15-minute period during the satellite passes chosen with high elevations. The tomographic receiving chain consisted of five receivers and was oriented approximately along the geomagnetic meridian close to the southward satellite passes. Three receiving sites – Kårvika (69.87° N, 18.93° E), Tromsø (69.59° N, 19.22° E), and Nordkjosbotn (69.22° N, 19.54° E) spaced about 50 km apart were located in northern Norway. Measurements at these receiving sites were used for detailed RT recovery of large-scale structures of the electron density in the ionospheric region close to the Tromsø heater. Another two remote tomographic receiving sites were arranged in Kiruna, Sweden (67.8° N, 20.4° E), and Oulu, Finland (65.0° N, 25.49° E). These sites are rather distant from the Norwegian group of stations. Tomographic measurements on such an extended chain make it possible to study the large-scale structure of the ionosphere within a wide spatial interval.

Plate 12c shows an image of the electron density distribution during the satellite pass at 14:31–14:49 UT on November 9, 1997 corresponding to the period of heating the ionosphere from 14:25–14:40 UT. The white lines outline the sector of maximal heating power. As seen from the figure, no manifestations of the heating effect are seen in the ionosphere. The F layer is horizontal and regular. Heating should be most effective in the ionosphere when the heating frequency is close to the plasma frequency. According to the estimates based on both the EISCAT IS radar data and radiotomography, the plasma frequency in the peak F layer was 3.1 to 3.4 MHz, which is lower than the heating frequency 4.04 MHz. Perhaps it was an inappropriate frequency of the heating wave that was the reason for the weak heating effect in the ionosphere during this period.

The RT reconstruction of the electron density restored from the data of the satellite pass from 17:52–18:10 UT on November 9 is shown in Plate 13a. This time interval corresponds to the heating of the ionosphere carried out from 17:45–18:01 UT. The observed structure differs noticeably from that for the previous time interval. A horizontal F layer with a maximum electron density of $0.9 \cdot 10^{11}\ \mathrm{m}^{-3}$ is seen in the south at a height of nearly 300 km. In the north, the electron density decreases to $(0.3\text{–}0.4) \cdot 10^{11}\ \mathrm{m}^{-3}$. The F layer is strongly tilted within the heating cone and slightly south of it. Within this region, two inclined plasma enhancements were encountered. The maximum plasma frequency was about 2.7 MHz in the south, which is lower than the daytime values discussed above, and therefore one might assume that the heating effect was weak. However, a most remarkable fact is that the tilted region overlapped the heating cone, and this strongly speaks for a heating effect. The shift of the tilted region southward of the heating cone could be caused by transport of the plasma away from the heating site.

An ionospheric cross section at 20:26–20:44 UT on November 1997 is depicted in Plate 13b. No heating was carried out during this satellite pass. The reconstruction indicates that the ionosphere relaxed to its regular state. The F layer is nearly horizontal and only slightly structured. No strong tilt and decrease in electron density similar to that in previous case from 17:46–18:01 UT is visible here.

Another image of the electron concentration distribution is shown in Plate 13c. This cross section was obtained for the RT data measured during the satellite pass from 22:23–22:40 UT on November 9, 1997. Heating the ionosphere was carried out from 22:18–22:33 UT. In general, the structure observed is very similar to that in Plate 13a with all features more pronounced, i.e., decreased electron density within the heating cone and a steep tilt extending from the F region in the south to E-region altitudes in the north. In this case, the tilt is more clearly restricted within the heating cone than in Plate 13a. There are also inclined plasma enhancements much like those in Plate 13a, and they seem to have field-aligned orientation. The plasma frequency in this case was close to the heating wave frequency; thus the observed structures are likely to be produced by the heating.

The above series of images prove that the RT method is a versatile tool for investigating not only natural ionospheric irregularities but also those of artificial origin. It should be mentioned here that the quality of the RT reconstructions depends on what a region of the ionosphere is reconstructed. Best reconstructed is the F layer ionosphere since its contribution to the measured data is usually considerably greater than the E layer contribution. Therefore in many RT reconstructions, when the E layer contribution is insignificant, a "reduced" ionosphere is recovered with a lower boundary at 150 km. However, if the E layer contribution to the measured data is noticeable amounting to units up to dozens percent of the F layer contribution, in case of good geometry of the experiment and large enough number of receivers and dense enough receiver chain, the reconstruction of the E region ionosphere is also possible. To reconstruct the D layer ionosphere from phase data seems not possible since the contribution of this region to the measured data is negligibly small as compared to the F layer.

It should be mentioned that in the course of experimental work, the RT results have been repeatedly compared with incoherent scattering data [Raymund et al., 1993; Kersley et al., 1993; Foster et al., 1994; Pakula et al., 1995; Heaton et al., 1995; Nygrén et al., 1996b; Walker et al., 1996]. On the whole, the comparison showed good agreement of the results within the accuracy of both methods. Also RT images have been compared with ionosond data [Kersley et al., 1993; Kunitsyn et al., 1994a] in electron density profiles and critical frequencies. Here also good agreement of the results is observed.

In particular, in [Franke et al., 2003] a comparison is carried out of a few hundred RT sections in the region of the equatorial anomaly with ionosond data in Manila and Chungli; statistical data are presented showing good agreement of the results.

In conclusion, let us give a brief summary of experimental work on ray ionospheric RT. Since 1990, the ray RT method was widely adopted in many countries for investigation the large-scale structures of the ionosphere:

- in Russia [Kunitsyn et al., 1990, 1994a, 1995b; Andreeva et al., 1990, 1992; Kunitsyn and Tereshchenko, 1991, 1992; Oraevsky et al., 1993; Tereshchenko et al., 2000a,b];
- in Great Britain [Pryse and Kersley, 1992; Kersley and E.Pryse, 1994; Raymund et al., 1993; Kersley et al., 1993, 1997; Pryse et al., 1993, 1995, 1997, 1998c, 1999; Heaton et al., 1993, 1995, 1996; Mitchell et al., 1995, 1997a,c, 1998; Walker et al., 1996, 1998; Jones et al., 1997; Idenden et al., 1998; Moen et al., 1998; Spenser et al., 1998];
- in the USA and Caribbean zone [Pakula et al., 1995; Foster et al., 1994; Bust et al., 1994, 2000, 2001; Cook and Close, 1995; Kronschnabl et al., 1995; Fougere, 1997; Bernhardt et al., 1997, 1998];
- in Finland and Scandinavia [Markkanen et al., 1995; Nygrén et al., 1996a,b, 1997];
- in Southeast Asia (China, Taiwan, the Philippines, from 1994–1996) [Huang et al., 1997, 1999; Andreeva et al., 2000b; Yeh et al., 2001];
- in Japan (1994) [Kunitake et al., 1995];
- in Holland and France (1995–1996) [Fehmers, 1996; Fehmers et al., 1998].

3. Inverse Scattering Problems

3.1 Statement of Inverse Scattering Problems in Ionospheric Radio Probing

As shown in Sect. 1.2, for high sounding frequencies, vector equation (1.11) splits up into three scalar equations, and it is sufficient to consider the equation for one component of the field

$$\Delta E + k^2 \varepsilon(\boldsymbol{r}, k)\, E = 0\,, \tag{3.1}$$

where

$$\varepsilon(\boldsymbol{r}, \omega) = 1 - \frac{4\pi r_e N(\boldsymbol{r})}{k^2\left(1 + \mathrm{i}\nu_{\text{eff}}(\boldsymbol{r})/\omega\right)}\,, \qquad k = \frac{2\pi f}{c} = \frac{\omega}{c}$$

is the wave number, $\nu_{\text{eff}}(\boldsymbol{r})$ is the efficient electron collision frequency, $N(\boldsymbol{r})$ is electron concentration, and the "ion" contribution to the dielectric permittivity may be neglected [Ginzburg, 1961]. The relation for ε has the same form both in the SI and cgs systems; only the expression for r_e varies.

Next, it will be convenient to introduce a complex function ,

$$q(\boldsymbol{r}, \omega) = k^2(1 - \varepsilon) = \frac{4\pi r_e N(\boldsymbol{r})}{1 + \mathrm{i}\nu_{\text{eff}}(\boldsymbol{r})/\omega}\,, \tag{3.2}$$

and divide it into two parts: $q \to q_0(z, \omega) + q(\boldsymbol{r}, \omega)$, one corresponding to the regular stratified ionosphere with dependences $N_0(z)$ and $\nu_0(z)$ and the other connected with three-dimensional irregularities against the stratified background medium $N(\boldsymbol{r})$ and $\nu(\boldsymbol{r})$. Performing this separation in a high-frequency ($\nu_{\text{eff}} \ll \omega$) approximation, we obtain

$$\begin{aligned} q_0(z, \omega) + q(\boldsymbol{r}, \omega) &\simeq 4\pi r_e \left[\frac{N_0(z)}{1 + \mathrm{i}\nu_0(z)/\omega}\right] \\ &+ 4\pi r_e \left[N(\boldsymbol{r})\left(1 - \frac{\mathrm{i}}{\omega}\left(\nu_0(z) + \nu(\boldsymbol{r})\right)\right) - \frac{\mathrm{i}\nu(\boldsymbol{r})}{\omega} N_0(z)\right]. \end{aligned} \tag{3.3}$$

The terms in square brackets correspond to the two parts q_0 and $q(\boldsymbol{r}, \omega)$. Here, separation of efficient collision frequency is also made: $\nu_{\text{eff}} = \nu_0(z, \omega) + \nu(\boldsymbol{r}, \omega)$. The general relation for the efficient frequency of electron collisions can be written as

$$\nu_{\mathrm{eff}} = \alpha_{\mathrm{en}} N_{\mathrm{n}} + \alpha_{\mathrm{ei}} N \,, \tag{3.4}$$

where the first summand is a contribution to the efficient collision frequency of electron collisions with neutral molecules having concentration N_{n} and the second summand is due to electron–ion collisions. The values α_{en} and α_{ei} are proportional to products of the corresponding collision cross sections and average thermal electron velocity; therefore, they are temperature-dependent; as well, α_{ei} slightly depends on the electron density via the "Coulomb logarithm" [Ginzburg, 1961]. If the temperature within three-dimensional irregularities differs little from that of the ambient medium, then $\alpha_{\mathrm{en}}(z)$ and $\alpha_{\mathrm{ei}}(z)$ will be functions of only the vertical coordinate z, since the temperature in the regular ionosphere varies appreciably with altitude. Separation, as before, of the concentration of neutral molecules $N_{\mathrm{n}} \to N_{\mathrm{n}0}(z) + N_{\mathrm{n}}(\boldsymbol{r})$ and electrons $N \to N_0(z) + N(\boldsymbol{r})$ into a regular part and that associated with irregularities gives

$$\begin{aligned} \nu_{\mathrm{eff}} = \nu_0(z) + \nu(\boldsymbol{r}) = {} & [\alpha_{\mathrm{en}}(z)\, N_{\mathrm{n}0}(z) + \alpha_{\mathrm{ei}}(z)\, N_0(z)] \\ & + [\alpha_{\mathrm{en}}(z)\, N_{\mathrm{n}}(\boldsymbol{r}) + \alpha_{\mathrm{ei}}(z)\, N(\boldsymbol{r})] \,. \end{aligned} \tag{3.5}$$

In particular problems, the contributions of different summands to the efficient collision frequency (3.4) may differ considerably. For example, in the regular ionosphere at altitudes up to 300 km, the first summand predominates, whereas above 300 km the contribution of electron–ion collisions becomes much larger [Ginzburg and Rukhadze, 1975; Ratcliffe, 1972]. If the temperature of irregularities deviates noticeably from that in the ambient regular ionosphere, it is necessary to separate the disturbance in temperature from the regular background and, in accordance with this, to separate α_{en} and α_{ei}. Then the relation (3.5) for $\nu_0(z)$ and $\nu(\boldsymbol{r})$ can be easily modified according to this division. The examples given show that a particular form of relation (3.5) for $\nu_0(z)$ and $\nu(\boldsymbol{r})$ depends on the specificity of the problem.

In any case, however, the complex and real parts of $q(\boldsymbol{r},\omega)$ are defined by three-dimensional disturbances in electron density $N(\boldsymbol{r})$, neutral molecule concentration $N_{\mathrm{n}}(\boldsymbol{r})$, and collision frequency $\nu(\boldsymbol{r})$. By reconstructing the complex function $q(\boldsymbol{r},\omega)$, it is also possible to recover the structure of the mentioned three-dimensional disturbances, given the information about the regular ionosphere. Therefore, in the following discussion, a problem of reconstruction of just the function $q(\boldsymbol{r},\omega)$ will be set. The latter, in accordance with (3.3), looks like

$$q(\boldsymbol{r},\omega) = \alpha\, N(\boldsymbol{r}) - \mathrm{i}\, \beta(\omega)\; N(\boldsymbol{r})\; \tilde{\nu}(\boldsymbol{r}) \tag{3.6}$$

with known functions

$$\begin{aligned} &\alpha = 4\pi r_{\mathrm{e}} \,, \quad \beta(\omega) = 4\pi r_{\mathrm{e}}/\omega \\ &\tilde{\nu}(\boldsymbol{r}) = \nu_0(z) + \nu(\boldsymbol{r})\; (N_0(z)/N(r) + 1) \,. \end{aligned}$$

Introducing q is justified also by the fact that the quantity $q \sim 1 - \varepsilon$ refers to generalized susceptibilities [Landau and Lifshitz, 1984] describing the response of the medium to the action of the field. Moreover, the scalar Helmholtz equation assumes the form of the stationary Schrödinger equation where q is the complex potential:

$$\Delta E + k^2 E - q_0(z,\omega)\, E - q(\boldsymbol{r},\omega)\, E = \delta(\boldsymbol{r} - \boldsymbol{r}_0)\,. \tag{3.7}$$

The Delta-function in the right-hand part of (3.7) corresponds to the sounding point source used in most radio probing experiments, because within the main lobe of the radiation pattern of an arbitrary source, the wave can be assumed spherical. The dimensional constant ahead of the δ function in (3.7) is omitted so as not to be hauled into all subsequent formulas. The constant is expressed in terms of the total power $\mathfrak{P}$ of the point source, and in the dimensional equation (3.7), the right-hand part is of the form $A_0\, \delta(\boldsymbol{r} - \boldsymbol{r}_0)$, $A_0 = \sqrt{2\,\pi\,\mathfrak{P}/(c\,\varepsilon_0)}$. It can be readily seen that, in accordance with the definition of complex amplitude E, the total power is expressed in terms of the integral over the closed surface that contains the source:

$$\mathfrak{P} = \int 2\,[\boldsymbol{E}\boldsymbol{H}^*]\,\mathrm{d}\boldsymbol{S}\,.$$

Here, $\boldsymbol{H}$ is magnetic field, $\boldsymbol{E}$ is electric field, and A_0 corresponds to the amplitude value of the point source field, i.e., $E_0(\boldsymbol{r}) = -A_0 \exp(\mathrm{i}kr)\,/(4\pi r)$. If necessary, it is always possible in further considerations to regain the dimensional field $E \rightarrow A_0 E$.

The inverse problem for equation (3.7) can be formulated as follows: to recover the structure of irregularities from the field measured within a certain limited region of a given surface over a limited frequency range and limited set of probing source positions. Knowledge of the complex function $q(\boldsymbol{r},\omega)$ will make it possible to reconstruct both $N(\boldsymbol{r})$ and $\nu(\boldsymbol{r})$. Note that $q(\boldsymbol{r},\omega)$ is a finite function since the irregularities $N(\boldsymbol{r})$ are spatially limited. Such an inverse problem (IP) of reconstructing the irregularities refers to dynamic inverse problems. Also the term "inverse scattering problem" (ISP) is used. In this case, the regular stratified ionosphere is assumed known, that is, either the inverse problem of reconstructing the stratified ionosphere has been solved by other well-developed methods or its influence can be neglected as, for example, in VHF probing. Of course, instead of measuring the field, it is possible to measure other parameters of radio signals (such as phase, Doppler frequency, group delay, and so on). Such IPs based on "kinematic" parameters of radio signals are special cases of the problem stated since all of these quantities are defined by the field. Note that (3.7) also describes scattering of HF radio waves quite well, with polarization effects neglected. In other words, it is possible to state an IP for (3.7) met by polarization components of the field, in HF-radio sounding as well, if the correction for polarization is not large. Variations in the polarization components will be smaller than the components themselves even if the sounding frequency is only 1.5–2 times as high as the critical frequency.

No exact solution of the IP (3.7) for a complex potential is known. It should be pointed out that in waves of a different nature propagating in real physical media, the complexity of the potential q is essential, and it is possible to derive appropriate dispersion relations that link the real and complex parts of the potential [Landau and Lifshitz, 1984]. For example, for cold ionospheric plasma

$$\operatorname{Re} q(\boldsymbol{r},\omega)/\omega^2 = \frac{1}{\pi} \fint \frac{\operatorname{Im} q(\boldsymbol{r},\Omega)}{\Omega^2(\Omega-\omega)} \, \mathrm{d}\Omega \,, \tag{3.8}$$

$$\operatorname{Im} q(\boldsymbol{r},\omega)/\omega^2 = -\frac{1}{\pi} \fint \frac{\operatorname{Re} q(\boldsymbol{r},\Omega)}{\Omega^2(\Omega-\omega)} \, \mathrm{d}\Omega - \frac{4\pi r_{\mathrm{e}} N(\boldsymbol{r})}{\omega \nu} \,.$$

There is an extensive literature of inverse problems for remote sensing and ISP; let us only mention here a series of monographs [Newton, 1990; Lavrentiev et al., 1986; Romanov, 1987; Louis, 1989; Anger, 1990; Ramm, 1992; Colton and Kress, 1992; Isakov, 1998]. Note that basic results for the solution of the 3-D inverse problem for the Schrödinger equation with a real, frequency independent potential $q(\boldsymbol{r})$ (with $q_0(\boldsymbol{r},\omega) = 0$) were obtained in [Faddeev, 1976; Newton, 1990]. As regards complex potentials, no rigorous results have been obtained even for the 1-D inverse problem, that is, no generalization is known of the well-known Gelfand–Levitan–Marchenko algorithm for this case, more so as to the arbitrary form of energy dependency (frequency ω) of the potential [Chadan and Sabatier, 1989]. Also, no exact solution is known for the inverse problem of $q(\boldsymbol{r},\omega)$ reconstruction in the presence of a stratified background.

In this connection, of certain interest are approximate approaches to solving the ISP. Approximate consideration is justified by the properties of the regular ionosphere: the typical scale sizes here are much greater than the probing wavelength, and the irregularities are often rather weak. Therefore, a geometric optics approximation is applicable for describing wave propagation in the ionosphere in the presence of a regular background. Scatter by weak irregularities is covered by such well-known approaches as the Born approximation (BA) and the Rytov approximation (RA). To calculate scattering by strong irregularities, some special methods will be used.

3.2 Inverse Scattering Problem for Weak Irregularities

It is worthwhile at first to consider the inverse scattering problem (ISP) within a high frequency limit, where the sounding frequencies are much higher than the critical frequency of the regular ionosphere. In this case, the refractive index is close to unity, and the influence of the regular ionosphere can be neglected. Therefore, instead of (3.7), the following equation should be used:

$$\Delta E + k^2 E - q(\boldsymbol{r},\omega)\, E = \delta(\boldsymbol{r}-\boldsymbol{r}_0)\,. \tag{3.9}$$

The differential equation given is equivalent to the Lippmann–Shwinger integral equation,

$$E(\boldsymbol{r}) = E_0(\boldsymbol{r}) + \int G(\boldsymbol{r} - \boldsymbol{r_1})\ E(\boldsymbol{r_1})\ q(\boldsymbol{r}_1, \omega)\ \mathrm{d}^3\boldsymbol{r}_1 \,. \tag{3.10}$$

Here $G(\boldsymbol{r}) = -\exp(ikr)/(4\pi r)$ is the free space Green function; $E_0(\boldsymbol{r}) = G(\boldsymbol{r} - \boldsymbol{r}_0)$ is the sounding source field.

Now let the operator $(\Delta + k^2)$ act upon the left-hand and right-hand sides of (3.10). Owing to the finiteness of $q(\boldsymbol{r}, \omega)$ over $\boldsymbol{r}$ and the uniform convergence of the integral, it is possible to differentiate under the integral sign. By definition, the Green function $(\Delta + k^2)\, G(\boldsymbol{r}) = \delta(\boldsymbol{r})$; hence the Lippmann–Shwinger integral equation (3.10) and differential equation (3.9) are equivalent, naturally under the condition that convolution exists for the Green function. From representation (3.10), denoting the corresponding integral operator by $\hat{G}$, we obtain the solution of (3.9), (3.10) in the form of the Born–Neumann series:

$$\begin{aligned} E &= E_0 + \hat{G}qE_0 + \hat{G}q\hat{G}qE_0 + \dots \,, \\ \hat{G}f &\equiv \int G(r - r_1)\ f(r_1)\ d^3r_1 \,. \end{aligned} \tag{3.11}$$

The Born series will have a uniform convergence under the condition of a restricted norm of the operator $\hat{G}$ [Kreyn, 1964]:

$$\left|\hat{G}q\right| \leq \max_{\boldsymbol{r}} \int \frac{|q(\boldsymbol{r})|}{4\pi\, |\boldsymbol{r} - \boldsymbol{r}'|}\ \mathrm{d}^3 r' < 1 \,.$$

For further estimations, let us use the following condition for the maximum size r_m of the region with irregularities and typical values q of irregularities $q\, r_\mathrm{m}^2 \leq 1$ that stems from the preceding inequality. In practice, a region containing scattering irregularities always has limited dimensions, since the sounding source "illuminates" only a limited group of irregularities. Typical values of N for natural irregularities $N \sim (10^2\text{–}10^4)\ \mathrm{cm}^{-3}$, i.e., $N/N_0 \sim (10^{-2}\text{–}10^{-3})$, $q \sim 0.5(10^{-6}\text{–}10^{-4})\,\mathrm{m}^{-2}$; in this case the uniform convergence of the Born series is valid for irregularities with the dimensions $r_\mathrm{m} \sim (10^2\text{–}10^3)\,\mathrm{m}$. The above condition is sufficient and provides, of course, a rough underrated estimate of r_m. In Sect. 3.5, a more exact condition for the applicability of the Born series is given, showing that this approximation is valid up to r_m equal to tens of kilometers and more. Besides, the first term of the asymptotic series can approximate the solution quite well, even if the whole series diverges. It is natural, therefore, that in many cases, the Born series is applicable to regions of greater dimensions.

In the case under consideration ($\lambda \ll r_\mathrm{m}$), the condition $q \leq 1/r_\mathrm{m}^2 = (\lambda/r_\mathrm{m})^2 \frac{1}{\lambda^2}$ may turn out to be violated even at small $q \ll k^2$ for quite "large" irregularities. In studying wave scattering with such "large" irregularities, the Rytov approximation is usually applied. A series of theoretical and experimental evidence [Tatarskii, 1971; Rytov et al., 1989; Ishimaru, 1978; Nayfe, 1976] exists for higher exactness and wider applicability of the Rytov approximation compared with the Born approximation. The Rytov series results from renormalizing the Born series and, hence, as for other renormalizing methods, can have a wider range of uniform applicability than the Born series. This conclusion, however, has been repeatedly challenged [Nayfe, 1976]. In some applications, the Rytov method includes both the Born approximation and geometric optics as limiting cases [Rytov et al., 1989; Ishimaru, 1978]. Applicability conditions for the Rytov approximation are the following: $|\lambda\nabla_\perp\Phi_1|^2 \ll |q|/k^2 < 1$, where Φ_1 is the first approximation of the complex phase of field $E = \exp(\Phi)$, $\Phi = \Phi_0 + \Phi_1 + \ldots$. Making use of such a representation of the field for the zeroth Φ_0 and the first approximation of the complex phase of the field, we obtain the following equation from (3.9):

$$\begin{aligned} &\exp\Phi_0\left[\Delta\Phi_0 + (\nabla\Phi_0)^2 + k^2\right] = \delta(\boldsymbol{r}-\boldsymbol{r_0})\,,\\ &\Delta\Phi_1 + 2\nabla\Phi_0\nabla\Phi_1 = q(\boldsymbol{r},k)\,. \end{aligned} \tag{3.12}$$

Now, within the framework of the above mentioned approximations, we shall consider several schemes for solutions of the inverse problem of reconstructing the complex potential $q(\boldsymbol{r})$ as a function of three spatial variables. Wherever possible, solution of the inverse problem will be generalized to the case of reconstructing the function of four variables $q(\boldsymbol{r},\omega)$. The first Born approximation of the field is

$$E_1(\boldsymbol{r}) \simeq -\frac{1}{4\pi}\int \frac{\exp(\mathrm{i}k\,|\boldsymbol{r}-\boldsymbol{r}_1|)}{|\boldsymbol{r}-\boldsymbol{r}_1|}\,q(\boldsymbol{r}_1)\,E_0(\boldsymbol{r}_1)\,\mathrm{d}^3r_1\,. \tag{3.13}$$

Within the Rytov approach, the first approximation of the complex phase of the field is expressed by the formula

$$\Phi_1(\boldsymbol{r}) \simeq \frac{\exp[-\Phi_0(\boldsymbol{r})]}{-4\pi}\int \frac{\exp(\mathrm{i}k\,|\boldsymbol{r}-\boldsymbol{r}_1|)}{|\boldsymbol{r}-\boldsymbol{r}_1|}\exp[\Phi_0(\boldsymbol{r}_1)]\,q(\boldsymbol{r}_1)\,\mathrm{d}^3r_1\,. \tag{3.14}$$

Relations (3.13) and (3.14) are integral equations of the first order with respect to $q(\boldsymbol{r})$. However, at a fixed sounding frequency, the given integral operators, generally, do not have inverse matches since finite functions $q(\boldsymbol{r})$ exist reducing integrals (3.13), (3.14) to zero [Devaney, 1978].

Let us consider the solutions of (3.13), (3.14) when the field is measured in the Fraunhofer zone and the curvature of sounding waves at the scattering irregularity can be neglected, i.e., if the following conditions are satisfied:

$$R, R_0 \gg r_\mathrm{m}\,, \qquad R, R_0 \gg \frac{r_\mathrm{m}^2}{\lambda}\,, \tag{3.15}$$

where $\boldsymbol{R} = \boldsymbol{r}_\mathrm{R} - \boldsymbol{r}_\mathrm{s}$, $\boldsymbol{R}_0 = \boldsymbol{r}_0 - \boldsymbol{r}_\mathrm{s}$, $\boldsymbol{r}_\mathrm{R}$ are the coordinates of the receivers; $\boldsymbol{r}_0$ are the coordinates of the sounding source; $\boldsymbol{r}_\mathrm{s}$ are approximate coordinates of the scattering irregularity; r_m is, as before, the maximum dimension of the object (the scattering irregularity). Then it is possible to make the expansion in the exponent indexes of the kind $k\,|\boldsymbol{r}_{\mathrm{R}} - \boldsymbol{r}_1| \simeq kR - k\boldsymbol{R}\boldsymbol{r}'/R$; $\boldsymbol{r}' = \boldsymbol{r}_1 - \boldsymbol{r}_{\mathrm{s}}$. After that, E_1 and Φ_1 become proportional to the Fourier transform of $q(\boldsymbol{r})$. The transform variable $\boldsymbol{p}$ contains the coordinates of the sources and receivers $\boldsymbol{p} = k(\boldsymbol{R}/R + \boldsymbol{R}_0/R_0)$. It is possible to invert the above Fourier transform and to solve the inverse problem using the data on the field in the Fraunhofer zone,

$$q(\boldsymbol{r}) = \int W(\boldsymbol{p}) \exp(\mathrm{i}\boldsymbol{p}\boldsymbol{r})\, \mathrm{d}^3 p\,, \tag{3.16}$$

where

$$W(\boldsymbol{p}) = \frac{2E_1 R R_0}{\pi} \exp[-\mathrm{i}k(R + R_0)] \text{ with the Born approximation;}$$

$$W(\boldsymbol{p}) = -\frac{\Phi_1 R R_0}{(2\pi)^2 |R - R_0|} \exp[\mathrm{i}k\,(|R - R_0| - R + R_0)]$$

with the Rytov approximation.

The direct and inverse Fourier transforms are determined by the Plancherel theorem [Reed and Simon, 1979], since $q(\boldsymbol{r})$ is a finite piecewise continuous function and, hence, it is quadratically integrable $q \in L^2$. Moreover, according to the Plancherel–Polya theorem [Fuks, 1963], $W(\boldsymbol{p})$ is the integral function of three variables. Therefore, an unambiguous (unique) reconstruction of $q(\boldsymbol{r})$ is possible if the data on the field within some elementary volume of $\boldsymbol{p}$-space are known. This is feasible, for example, when the field is measured at a piece of the surface within a certain frequency range or when the field is measured at a piece of the surface with different positions of the source varying continuously within a certain interval.

The results of solving the inverse problem of reconstructing $q(\boldsymbol{r})$ from the field in the far zone within the first Born approximation are well known and were reported repeatedly by various authors [Fadeev, 1956; Wyatt, 1968; Schmidt-Weinmar et al., 1975; Lam et al., 1976; Devaney, 1978]. However, for the purpose of possible ionospheric applications, it is useful to make the following generalization of the given results as follows.

It should be noted that the complex potential q is reconstructed not only when q is independent on ω, as in (3.16), but also when $q(\boldsymbol{r}, \omega)$ is of the form $q(\boldsymbol{r})\,\psi(\omega)$ with the known $\psi(\omega)$, or if the real and imaginary parts of the complex integral (3.6) take the above form, as occurs in the ionospheric plasma (with $\omega \gg \nu$) $q(\boldsymbol{r}, \omega) = \alpha\, N(\boldsymbol{r}) - \beta(\omega)\, N(\boldsymbol{r})\, \tilde{\nu}(\boldsymbol{r})$. The reconstruction formulas for the real and complex parts of the potential become more complicated:

$$\begin{aligned} N(\boldsymbol{r}) &= \int \left[\frac{W(\boldsymbol{p}) + W^*(-\boldsymbol{p})}{2\alpha}\right] \exp(\mathrm{i}\boldsymbol{p}\boldsymbol{r})\, \mathrm{d}^3 p\,, \\ N(\boldsymbol{r})\, \tilde{\nu}(\boldsymbol{r}) &= \int \left[\frac{W(\boldsymbol{p}) + W^*(-\boldsymbol{p})}{2\mathrm{i}\beta(\omega)}\right] \exp(\mathrm{i}\boldsymbol{p}\boldsymbol{r})\, \mathrm{d}^3 p\,. \end{aligned} \tag{3.17}$$

Generally speaking, asymptotic approximations of function $\boldsymbol{p}$ here are not W but the integrands in square brackets. Therefore, the designation $W(-\boldsymbol{p})$ only symbolizes an "inversion" of the source and the receiver with respect to the object (the coordinates $\boldsymbol{r}_{\mathrm{s}}$) $\boldsymbol{r}_{\mathrm{s}}$) $\boldsymbol{R} \to -\boldsymbol{R}$, $\boldsymbol{R}_0 \to -\boldsymbol{R}_0$.

In this section, to derive integral transformations, we use the information on approximate coordinates of the scatterer $\boldsymbol{r}_{\mathrm{s}}$ and its maximum dimension. Location of the scattering object will be considered in Sect. 3.3.

Unfortunately, in most cases, the conditions (3.15) are violated when radio waves are scattered by large-scale irregularities of the ionosphere. Inequality $R/r_{\mathrm{m}} \gg r_{\mathrm{m}}/\lambda$ is rather rigid in view of the necessity to resolve the internal structure of an object, i.e., the necessity for condition $r_{\mathrm{m}} \gg \lambda$. In this connection, of particular interest for ionosphere radio probing are inverse problems of reconstructing the irregularities from the data on the field in the Fresnel zone where opposite inequalities R, $R_0 \leq r_{\mathrm{m}}^2/\lambda$ are possible. Of course, the field measured in the Fresnel zone can be (analytically) continued into the Fraunhofer zone, but from a practical point of view, it is useful to consider the inverse problem of reconstructing the irregularities from the data measured in the Fresnel zone. As before, initial relations for solving the inverse problem are (3.13) and (3.14). Due to the smallness of the aperture angles, the Fresnel paraxial approximation is valid. In this approximation, two different cases should be distinguished of mutual disposition of the site of measurements and the probing source. In the first case, the source and the receiving site are located on approximately opposite sides of the scatterer, i.e., the field is measured scattered almost in the forward direction. In the other case, the source and the receivers lie approximately on the same side of the scatterer and the nearly backscattered field is measured.

At first, let us consider the case of "forward" scattering for $q(\boldsymbol{r}, \omega)$ of an arbitrary form. Hereinafter we will use the Cartesian coordinates with particular axis z pointing close to the direction of the sounding wave propagation and a transverse coordinate $\boldsymbol{\rho}(x, y)$ in the plane orthogonal to z. Let the coordinates of the receivers and the sources be equal to $\boldsymbol{r}_{\mathrm{R}} = (\boldsymbol{\rho}_{\mathrm{R}}, z_{\mathrm{R}})$, $\boldsymbol{r}_0 = (\boldsymbol{\rho}_0, z_0)$, respectively. Now we make the "Fresnel expansion" of the exponent indexes of the type

$$|\boldsymbol{r}_{\mathrm{R}} - \boldsymbol{r}_1| \simeq z_{\mathrm{R}} - z_1 + \frac{(\boldsymbol{\rho}_{\mathrm{R}} - \boldsymbol{\rho}_1)^2}{2z_{\mathrm{R}} - 2z_{\mathrm{s}}}, \tag{3.18}$$

that are valid when these conditions [Kunitsyn, 1986a; Gusev and Kunitsyn, 1986] are satisfied:

$$|z_{\mathrm{R}} - z_{\mathrm{s}}|, |z_0 - z_{\mathrm{s}}| \gg \rho_{\mathrm{R}}, \rho_0, z_{\mathrm{m}}, \sqrt{\frac{r_{\mathrm{m}}^3}{\lambda}}, \sqrt{\frac{z_{\mathrm{m}}\rho_{\mathrm{R}}^2}{\lambda}}, \sqrt[1/3]{\frac{\rho_{\mathrm{R}}^4}{\lambda}}, \sqrt[1/3]{\frac{\rho_0^4}{\lambda}}. \tag{3.19}$$

Then, in accordance with (3.13), (3.14), the field and its complex phase will be proportional to Fresnel transforms of $q(\boldsymbol{r}, \omega)$:

$$V(\boldsymbol{s}, \omega) = \int \mathrm{d}^2\rho \, \mathrm{d}z \, q(\boldsymbol{r}, \omega) \exp\left[\frac{\mathrm{i}k(\boldsymbol{s} - \boldsymbol{\rho})^2}{2\zeta}\right]. \tag{3.20}$$

Here,

$$V(\boldsymbol{s},\omega) = -\frac{4\pi\zeta E_1(\boldsymbol{r}_\mathrm{R})}{E_0(\boldsymbol{r}_\mathrm{R})} \text{ for the Born approximation,}$$
$$V(\boldsymbol{s},\omega) = -4\pi\zeta\Phi_1(\boldsymbol{r}_\mathbf{R}) \text{ for the Rytov approximation.}$$

The derivative $\boldsymbol{s}$ contains the coordinates of both the receivers and the source; ζ is the reduced distance:

$$\boldsymbol{s} = \frac{z_0 - z_\mathrm{s}}{z_0 - z_\mathrm{R}}\boldsymbol{\rho} + \frac{z_\mathrm{s} - z_\mathrm{R}}{z_0 - z_\mathrm{R}}\boldsymbol{\rho}_0 \,,$$

$$\zeta = \frac{(z_0 - z_\mathrm{s})(z_\mathrm{s} - z_\mathrm{R})}{z_0 - z_\mathrm{R}} \,.$$

Geometrically, the vector $\boldsymbol{s}$ is a segment in a plane containing the object ($z = z_\mathrm{s}$) that connects the origin of the coordinates $\boldsymbol{\rho} = 0$ with the intersection of the object plane with the line of sight (the line from the source to the receiver). This can be easily seen from the equation of the straight line passing through the points of intersection $(z_\mathrm{R}, \boldsymbol{\rho}_\mathrm{R})$ and $(z_0, \boldsymbol{\rho}_0)$:

$$\frac{x - x_\mathrm{R}}{x_0 - x_\mathrm{R}} = \frac{y - y_\mathrm{R}}{y_0 - y_\mathrm{R}} = \frac{z - z_\mathrm{R}}{z_0 - z_\mathrm{R}} \,.$$

This yields the relation for the coordinates (s_x, s_y) of the vector $\boldsymbol{s}$ in the plane $(z = z_\mathrm{s})$: $\boldsymbol{s} = \boldsymbol{\rho}_\mathrm{R} + (\boldsymbol{\rho}_0 - \boldsymbol{\rho}_\mathrm{R})(z_\mathrm{s} - z_\mathrm{R})/(z_0 - z_\mathrm{R})$, which is equivalent to that introduced above. Note that the receivers (as well as the sources) are not necessarily located in the same plane z_R. If one of the receivers is located in a different plane $z_\mathrm{R}{}'$ and has the coordinates $\rho_\mathrm{R}{}'$, then its coordinates can readily be recalculated for the plane z_R without varying $\boldsymbol{s}$: $\boldsymbol{\rho}_\mathrm{R} = (z_0 - z_\mathrm{R})/(z_0 - z_\mathrm{R}{}')\boldsymbol{\rho}_\mathrm{R}{}'$. The fields at the points with equal $\boldsymbol{s}$ do coincide, as one can see from (3.20).

Making the inverse Fresnel transform of (3.20), we obtain [Kunitsyn, 1986a; Kunitsyn and Tereshchenko, 1991] the solution of the inverse problem:

$$q_z(\boldsymbol{\rho},\omega) \equiv \int q(\boldsymbol{r},\omega)\,\mathrm{d}z = \left(\frac{k}{2\pi\zeta}\right)^2 \int V(\boldsymbol{s},\omega)\,\exp\left[-\frac{\mathrm{i}k\,(\boldsymbol{s}-\boldsymbol{\rho})^2}{2\zeta}\right]\,\mathrm{d}^2 s\,. \tag{3.21}$$

The relation (3.21) shows that it is possible to solve the inverse problem of reconstructing the integral $q_z(\boldsymbol{\rho},\omega)$ in the frameworks of both the Born and Rytov approximations from the field measured in the Fresnel zone in nearly forward scattering. In this case, the given solution of the inverse problem is unique. To show this, let us consider the function $V(\boldsymbol{\sigma},\omega)$ that is an extension of the function $V(\boldsymbol{s},\omega)$ (3.20) given in the space of real variables s_x, s_y to the whole space C^2 of complex variables $\sigma_x = s_x + \mathrm{i}\,s_x{}'$, $\sigma_y = s_y + \mathrm{i}\,s_y{}'$. Since the piecewise-continuous finite (over $\boldsymbol{\rho}$) function $q_z(\boldsymbol{\rho},\omega)$ belongs to L^1, the integral in (3.11) may be considered in the ordinary sense. The region covered by the integral is limited. Whence it follows that this integral absolutely and uniformly converges in any limited region of the C^2 space; moreover, the integrand is an integral function of the variables σ_x and σ_y. Therefore $V(\boldsymbol{\sigma}, \omega)$ is an integral function of two variables. Put another way, by virtue of the main theorem of Hartogs [Shabat, 1976], the function is analytical at the point, if it is analytical in each variable taken separately. The analyticity of $V(\boldsymbol{\sigma}, \omega)$ with respect to the variables σ_x, σ_y on the whole plane σ for finite piecewise-continuous $q(\boldsymbol{r}, \omega)$ is evident. Consequently, for the unique reconstruction of $q_z(\boldsymbol{r}, \omega)$, we need the data on $V(\boldsymbol{s}, \omega)$ at a limited elementary surface of the plane $\boldsymbol{s}$, wherefrom the analytical continuation of $V(\boldsymbol{s}, \omega)$ is possible onto the whole space. The coefficients of the power series of the analytical function $V(\boldsymbol{\sigma}, \omega)$ can be obtained by calculating the derivatives in the directions of the real axes.

Thereby the analytical function $V(\boldsymbol{s}, \omega)$ is determined in the whole complex neighborhood of any point σ_x, σ_y from its values in the real neighborhood, i.e., on the open set of variables s_x, s_y that is obtained as a result of varying only the real parts of the complex variables.

It should be noted that if the Born and Rytov approximations of weak scattering are not applied and the small-angle forward scattering is treated in the Fresnel approximation, then instead of (3.20), we obtain the following integral equation:

$$E(\boldsymbol{r}) = E_0(\boldsymbol{r}) - \frac{E_0(\boldsymbol{r})}{4\pi} \int \mathrm{d}^3 r_1 \frac{q(\boldsymbol{r}_1)}{\zeta} \frac{E(\boldsymbol{r}_1)}{E_0(\boldsymbol{r}_1)} \exp\left[\frac{\mathrm{i}k}{2\zeta} (\boldsymbol{s} - \boldsymbol{\rho}_1)^2\right] . \quad (3.22)$$

In the general case, $\boldsymbol{s}$ and ζ contain, in contrast to (3.20), the integration variable z_1 instead of z_{s}, and ζ can be factored outside the integral sign only under conditions (3.10). Equation (3.22) is a paraxial approximation of the Lippmann–Schwinger equation (3.10). Similar relations will be used in further considerations of strong scatterers and statistical inverse problems. Equation (3.22) is obtained using the Fresnel small-angle expansion of the Green function

$$\begin{aligned} \frac{\exp(\mathrm{i}k\,|\boldsymbol{r} - \boldsymbol{r}_1|)}{|\boldsymbol{r} - \boldsymbol{r}_1|} &\sim \frac{\Theta(z - z_1)}{z - z_1} \exp\left\{ \mathrm{i}k \left[z - z_1 + \frac{(\boldsymbol{\rho} - \boldsymbol{\rho}_1)^2}{2(z - z_1)} \right] \right\} \\ &+ \frac{\Theta(z_1 - z)}{z_1 - z} \exp\left\{ \mathrm{i}k \left[z_1 - z + \frac{(\boldsymbol{\rho} - \boldsymbol{\rho}_1)^2}{2(z_1 - z)} \right] \right\} . \end{aligned}$$

The first term corresponds to the wave from the point source propagating in the positive direction of the z-axis; the second term corresponds to the wave propagating in the opposite direction. The latter refers to forward scattering, and it was used in (3.22), where the presence of the unit step $\Theta(z_1 - z)$ in the integrand was implied from the derivation conditions.

The Fresnel transform applied here is a pair of superexact integral transforms of the shape

$$V(\boldsymbol{s}) = \int q_z(\boldsymbol{\rho}) \exp\left[\frac{\mathrm{i}a(\boldsymbol{s}-\boldsymbol{\rho})^2}{2}\right] \mathrm{d}^2\rho\,,$$

$$q_z(\boldsymbol{\rho}) = \left(\frac{a}{2\pi}\right)^2 \int V(\boldsymbol{s}) \exp\left[-\frac{\mathrm{i}a(\boldsymbol{s}-\boldsymbol{\rho})^2}{2}\right] \mathrm{d}^2 s\,,$$

that are reduced to two-dimensional Fourier transforms by redesignating the functions $V \to V \exp(-\mathrm{i}a\, s^2/2)$ and $q_z \to q_z \exp(-\mathrm{i}a\rho^2/2)$. In our opinion, however, it is the Fresnel transform that should be applied in this case because it has a clear physical sense. The transform kernel is proportional to the Green function of the parabolic equation to which (3.7) is reduced within the paraxial approximation used in (3.18), (3.19). With this approximation, that is just the transformation of the field from one plane to another, same as the Fresnel transform is.

From the forward-scattered field it is possible to reconstruct only the two-dimensional structure of the object. This results from the great difference between the longitudinal (in the direction of the propagating sounding wave) and transversal resolutions at small-angle receiving apertures $\Theta \ll 1$. The transversal resolution is known to be proportional to λ/Θ, whereas the longitudinal is proportional to λ/Θ^2. Therefore, with the limitations (3.19) on the longitudinal scale of the object, only a two-dimensional structure can be reconstructed integrated in the direction of propagation of the probing wave.

It is worthwhile to consider the case of "backward" scattering ($z_0, z_\mathrm{R} < z_\mathrm{s}$) in the framework of the Born approximation only. Making expansions of the exponent indexes in a way similar to (3.18), we obtain from (3.13) the following:

$$\begin{aligned} U(\boldsymbol{s},\omega) &\equiv (4\pi)^2 Z_\mathrm{R} Z_0\, E_1(\boldsymbol{r}_\mathrm{R}) \exp\left\{-\mathrm{i}k\left[Z + \frac{(\boldsymbol{\rho}_\mathrm{R}-\boldsymbol{\rho}_0)^2}{2Z}\right]\right\} \\ &\simeq \int q(\boldsymbol{r},\omega) \exp\left[2\mathrm{i}kz_1 + \frac{\mathrm{i}k(\boldsymbol{s}-\boldsymbol{\rho}_1)^2}{2\zeta}\right] \mathrm{d}^3 r_1\,. \end{aligned} \tag{3.23}$$

Here $\boldsymbol{s} = (z_\mathrm{s}-z_0)\boldsymbol{\rho}_\mathrm{R}/Z + (z_\mathrm{s}-z_\mathrm{R})\boldsymbol{\rho}_0/Z$, $\zeta = Z_\mathrm{R}Z_0/Z$, $Z_\mathrm{R} = z_\mathrm{s} - z_\mathrm{R}$, $Z_0 = z_\mathrm{s} - z_0$, $Z = 2z_\mathrm{s} - z_0 - z_\mathrm{R}$. Performing the inverse Fresnel transform with respect to the variable $\boldsymbol{s}$, we get

$$\begin{aligned} \tilde{q}(\boldsymbol{\rho},\omega) &= \int q(\boldsymbol{r},\omega) \exp(2\mathrm{i}kz)\, \mathrm{d}z \\ &\simeq \left(\frac{k}{2\pi\zeta}\right)^2 \int U(\boldsymbol{s},\omega) \exp\left[-\frac{\mathrm{i}k(\boldsymbol{s}-\boldsymbol{\rho})^2}{2\zeta}\right] \mathrm{d}^2 s\,, \end{aligned} \tag{3.24}$$

i.e., the integral $\tilde{q}(\boldsymbol{\rho},\omega)$ is reconstructed from the data on the field in the Fresnel zone in backward scattering. The uniqueness of this reconstruction can be shown by the same reasoning as given above for the uniqueness of reconstructing $q_z(\boldsymbol{\rho},\omega)$. If $q(\boldsymbol{r},\omega)$ is of the form $q(\boldsymbol{r})\,\alpha(\omega)$ with real q and α, which for an isotropic ionospheric plasma corresponds to the neglect of absorption, then inversion of (3.14) is possible, naturally, if the data at various frequencies are available. Then the solution of the inverse scattering problem becomes a combination of (3.23) and (3.24) [Kunitsyn, 1986a; Kunitsyn and Tereshchenko, 1991]:

$$q(\boldsymbol{r}) = \frac{1}{\pi}\int \frac{\exp(-2\mathrm{i}kz)}{\alpha(k)}\,\mathrm{d}k\,\left(\frac{k}{2\pi\zeta}\right)^2 \int U(\boldsymbol{s},\omega)\,\exp\left[-\frac{\mathrm{i}k(\boldsymbol{s}-\boldsymbol{\rho})^2}{2\zeta}\right]\,\mathrm{d}^2 s\,. \tag{3.25}$$

An explicit inversion is also possible in the more general case when the known frequency dependence is included as a multiplier in the real and complex parts of (3.6), i.e., as in the ionospheric plasma, $q(\boldsymbol{r},\omega) = \alpha\,N(\boldsymbol{r}) - \mathrm{i}\beta(\omega)\,N(\boldsymbol{r})\,\tilde{\nu}(\boldsymbol{r})$. But here, to make inversion in the same way as for reception in the Fraunhofer zone, it is necessary to measure the field while sounding the object from "another side" (as before, the backward scattering is recorded), i.e., data are required on $V_{\mathrm{I}}(\boldsymbol{s},\omega)$ and, accordingly, $\tilde{q}_{\mathrm{I}}(\boldsymbol{\rho},\omega)$ (3.24) with the "inverted" position of the receiver-transmitter system with respect to the object:

$$N(\boldsymbol{r}) = \frac{1}{\pi}\int \left(\frac{\tilde{q}_{\mathrm{I}}+\tilde{q}^*}{2\alpha}\right)\,\exp(-2\mathrm{i}kz)\,\mathrm{d}k\,,$$

$$N(\boldsymbol{r})\,\tilde{\nu}(\boldsymbol{r}) = \frac{1}{\pi}\int \left(\frac{\tilde{q}_{\mathrm{I}}-\tilde{q}^*}{2\mathrm{i}\beta(\omega)}\right)\,\exp(-2\mathrm{i}kz)\,\mathrm{d}k\,, \tag{3.26}$$

where $\tilde{q}$ and $\tilde{q}_{\mathrm{I}}$ are determined after the Fresnel transform (3.24) of the experimental data $V(\boldsymbol{s},\omega)$ and $V_{\mathrm{I}}(\boldsymbol{s},\omega)$, respectively.

It is quite easy to justify the validity of the transition from (3.24) to (3.25) or (3.26). By virtue of its piecewise continuity and finiteness (with respect to $\boldsymbol{r}$), the function $q(\boldsymbol{r},\omega) \in L^1 \cap L^2$; therefore it is possible to derive the direct and inverse Fourier transforms [Reed and Simon, 1979]. In this case, $\tilde{q}(\boldsymbol{\rho},\omega)$ [for $q(\boldsymbol{r})$ independent of ω] or the expressions in square brackets included in (3.26), will be integral functions in the plane of the complex variable $\Omega = \omega + \mathrm{i}\omega'$. From this also, the uniqueness follows of reconstructing $q(\boldsymbol{r})$ (3.25) or $N(\boldsymbol{r})$, $\tilde{\nu(\boldsymbol{r})}$ (3.26) from the data on $\tilde{q}(\boldsymbol{\rho},\omega)$ or $[(\tilde{q}+\tilde{q}_{\mathrm{I}}^*)/(2\alpha)]$ and $[(\tilde{q}-\tilde{q}_{\mathrm{I}}^*)/(2\mathrm{i}\beta)]$ at a given interval of the frequency axis. These data are uniquely analytically continued onto the whole complex space Ω.

Now we formulate the conditions sufficient for the existence and uniqueness of the transformations (3.20) $\to$ (3.21) and (3.23) $\to$ (3.24). As noted above, the transformations mentioned are carried out by means of the Fresnel transform. If $q(\boldsymbol{r},\omega)$ is a finite (with respect to $\boldsymbol{r}$) piecewise continuous and, consequently, quadratically integrable function ($q \in L^2$), then the functions $\tilde{q}(\boldsymbol{\rho},\omega)$ and $q_z(\boldsymbol{\rho},\omega)$ will also be quadratically integrable. Therefore, it is possible to obtain the direct and inverse two-dimensional Fourier transforms of these function, and convolutions (3.20) and (3.23) can be represented in the form of a Fourier transform of the product of $\hat{q}(\boldsymbol{v},\omega)$ – the Fourier transform of $\tilde{q}(\boldsymbol{\rho},\omega)$ [or $q_z(\boldsymbol{\rho},\omega)$] and $\hat{f}(\boldsymbol{v},\omega) = 2\mathrm{i}\zeta \exp(\mathrm{i}\zeta v^2/2k)/k$ – the Fourier transform of the function $\exp(\mathrm{i}k\rho^2/2\zeta)$. Since $\hat{q} \in L^2$ and $\hat{q}\hat{f} \in L^2$, it is possible to determine the one-to-one direct and inverse transformations of the kind of (3.20)→ (3.21) and (3.23)→ (3.24).

Relations (3.16) or (3.17), (3.21) and (3.25) or (3.26) yield solutions of the inverse scattering problem for weak irregularities from the data on the field in the Fraunhofer and Fresnel zones within the Born and the Rytov approximations. Besides, it is possible, in principle, to solve the inverse scattering problem if the data are available over a limited frequency range and within a limited volume of s-space. The given volume of s-space can be "filled" by varying the coordinates of either receivers only or sources only, or by varying them together. Then, by performing an analytical continuation from the given regions onto the whole s- and ω-spaces, the inverse scattering problem can be solved by making use of the above mentioned relations. However, the procedure of analytical continuation is highly unstable; therefore, it is worthwhile to consider the relationship between the solution of the inverse problem and the functions obtained from transformations like (3.16), (3.21), and (3.25) of the field measured within a limited space volume and over a limited frequency range [for (3.16) and (3.25)], i.e., from limited data. These questions are discussed in Sect. 4.2.

3.3 Determining the Location and the Parameters of Irregularities

The solutions of the inverse scattering problem obtained in the previous section have the form of linear integral Fourier and Fresnel transforms. To perform such transformations of the experimental data, approximate coordinates of the scatterer must be known since they appear in the parameters of integral transformations. The major problem here is to determine the distance to the scatterer along the z-coordinate; the coordinate $\boldsymbol{\rho}_{\mathrm{s}}$ of the scatterer can be taken equal to zero. As mentioned above, due to the smallness of the receiving apertures, only rather large irregularities with dimensions exceeding the wavelength can be reconstructed. Such irregularities scatter the waves into a narrow zone along the source–receiver line. Therefore, a noticeable scattered field arises only if the scatterer appears close to this line ($\boldsymbol{\rho}_{\mathrm{s}} \approx 0$). If only the approximate distance to the scatterer is known, $\boldsymbol{s}$, ζ are determined incorrectly and, consequently, the function is reconstructed with distortions.

Let us illustrate the distortions by an example of probing the ionosphere with a moving satellite-borne point source. The distance from the satellite to the receivers ($z_0 - z_{\mathrm{R}}$) is known quite exactly. Assume that a systematic error $\varDelta$ is contained in the estimate of the distances from the source to the scatterer $(z_0 - z_s)$ and from the scatterer to the receivers $(z_s - z_R)$ so that $z_s{}' = z_s - \Delta$. It can be easily seen that if the error Δ in the modulus is smaller than the value $2\lambda\zeta^2/(\max s)^2$, the function being reconstructed $q_z{}'(\boldsymbol{\rho})$ takes the form

$$q_z{}'(\boldsymbol{\rho}) \approx q_z\left(\frac{x}{u_x}, \frac{y}{u_y}\right) \exp\left[\frac{\mathrm{i}k(y^2 - x^2)\dfrac{\Delta}{\zeta}}{2\zeta u_x u_y}\right], \tag{3.27}$$

where $u_x = 1 - \Delta/(z_s - z_R)$, $u_y = 1 + \Delta/(z_0 - z_s)$. The relation given makes it possible to determine the distance to the scatterer by recovering the structure with various Δ. When passing through the true position of the scatterer $(\Delta = 0)$, the "phase front" curvature of the function to be reconstructed is reversed with respect to each coordinate. It can be shown that the scale size of the distance Δ, where reconstruction errors are small, corresponds to longitudinal resolution of the receiving system. This conclusion is confirmed by numerical simulation of reconstruction and distance determination (see Sects. 4.2 and 4.4).

The coordinates x, y do not enter in (3.27) symmetrically. This is connected with the fact that the receivers are arranged along the y-axis, while the transmitter moves along the x-axis. When passing from negative to positive Δ, the distance is decreasing to the receivers and increasing to the transmitter. If a two-dimensional network and one stationary transmitter are used for reception, the relation of distortions of the function q' being reconstructed to the true q will be symmetrical with respect to the coordinates (x, y).

The solutions of the inverse scattering problem obtained above make it possible to suggest a method for determination of the parameters of scattering irregularities by recording the field at a few receiving sites. Quite often in practical applications of particular interest are such parameters of the scatterer as the total "mass" $\mathfrak{M}$, "center of mass" $\boldsymbol{\mathfrak{R}}$, "mean-square dimension" $\mathfrak{R}_0$, and other moments of the function $q(\boldsymbol{r})$:

$$\begin{aligned} &\mathfrak{M} \equiv \int q(\boldsymbol{r})\,\mathrm{d}^3 r\,, \boldsymbol{\mathfrak{R}} \equiv (\mathfrak{M})^{-1} \int \boldsymbol{r}\, q(\boldsymbol{r})\,\mathrm{d}^3 r\,, \\ &\mathfrak{R}_0^2 \equiv (\mathfrak{M})^{-1} \int (\boldsymbol{r} - \boldsymbol{\mathfrak{R}})^2\, q(\boldsymbol{r})\,\mathrm{d}^3 r\,, \ \ldots \end{aligned} \tag{3.28}$$

The parameters mentioned can be found from the first derivatives of the function $W(\boldsymbol{p})$ (3.10) [Gusev and Kunitsyn, 1986]:

$$\begin{aligned} &\mathfrak{M} = (2\pi)^3\, W(\boldsymbol{p} = 0)\,, \ \boldsymbol{\mathfrak{R}} = (2\pi)^3 \frac{\partial}{\partial \boldsymbol{p}} \ln W(\boldsymbol{p}) \bigg|_{\boldsymbol{p}=0}\,, \\ &\mathfrak{R}_0^2 = -(2\pi)^3 \frac{\Delta_p W(\boldsymbol{p})}{W(\boldsymbol{p})} + \boldsymbol{\mathfrak{R}}^2 \bigg|_{\boldsymbol{p}=0}\,. \end{aligned} \tag{3.29}$$

The formulas given follow from the definition of the Fourier transform $W(\boldsymbol{p})$ via $q(\boldsymbol{r})$; Δ_p is the Laplacian with respect to the variable $\boldsymbol{p}$. All of these quantities can by determined from an experiment by measuring the field scattered forward and its first derivatives, which requires rather few receivers. For example to determine the dimensions in the plane transverse to the direction of the sounding wave propagation, it is quite sufficient to have five receivers (three receivers at each axis to measure the second derivative, one of the receivers possibly shared). Consider the following example: a scatterer as a uniform sphere of the radius R_s, $q = q_\mathrm{s}$, the center at the point $\boldsymbol{r}_\mathrm{s}$. Then

$$\begin{aligned} W(p) &= (2\pi)^{-3} \exp(-\mathrm{i}p r_\mathrm{s})\, q_\mathrm{s}\, \frac{4\pi}{p^3} \left(\sin pR_\mathrm{s} - pR_\mathrm{s} \cos pR_\mathrm{s}\right) \\ &\simeq 4\pi q_\mathrm{s} \left[\frac{R_\mathrm{s}^3}{3} - p^2 \frac{R_\mathrm{s}^5}{30} - \mathrm{i}R_\mathrm{s}^3 \frac{\boldsymbol{p}\boldsymbol{r}_\mathrm{s}}{3} - R_\mathrm{s}^3 \frac{(\boldsymbol{p}\boldsymbol{r}_\mathrm{s})^2}{6} + \dots \right] . \end{aligned}$$

Whence, by (3.29),

$$\mathfrak{M} = 4\pi q_\mathrm{s} \frac{R_\mathrm{s}^3}{3} , \quad \boldsymbol{\mathfrak{R}} = \boldsymbol{r}_\mathrm{s} ; \; \mathfrak{R}_0^2 = \frac{R_\mathrm{s}^2}{5} .$$

The data on the field in the Fresnel zone (forward scattering) make it possible to reconstruct the integral q_z (3.21) in the direction of the sounding wave propagation; therefore, the parameters of this two-dimensional function can also be expressed via the first derivatives of the two-dimensional Fourier transform $\hat{V}(\boldsymbol{v}) = \int V(\boldsymbol{s}) = \exp(\mathrm{i}\boldsymbol{v}\boldsymbol{s})\; \mathrm{d}^2 s$ of the data $V(\boldsymbol{s})$:

$$\begin{aligned} \mathfrak{M} &= -\mathrm{i}\frac{k}{2\pi\zeta}\, \hat{V}(\boldsymbol{v}=0) \; ; \; \boldsymbol{\mathfrak{R}} = -\mathrm{i}\frac{\partial}{\partial \boldsymbol{v}} \ln \hat{V} \bigg|_{\boldsymbol{v}=0} , \\ \mathfrak{R}_{0x}^2 &= -\left(\frac{1}{\hat{v}} \frac{\partial^2 \hat{V}}{\partial v_x^2} + \mathrm{i}\frac{\zeta c}{\omega} + \mathfrak{R}_x^2 \right) \bigg|_{\boldsymbol{v}=0} , \\ \mathfrak{R}_{0y}^2 &= -\left(\frac{1}{\hat{v}} \frac{\partial^2 \hat{V}}{\partial v_y^2} + \mathrm{i}\frac{\zeta c}{\omega} + \mathfrak{R}_y^2 \right) \bigg|_{\boldsymbol{v}=0} . \end{aligned} \tag{3.30}$$

It is, however, impossible to determine these parameters by means of several receivers, because the derivatives of the Fourier transform $\hat{V}$ are proportional to the moments of $V(\boldsymbol{s})$:

$$\frac{\partial^m \hat{V}}{\partial v_x^m} \bigg|_{\boldsymbol{v}=0} = \int (\mathrm{i}s_x)^m \, V(\boldsymbol{s}) \; \mathrm{d}^2 s \, ;$$

determination of the latter requires the field $V(\boldsymbol{s})$ to be known over a wide range of $\boldsymbol{s}$. Only an approximate estimation of the parameters is possible if $V(\boldsymbol{s})$ at several points is fitted by some given functions. Determination of the scatterer parameters from the Fresnel data can also be made without using the Fourier transform data, if the field is extended into the Fraunhofer zone where the relations for the parameters have the very simple form (3.29).

Let us consider in more detail the continuation of the field from the small-angle recording aperture since these relations will be necessary in the further analysis. It is known that the Helmholtz equation in free space directly yields the expression for the field $E(\boldsymbol{r})$ through the values of the field and its derivative on some closed surface S_{R} surrounding the point $\boldsymbol{r}$:

$$E(\boldsymbol{r}) = \int \mathrm{d}S_{\mathrm{R}} \left[E(\boldsymbol{r}_{\mathrm{R}}) \frac{\partial}{\partial n} G(\boldsymbol{r} - \boldsymbol{r}_{\mathrm{R}}) - G \frac{\partial}{\partial n} E(\boldsymbol{r}_{\mathrm{R}}) \right] . \tag{3.31}$$

This is the so-called integral theorem of Helmholtz–Kirchhoff [Born and Wolf, 1970]. Here, G is the free-space Green function; $\partial/\partial n$ denotes differentiation with respect to the external normal to S_{R}. It turns out that even if the field is not detected at a closed surface S_{R}, its "extension" from this surface, in accordance with (3.31), is a field close to the true one. Needless to say in this case that the surface must be large enough in units of wavelengths. Moreover, the extended field will be close to the true one at those points of observation where the object is seen through the "window" cut out by the detecting system.

Consider now such a continuation of the field measured at a small-angle aperture that is a receiving grid in the z_{R} plane. Let us derive an expression for the field $E(\boldsymbol{r})$ within the small-angle region of the z-plane. The derivative with respect to the normal of the Green function reduces to multiplying by $\mathrm{i}k$:

$$\frac{\partial}{\partial n} G(\boldsymbol{r} - \boldsymbol{r}_{\mathrm{R}}) = \frac{\partial G}{\partial z_{\mathrm{R}}} = \frac{\mathrm{i}k - 1}{|\boldsymbol{r}_{\mathrm{R}} - \boldsymbol{r}|} G \frac{\partial |\boldsymbol{r}_{\mathrm{R}} - \boldsymbol{r}|}{\partial z_{\mathrm{R}}} \simeq \mathrm{i}kG \frac{z_{\mathrm{R}} - z}{|\boldsymbol{r}_{\mathrm{R}} - \boldsymbol{r}|} \simeq \mathrm{i}kG . \tag{3.32}$$

Here the paraxial approximation and the condition $|\boldsymbol{r}_{\mathrm{R}} - \boldsymbol{r}| \gg 1/k$ are used that is sometimes referred to as an optical approximation. For the ionospheric problems under consideration, these approximations are valid with a high accuracy. Similarly, taking into account the fact that

$$\frac{\partial |\boldsymbol{r}_0 - \boldsymbol{r}_{\mathrm{R}}|}{\partial z_{\mathrm{R}}} \simeq \frac{\partial |\boldsymbol{r}' - \boldsymbol{r}_{\mathrm{R}}|}{\partial z_{\mathrm{R}}} \simeq -1 ,$$

the asymptotics of the normal derivative of the field is obtained from (3.10):

$$\frac{\partial E(\boldsymbol{r}_{\mathrm{R}})}{\partial z_{\mathrm{R}}} = \frac{\partial E_0}{\partial z_{\mathrm{R}}} + \int E(\boldsymbol{r}')\, q(\boldsymbol{r}') \frac{\partial}{\partial z_{\mathrm{R}}} G(\boldsymbol{r}_{\mathrm{R}} - \boldsymbol{r}') \simeq \mathrm{i}kE(\boldsymbol{r}_{\mathrm{R}}) . \tag{3.33}$$

Substituting these paraxial approximations of the derivatives of the Green function and the field in (3.31) gives the formula for "extending" the field from a small-angle aperture into the paraxial region

$$E_{\mathrm{c}}(\boldsymbol{r}) \simeq \int \mathrm{d}^2\rho_{\mathrm{R}}\, 2\mathrm{i}kE(\boldsymbol{r}_{\mathrm{R}})\, G(\boldsymbol{r}_{\mathrm{R}} - \boldsymbol{r}) . \tag{3.34}$$

Next, making use of the expression for the field $E_{\mathrm{R}}(\boldsymbol{r}_{\mathrm{R}})$ within the Fresnel approximation (3.20) and integrating over the whole surface ρ_{R}, we arrive at a relation similar to (3.21):

$$E_c(\boldsymbol{r}) = \frac{E_0(\boldsymbol{r})}{4\pi\zeta'}\int d^2r'\,q(\boldsymbol{r}')\exp\left[\frac{ik(\boldsymbol{s}'-\boldsymbol{\rho}')^2}{2\zeta'}\right]. \tag{3.35}$$

The intermediate computations omitted are quite simple though rather lengthy. In the derivation of (3.35), the following equality was used:

$$\frac{(\boldsymbol{\rho}-\boldsymbol{\rho}')^2}{z_s - z} + \frac{(\boldsymbol{\rho}_0-\boldsymbol{\rho}')^2}{z_0 - z_s} = \frac{(\boldsymbol{\rho}-\boldsymbol{\rho}_0)^2}{z_0 - z} + \frac{(\boldsymbol{s}'-\boldsymbol{\rho}')^2}{\zeta'^2},$$

that defines $\boldsymbol{s}'$ and ζ':

$$\boldsymbol{s}' = \frac{z_0 - z_s}{z_0 - z}\boldsymbol{\rho} + \frac{z_s - z}{z_0 - z}\boldsymbol{\rho}_0,$$
$$\zeta' = \frac{(z_0 - z_s)(z_s - z)}{z_0 - z}.$$

Relation (3.35) shows that the form of the relationship to field E_c extended from the given aperture remains the same for corresponding changes in $\boldsymbol{s}$ and ζ. Passing from integration with respect to the finite aperture to integration over the whole plane $\boldsymbol{\rho}_R$ is possible when the dimension of the detecting aperture is much greater than the dimension $\sqrt{\lambda\zeta}$ of the Fresnel zone.

The possibility of extending the field from the Fresnel zone allows us to obtain rather simple expressions for the parameters of irregularites. For this purpose it is sufficient, in accordance with (3.35), to pass into the Fraunhofer zone $(\boldsymbol{s},\zeta) \to (\boldsymbol{s}',\zeta')$, where $kr_m^2/\zeta' \ll \pi$; then,

$$\begin{aligned}
\mathfrak{M} &= V_c\,(\boldsymbol{s}'=0)\ ;\ \boldsymbol{\mathfrak{R}} = \frac{1}{k}\frac{\partial}{\partial \boldsymbol{s}'}\left[V_c\exp\left(-\frac{iks^2}{2\zeta^2}\right)\right]\Bigg|_{\boldsymbol{s}'=0},\\
\mathfrak{R}_{0x}^2 &= -\frac{1}{k^2}\frac{\partial^2}{\partial s_x'^2}\left[V_c\exp\left(-\frac{iks^2}{2\zeta^2}\right)\right]\Bigg|_{\boldsymbol{s}'=0},\\
\mathfrak{R}_{0y}^2 &= -\frac{1}{k^2}\frac{\partial^2}{\partial s_y'^2}\left[V_c\exp\left(-\frac{iks^2}{2\zeta^2}\right)\right]\Bigg|_{\boldsymbol{s}'=0}.
\end{aligned} \tag{3.36}$$

Here the function V_c is defined as in (3.20) via extended field E_c or its complex amplitude. Only the transverse coordinates of the center of mass and the transverse dimensions of the irregularity can be found from (3.36). The longitudinal coordinates and dimensions of the irregularity cannot, naturally, be determined in the framework of this paraxial approximation.

3.4 Inverse Scattering Problem in a Stratified Medium

If the sounding frequencies are comparable with the critical frequency of the ionosphere, a regular stratified structure significantly affects the propagation of radio waves; therefore, (3.7) should be solved without passing to (3.9). Since the changes in a regular ionosphere are smooth and their typical scales are large, wave propagation in the presence of a background stratified ionosphere can be described in the context of geometric optics. It turns out that for practically important cases of "forward" and "backward" scattering in the direction perpendicular to the ionospheric layers, i.e., in the vertical direction, one can obtain the following representation for the Green function, within the paraxial approximation, in terms of the known phase and group paths:

$$G(\boldsymbol{r},\boldsymbol{r}_0) \simeq -\frac{1}{4\pi\sqrt{n(z,\omega)\,n(z_0,\omega)}}\frac{1}{g(z_0,z)} \times \exp\left[\mathrm{i}k\int n\,\mathrm{d}z + \frac{\mathrm{i}k(\boldsymbol{\rho}-\boldsymbol{\rho}_0)^2}{2\,g(z_0,z)}\right]. \tag{3.37}$$

Here, $n(z,\omega) = [1 - q_0(z)/k^2]^{1/2}$ is the refractive index in the regular ionosphere; $g(z,z_0) = \int_{z_0}^{z}\frac{\mathrm{d}z}{n}$ is the group path. Derivation of (3.37) is given in [Kunitsyn, 1985, 1986a]. Remember that in stating the inverse problem of reconstructing three-dimensional irregularities in the presence of a stratified background, the regular ionosphere is assumed known, i.e., either $n(z)$ or the group and phase paths are known. The phase path – $\int n\,\mathrm{d}z$ – can be found from the transionospheric sounding data [Kunitsyn and Smorodinov, 1986] measured over a wide frequency range.

A paraxial approximation for the Green function of the form (3.37) can be derived in a simpler way by a standard transition to a parabolic equation [Kunitsyn and Tereshchenko, 1991]. Assuming $E = \frac{F}{\sqrt{n}}\exp\left(\mathrm{i}k\int n\,\mathrm{d}z\right)$ and neglecting the ordinary method of the second derivative, let us pass from (3.7) to the parabolic equation with the transverse [in coordinates $\boldsymbol{\rho} = (x,y)$] Laplacian $\Delta_\perp$:

$$2\mathrm{i}kn\frac{\partial F}{\partial z} + \Delta_\perp F = \sqrt{n(z_0)}\,\delta(\boldsymbol{r}-\boldsymbol{r}_0)\,.$$

Then, introducing the group path variable g: $\mathrm{d}g = \frac{\mathrm{d}z}{n}$, $\frac{\partial}{\partial g} = n\frac{\partial}{\partial z}$, and $\delta(g-g_0) = \delta\left[\frac{\mathrm{d}g}{\mathrm{d}z}\,(z-z_0)\right] = n(z_0)\,\delta(z-z_0)$, we arrive at the following equation:

$$2\mathrm{i}k\frac{\partial F}{\partial g} + \Delta_\perp F = \frac{1}{\sqrt{n(z_0)}}\,\delta(g-g_0)\,\delta(\boldsymbol{\rho}-\boldsymbol{\rho}_0)\,. \tag{3.38}$$

Writing the fundamental solution of the given equation for F [Vladimirov, 1988], we come to (3.37).

With this representation of the Green function used in (3.20) or (3.23) instead of the Green function of free space, we obtain the solution of the inverse scattering problem in a stratified medium. For forward scattering, the solution coincides in its form with (3.21):

$$q_z(\boldsymbol{\rho}) = \left(\frac{k}{2\pi\zeta}\right)^2 \int V(\boldsymbol{s}) \exp\left[-\frac{\mathrm{i}k(\boldsymbol{s}-\boldsymbol{\rho})^2}{2\zeta}\right] \mathrm{d}^2\boldsymbol{s}\,, \tag{3.39}$$

where $V(\boldsymbol{s}) = -4\pi\zeta n(z_\mathrm{s})\, E_1(\boldsymbol{r}_\mathrm{R})\,/E_0(\boldsymbol{r}_\mathrm{R})$ in the Born approximation and $V(\boldsymbol{s}) = -4\pi\zeta n(z_\mathrm{s})\, \Phi_1(\boldsymbol{r}_\mathrm{R})$ in the Rytov approximation.

However, the meaning of the variables $\boldsymbol{s}$ and ζ is somewhat different:

$$\boldsymbol{s} = \frac{g_0}{g_0+g_\mathrm{R}}\,\boldsymbol{\rho}_\mathrm{R} + \frac{g_\mathrm{R}}{g_0+g_\mathrm{R}}\,\boldsymbol{\rho}_0\,, \quad \zeta = \frac{g_0 g_\mathrm{R}}{g_0+g_\mathrm{R}}\,,$$

$$g_\mathrm{R} = \int_{z_\mathrm{R}}^{z_\mathrm{s}} \frac{\mathrm{d}z}{n}\,, \quad g_0 = \int_{z_\mathrm{s}}^{z_0} \frac{\mathrm{d}z}{n}\,.$$

In backward scattering, the explicit formula for reconstruction, analogous to (3.25), can also be derived, if the condition is met of small variations in the refractive coefficient at the irregularity within the frequency range $\Delta\omega$ used:

$$\left|z_\mathrm{m}\frac{\partial n}{\partial z}\right| \quad \left|\Delta\omega\frac{\partial n}{\partial \omega}\right| \ll n(z=z_\mathrm{s},\omega_0)\,; \quad \left|\frac{\partial n}{\partial \omega}\Delta\omega\frac{z_\mathrm{m}}{\lambda}\right|\,, \quad \left|\frac{z_\mathrm{m}^2}{\lambda}\frac{\partial n}{\partial z}\right| \ll 1\,.$$

Then,

$$\begin{aligned} q(\boldsymbol{r}) = \int & \frac{\exp\left[\dfrac{2\mathrm{i}\omega z\, n(z_\mathrm{s})}{c}\right]}{\pi c}\,\mathrm{d}\omega \\ & \times \left(\frac{k}{2\pi\zeta}\right)^2 \int U(\boldsymbol{s},\omega)\, \exp\left[-\frac{\mathrm{i}k(\boldsymbol{s}-\boldsymbol{\rho})^2}{2\zeta}\right] \mathrm{d}^2 s\,. \end{aligned} \tag{3.40}$$

Here,

$$\begin{aligned} U(\boldsymbol{s},\omega) &= (4\pi)^2 n(z_\mathrm{s})\,\sqrt{n(z_0)\,n(z_\mathrm{R})}\,E_1(r_\mathrm{R}) \\ &\times \exp\left\{-\mathrm{i}k\left[\int_{z_\mathrm{R}}^{z_\mathrm{S}} n\,\mathrm{d}z + \int_{z_0}^{z_\mathrm{S}} n\,\mathrm{d}z + \frac{(\boldsymbol{\rho}_\mathrm{R}-\boldsymbol{\rho}_0)^2}{2(g_0+g_\mathrm{R})}\right]\right\}\,. \end{aligned}$$

Generalization of (3.40) to the known shape of frequency dependence is similar to generalization (3.25) → (3.26).

Note that it is possible to describe inverse scattering problems using the parabolic equation method applied here in deriving the expression for the Green function in a stratified medium. The parabolic equation suggested by Leontovich and Fock is a differential treatment of the Fresnel small-angle diffraction [Fok, 1970; Vaynshteyn, 1988]. Later on, the parabolic equation method was extensively developed and widely applied in various wave problems, including those of wave propagation in randomly inhomogeneous media, nonlinear interaction of wave packets and beams, diffraction in anisotropic media, and so on. At present, the parabolic equation method or Fresnel small-angle approximation are well developed theoretical tools of particular importance as a basis for studying a wide range of phenomena related to propagation and diffraction of waves.

3.5 Inverse Scattering Problem for Strong Irregularities

In many practically important cases of rather strong and extended irregularities, the weak scattering approximation is not valid. Particular numerical estimates of intensities and dimensions of such irregularities will be given at the end of the section. A number of studies [Prosser, 1969, 1976; Johnson and Tracy, 1983; Ney et al., 1984; Lesselier et al., 1985; Datta and Bandyopadhyay, 1986; Caorsi et al., 1988; Burov et al., 1989] are known on ISP of strong scatterers. The studies mentioned are based on different iterative schemes. The iteration procedures are constructed either on the basis of expansion into a series of the type (3.11) [Prosser, 1969, 1976; Burov et al., 1989] or on iterative solution of integral equation (3.10). The latter can be implemented in the following way. After a certain zero approximation for the potential q is given (by either inserting a priori information or inversion assuming a weak scattering approximation), linear integral equation (3.10) is solved for the field inside the scatterer. Then (3.10) is solved as a first kind of linear integral equation with respect to the potential with a known kernel that includes the approximation of the field found inside the scatterer. The potential approximation obtained is used at the next iterative step and so on. To solve integral equations, the methods of moments and conjugated gradients were applied [Ney et al., 1984; Lesselier et al., 1985; Datta and Bandyopadhyay, 1986; Caorsi et al., 1988]. However, in the case of large scatterers like ionospheric irregularities, when the size of the scatterer can be 10^3–10^4 times the wavelength, such iterative methods for ISP solving are inapplicable.

Here, an approach is described to reconstruct strongly scattering large irregularities from small-angle scatter data. This approach is based on asymptotic representation of the field in large scatterers and on analytical inversion [Kunitsyn, 1992] in solving the ISP. As done previously, consideration of the ISP will be based on Lippmann–Schwinger integral equation (3.10). Generalization to a stratified background can be carried out using the expressions for the Green function in the stratified medium, as obtained in Sect. 3.4. Next, to reduce calculations, it is convenient to superimpose the origin of the Cartesian coordinate system $\boldsymbol{r} = (\boldsymbol{\rho}, z)$ on a limited irregularity. Consider a field E using a paraxial approximation in the far zone ($z \gg r_{\mathrm{m}}, \rho$; $kr_{\mathrm{m}}^2/z \ll 1$) at small angles of "forward" scattering with respect to the direction of propagation (the z-axis) of the probing wave $E_0 = \exp ikz$. Expanding the exponent in the Green function (3.10),

$$k\left|\boldsymbol{r} - \boldsymbol{r}_1\right| \simeq kr - \frac{k\boldsymbol{r}\boldsymbol{r}_1}{r} + O\left(\frac{kr_{\mathrm{m}}^2}{r}\right)$$

$$\simeq k(r - z_1) - \frac{k\boldsymbol{\rho}\boldsymbol{\rho}_1}{z} + O\left(\frac{kr_{\mathrm{m}}^2}{r} + \frac{kr_{\mathrm{m}}\rho^2}{z^2}\right),$$

one obtains the following relation from (3.10):

$$A\left(\frac{k\boldsymbol{\rho}}{z}, k\right) \equiv (E - E_0)(-4\pi r) \exp(-\mathrm{i}kr)$$
$$= \int \exp\left(-\frac{\mathrm{i}\boldsymbol{\rho}\boldsymbol{\rho}_1}{z} - \mathrm{i}kz_1\right) q(\boldsymbol{r}_1, k)\, E(\boldsymbol{r}_1, k)\ \mathrm{d}^3 r_1\,. \tag{3.41}$$

Denote $U_{\mathbf{i}} = E(\boldsymbol{r}, k)\, E_0^{-1}(\boldsymbol{r}, k)$, $Q(\boldsymbol{r}, k) \equiv q(\boldsymbol{r}, k)\, U_{\mathbf{i}}$ and $Q_z(\boldsymbol{\rho}, k) \equiv \int Q\, \mathrm{d}z$. Index $\mathbf{i}$ of the normalized field indicates the dependence of the latter on the incidence direction $\mathbf{i}$ (the unit vector) of the sounding wave. Here, formula (3.41) is a two-dimensional Fourier transform with respect to the coordinate ρ_1:

$$\int \exp\left(-\frac{\mathrm{i}k\boldsymbol{\rho}\boldsymbol{\rho}_1}{z}\right) Q_z(\boldsymbol{\rho}_1, k)\ \mathrm{d}^2\rho_1 = A\left(\frac{k\boldsymbol{\rho}}{z}, k\right)\,. \tag{3.42}$$

Inversion of this transform yields the relation

$$Q_z(\boldsymbol{\rho}, k) = \left(\frac{k}{2\pi z}\right)^2 \int \exp\left(\frac{\mathrm{i}k\boldsymbol{\rho}\boldsymbol{\rho}'}{z}\right) A\left(\frac{k\boldsymbol{\rho}'}{z}, k\right)\ \mathrm{d}^2\rho'\,. \tag{3.43}$$

Let us introduce the Fourier transform $\hat{Q}$ of the function Q:

$$\hat{Q}(\boldsymbol{p}, k) = \int Q(\boldsymbol{r}, k) \exp(-\mathrm{i}\boldsymbol{p}\boldsymbol{r})\, \mathrm{d}^3 r\,.$$

According to the definition of Q_z,

$$\hat{Q}(\boldsymbol{s}, p_z = 0) = \int Q_z(\boldsymbol{\rho}, k) \exp(-\mathrm{i}\boldsymbol{s}\boldsymbol{\rho})\ \mathrm{d}^2\rho\,,$$

where $\boldsymbol{s} = (p_x, p_y)$ is a two-dimensional vector, $\boldsymbol{p} = (\boldsymbol{s}, p_z)$. Comparing this equality with (3.42), one arrives at the formula

$$\hat{Q}(\boldsymbol{s}, 0, k) = A\left(\frac{k\boldsymbol{\rho}}{z}, k\right)\,. \tag{3.44}$$

Relation (3.44) makes it possible to determine two-dimensional (orthogonal to z) cross section of the Fourier image of function Q from measurements of field E in a paraxial region adjacent to the z-axis. The dimensions of this section are defined by the dimensions (in variable $k\boldsymbol{\rho}/z$) of the region of field measurement. Varying the directions of the probing wave propagation, one can obtain other two-dimensional sections of the Fourier image $\hat{Q}(\boldsymbol{p}, k)$ or corresponding projections $Q_z(\boldsymbol{\rho}, k)$. It should be noted that, generally, the function $Q = qU_{\mathbf{i}}$ changes with variations in the direction of incidence of the sounding wave, since the field inside the scatterer becomes different. If the field inside the scatterer is somehow determined, then reconstruction of $q(\boldsymbol{r}, k)$ becomes possible from sets (3.43) of the projections Q_z with the given field $U_{\mathbf{i}}$. This is the usual tomographic problem of reconstructing a three-dimensional $q(\boldsymbol{r}, k)$ with a fixed k from a set of two-dimensional projections

$Q_z(\boldsymbol{\rho}, k)$ (3.43) with known weight $U_\mathbf{i}$. Of course, exact determination of the field inside the scatterer is equivalent to solving the ISP; therefore, at first sight, the approach described seems to be of little value. However, this can be used as a basis for constructing an iterative procedure: the approximation of the potential q defines the field $U_\mathbf{i}$ (forward problem) that makes it possible to find from the projections Q_z the next approximation (the inverse problem) for q. Although this approach greatly simplifies the inverse problem owing to inversions (3.43), (3.44), solving the direct problem still entails considerable difficulty in the case of large scatterers. In such situations, it is necessary to apply an asymptotic representation of the field already at the step of the forward problem.

Before we proceed to derivation of appropriate asymptotic representations, let us answer the question: what would result from reconstructing the potential within the weak scattering approximation (Sect. 3.2) if the potential is not weak? In the case of a weak scatterer, the product qE can be substituted by qE_0 in the integrands of (3.13), (3.14). Such a determination of the field E_0 underlies further solutions of the inverse problem. If this equality is not valid, it can be easily seen that the mentioned solutions of the inverse problem are not reconstructions of the potential q sought but those of the function Q (or Q_z), since the identity $qE = QE_0$ is valid.

Let us find an asymptotic representation for the field inside the scatterer in the practically important case of high frequencies when the dimensions of the scatterer and its typical details noticeably exceed the wavelength. Hereinafter, ISP solutions at a fixed frequency will be considered; therefore the argument k in the dependence $q(\boldsymbol{r}, k)$ will be omitted in further discussion, but it will be retained in the function $Q(\boldsymbol{r}, k)$, because here it signifies a large parameter with respect to which the asymptotic expansion is performed:

$$
\begin{aligned}
Q(\boldsymbol{r}_1, k) = q(\boldsymbol{r}_1)\exp(-\mathrm{i}kz_1)\Bigg[&\exp(\mathrm{i}kz_1) + \beta\int \frac{\exp(\mathrm{i}k\,|\boldsymbol{r}_2 - \boldsymbol{r}_1| + \mathrm{i}kz_2)}{|\boldsymbol{r}_2 - \boldsymbol{r}_1|} \\
&\times q(\boldsymbol{r}_2)\ \mathrm{d}^3 r_2 + \beta^2 \int \frac{\exp(\mathrm{i}k\,|\boldsymbol{r}_2 - \boldsymbol{r}_1|)}{|\boldsymbol{r}_2 - \boldsymbol{r}_1|}\, q(\boldsymbol{r}_2)\ \mathrm{d}^3 r_2 \\
&\times \int \frac{\exp(\mathrm{i}k\,|\boldsymbol{r}_3 - \boldsymbol{r}_2| + \mathrm{i}kz_3)}{|\boldsymbol{r}_3 - \boldsymbol{r}_2|}\, q(\boldsymbol{r}_3)\ \mathrm{d}^3 r_3 + \dots\Bigg]\,.
\end{aligned} \tag{3.45}
$$

The terms in square brackets represent the field E as a Born–Neumann series, $\beta = -(4\pi)^{-1}$. Let us convert to the coordinates $\boldsymbol{R}_n = \boldsymbol{r}_{n+1} - \boldsymbol{r}_n = (\boldsymbol{P}_n, Z_n)$, $\boldsymbol{P}_n = (X_n, Y_n)$:

$$
\begin{aligned}
Q(\boldsymbol{r}_1, k) = q(\boldsymbol{r}_1)\Bigg[&1 + \beta\int \frac{\exp[\mathrm{i}k(R_1 + Z_1)]}{R_1}\, q(\boldsymbol{R}_1 + \boldsymbol{r}_1)\ \mathrm{d}^3 R_1 \\
&+ \beta^2 \int \frac{\exp[\mathrm{i}k(R_1 + Z_1)]}{R_1}\ \mathrm{d}^3 R_1 \int \frac{\exp[\mathrm{i}k(R_2 + Z_2)]}{R_2} \\
&\times q(\boldsymbol{R}_1 + \boldsymbol{r}_1)\, q(\boldsymbol{R}_1 + \boldsymbol{R}_2 + \boldsymbol{r}_1)\ D^3 R_2 + \dots\Bigg]\,.
\end{aligned} \tag{3.46}
$$

Let us obtain an asymptotic expansion in a power series of $1/k$ of the second summand taken as an example. It can be easily understood that the minimum of the phase function is achieved at the semiaxis $Z_1 < 0$. In view of this fact, it is appropriate to pass from $\boldsymbol{R}_1$ to parabolic coordinates (ξ, η, ϕ): $X_1 = \xi\eta \cos\phi$, $Y_1 = \xi\eta \sin\phi$, $Z_1 = (\eta^2 - \xi^2)/2$, after which the second summand assumes the form

$$\begin{aligned} &\int \frac{\exp[\mathrm{i}k(R_1 + Z_1)]}{R_1}\, q(\boldsymbol{R}_1 + \boldsymbol{r}_1)\ \mathrm{d}^3 R_1 \\ &\quad = 2\int_0^{\infty} \xi\ \mathrm{d}\xi \int_0^{2\pi} \mathrm{d}\phi \int_0^{\infty} \mathrm{d}\eta\, \eta \exp\left(\mathrm{i}\eta^2\right) q(\xi, \eta, \phi)\ . \end{aligned}$$

Here, it is taken into account that the Jacobian of passing to parabolic coordinates is $(\xi^2 + \eta^2)\xi\eta$ and $2R_1 = \xi^2 + \eta^2$.

In accordance with the Erdélyi lemma [Fedoryuk, 1987], we obtain the asymptotic expansion of the integral in η:

$$\begin{aligned} \int_0^{\infty} \mathrm{d}\eta\, \eta \exp\left(\mathrm{i}\eta^2\right) f(\eta) \sim \frac{1}{k} \sum_{n=0}^{\infty} k^{-\frac{n}{2}} \frac{\Gamma\left(\frac{n+2}{2}\right)}{2n!} \\ \times \exp\left[\frac{\mathrm{i}\pi(n+2)}{4}\right] \frac{\partial^n f}{\partial \eta^n}\bigg|_{\eta=0}\ , \end{aligned} \tag{3.47}$$

where $f(\eta) = \int_0^{\infty} \xi\, \mathrm{d}\xi \int_0^{2\pi} \mathrm{d}\phi\, q(\xi, \eta, \phi)$. The first term of this series is proportional to the value $f(0)$ that reduces to the integral with respect to ξ and, hence, to a of the potential q. The second term ($\sim k^{-2}$) contains the first derivative with respect to η that can be suitably expressed in terms of the second derivative with respect to $\boldsymbol{\rho}$. Similarly, by passing to sets of parabolic coordinates, the asymptotic expansions of other terms of the series (3.46) can be made with integration with respect to the coordinates $\mathrm{d}^3 R_1$, $\mathrm{d}^3 R_2$ and so on. The asymptotic expansions will have the form of products of a series like (3.47) with corresponding differential operators. Therefore, the principal term of expansion in terms of $1/k$ for the third summand in (3.46) will be proportional to (k^{-2}) and so on. Based on representations like (3.47), specific calculation is possible of further terms of the asymptotic series (3.46). Besides stationary points of exponential multipliers, also singular points of the off-exponential multiplier, i.e., the potential and its products, and the corner points of the boundary of the support q will contribute to the asymptotic expansion [Fedoryuk, 1987]. Apropos, note an interesting statement of the problem of determining such singular points by tomography. It can be shown that for a potential of infinite smoothness, the expansion in $(1/k)$ will contain no fractional powers. Only the general type of expansion and contributions of the first summands are of importance here. Omitting calculations briefly described above, we write the result in the following way:

$$
\begin{aligned}
Q(\boldsymbol{r}_1) \simeq q(\boldsymbol{r}_1) \Bigg[1 &- \frac{\mathrm{i}}{2k} \int_{-\infty}^{z_1} q(\boldsymbol{\rho}_1 + \boldsymbol{z}_2) \, \mathrm{d}z_2 \\
&+ \frac{1}{4k^2} \int_{-\infty}^{z_1} (z_1 - z_2) \frac{\partial^2 q}{\partial \rho^2} (\boldsymbol{\rho}_1 + \boldsymbol{z}_2) \, \mathrm{d}z_2 \\
&- \frac{\mathrm{i}}{4k^2} \int_{-\infty}^{z_1} q(\boldsymbol{\rho}_1 + \boldsymbol{z}_2) \, \mathrm{d}z_2 \int_{-\infty}^{z_1} q(\boldsymbol{\rho}_1 + \boldsymbol{z}_3) \, \mathrm{d}z_3 + o\left(\frac{1}{k^2}\right) \Bigg] .
\end{aligned} \tag{3.48}
$$

Vector notation $\boldsymbol{z}$ in the argument q implies that integrating over z is performed along straight lines parallel to the z-axis. By definition, $Q/q = E/E_0$; therefore, the expression in square brackets in (3.48) is the ratio of the true field to the zero approximation field.

If the Rytov approximation series is taken as a basis for the power expansion of $1/k$, the expression for $Q/q = E/E_0 = \exp(\Phi - \Phi_0)$ becomes somewhat different. Making use of (3.14) and representation [Tatarskii, 1971] for further terms of the Rytov approximation series allows us to perform asymptotic expansions similar to (3.48). To an accuracy of squared terms (k^{-2}), we obtain

$$
\begin{aligned}
\Phi(r_1) - \Phi_0 \simeq &- \frac{\mathrm{i}}{2k} \int_{-\infty}^{z_1} q(\boldsymbol{\rho}_1 + \boldsymbol{z}_2) \, \mathrm{d}z_2 \\
&+ \frac{1}{4k^2} \int_{-\infty}^{z_1} (z_1 - z_2) \frac{\partial^2 q}{\partial \rho^2} (\boldsymbol{\rho}_1 + \boldsymbol{z}_2) \, \mathrm{d}z_2 + O\left(\frac{1}{k^3}\right) .
\end{aligned} \tag{3.49}
$$

Note that summation of the terms of the series (3.48) containing no transverse derivatives yields the complex phase approximation, i.e., the first summand of series (3.49). In other words, summation of the given subsequence of the Born series gives the first approximation of the complex phase in the Rytov series. Therefore, series (3.49) appears to be preferable for applications.

Based on relations (3.48), (3.49), a simple iterative procedure for solving ISP can be built if one takes into account the small number of summands in series (3.48), (3.49). In contrast to the known iterative procedures, in this case there is no need to solve either a forward or an inverse scattering problem taking into account the diffraction effects. As mentioned above, reconstruction according to algorithm (3.21) means reconstructing the projections $Q_z(\boldsymbol{\rho}, \mathbf{i})$,

$$
\int q(\boldsymbol{r}) \, U_{\mathbf{i}}[q] \, \mathrm{d}z = Q_z(\boldsymbol{\rho}, \mathbf{i}) . \tag{3.50}
$$

The zero approximation q_0 of the potential defines the approximation of the functional $U_{\mathbf{i}}[q_0]$ of a normalized field, after which an ordinary tomographic problem of reconstructing $q(\boldsymbol{r})$ (3.50) with known weight $U_{\mathbf{i}}[q_0]$ is solved. Then, the iterative procedure is repeated using the approximation obtained for the potential. The zero approximation is given based on a priori information or assuming weak scatterer $U_{\mathbf{i}}[q_0] = 1$.

Representations like (3.48), (3.49) in the case of frequency independent potentials allow separation of the summands by powers of $1/k$. For determination of the first summand, even the ISP solution within the weak scattering approximation (3.21) can be used. However, separation of summands requires measurements of the field over a wide range of frequencies.

The applicability condition for the Born approximation will be smallness of the first ($\sim k^{-1}$) and all other subsequent terms of the expansion (3.48). From the constraint on the first summand, the known condition of applicability for the Born approximation follows:

$$\frac{q\,r_{\mathrm{m}}}{2k} \ll 1\,. \tag{3.51}$$

Here q is the typical (average) value of the potential; r_{m} is the maximum size of the irregularity.

For disturbances in ionospheric electron density ($q \approx 4\pi r_{\mathrm{e}} N$) of magnitude $N \sim 10^{11}$ el/m^3, the limitation $r_{\mathrm{m}} \ll 2\,\mathrm{km}$ follows from (3.51) if the probing wavelength is $\lambda = 2\,\mathrm{m}$. Accordingly, $N \sim 10^{10}\ \mathrm{m}^{-3}$ restricts the irregularity dimension to 20 km and $N \sim 10^{9}\,\mathrm{m}^{-3}$ to 200 km. Note that electron density perturbations that scatter radio waves should be counted against the background regular stratified ionosphere; therefore with the background value $N \approx 10^{11}\ \mathrm{m}^{-3}$ and sufficiently strong (10%) disturbances, the Born approximation is valid up to scales of tens of kilometers. However, in the region of the main ionospheric maximum, where $N \approx 10^{12}\,\mathrm{m}^{-3}$, limitations on the sizes of perturbations become significant, and it becomes necessary to take the term (k^{-1}) into account.

The second summand of series (3.48), (3.49) for simple "structureless" scatterers is always (for $kr_{\mathrm{m}} \gg 1$) smaller than the first one. However, if the localized scatterer has a "fine" structure and contains "internal details" on the scale $a \gg \lambda$ and typical potential value q', the second summand may exceed the first one. Hence, another necessary condition for applicability of the Born approximation is smallness of the second summand with a transverse derivative

$$\frac{q' r_{\mathrm{m}}}{4k^2 a} \ll 1\,. \tag{3.52}$$

Strong scattering, when the scattered field exceeds the incident field, would be observed under the condition

$$\exp[\mathrm{Re}(\Phi - \Phi_0)] > 2\cos[\mathrm{Im}(\Phi - \Phi_0)]\ .$$

The "focusing" ($\partial^2 q/\partial \boldsymbol{\rho}^2 > 0$) constituent of the scatterer with size $a \geq \sqrt{\lambda r_{\mathrm{m}}}$ may produce a strong scattered field according to the above condition, when inequality (3.51) is violated and $q' > 4k^2 a/r_{\mathrm{m}}$. The scatterer detail itself may be either a strong ($q' > k/a$) irregularity or a weak (if $r_{\mathrm{m}} \gg ka^2$) one.

The Rytov expansion is a result of renormalization of the Born expansion, and therefore, it generally has a wider domain of uniform applicability than the Born expansion. As has been already mentioned above, the sum of the terms of Born series (3.48) containing no transverse derivatives gives the first approximation of the Rytov expansion. Hence, the Rytov approximation is also applicable when the first term of series (3.49) is comparable to unity, i.e., condition (3.51) is violated.

In most cases of practical importance, it is sufficient to take into account only the first term of series (3.49). Here, the scatterer is no longer the Born one according to (3.51), but inequality (3.52) is satisfied. Such conditions make it possible to obtain an analytical formula relating the function Q_z reconstructed from the experimental data to the projection of the scattering potential q_z. Based on the above assumptions and (3.49), we obtain the following relationship:

$$Q(\boldsymbol{r}) \simeq q(\boldsymbol{r}) \exp\left[-\frac{\mathrm{i}}{2k} \int_{-\infty}^{z} q(\boldsymbol{\rho}, z') \, \mathrm{d}z'\right] . \tag{3.53}$$

Let us represent the exponent as a series and integrate (3.53) termwise. It can be easily shown that in this case one will obtain integrals of the form

$$I_n \equiv \int_{-\infty}^{z} \mathrm{d}z \, q(z) \left[\int_{-\infty}^{z} \mathrm{d}z \, q(z)\right]^{n-1} = \frac{q_z}{n} \equiv \frac{1}{n} \left[\int_{-\infty}^{z} \mathrm{d}z \, q(z)\right]^{n} .$$

The equality results from integration by parts. Summing the expansion (3.53) after term-by-term integration, we arrive at the formula

$$Q_z(\boldsymbol{\rho}) = 2\mathrm{i}k \left[\exp\left(-\frac{\mathrm{i}q_z(\boldsymbol{\rho})}{2k}\right) - 1\right] . \tag{3.54}$$

An inverse relation being a solution of the problem of obtaining projections has the form

$$q_z(\boldsymbol{\rho}) = 2k \arctan\left(\frac{\mathrm{Re}Q_z}{2k + \mathrm{Im}Q_z}\right) + \mathrm{i}k \ln\left[\left(\frac{1 + \mathrm{Im}Q_z}{2k}\right)^2 + \left(\frac{\mathrm{Re}Q_z}{2k}\right)^2\right]$$
$$+ 4k\pi m . \tag{3.55}$$

Here m is an integer and q_z is determined with an accuracy of the real term $4k\pi m$. One can see that representation (3.53) corresponds (since $Q/q = E/E_0$) to the geometric optics approximation with rectilinear trajectories in calculating the field inside the scatterer; therefore, changing the potential by a real constant $4k\pi m$ does not affect the phase shift, and relation (3.54) remains unchanged. However, when the potential q is finite and projections $Q_z(\boldsymbol{\rho})$ are reconstructed in a region that includes the support of the function $q_z(\boldsymbol{\rho})$, the dependence $q_z(\boldsymbol{\rho})$ can be reconstructed from (3.54), (3.55), taking into account the jumps of the phase $Q_z(\boldsymbol{\rho})$. In the case of a real potential, the argument of the logarithm reduces to unity and formula (3.55) becomes simpler:

$$q_z(\boldsymbol{\rho}) = \pm 4k \arcsin\left[\frac{|Q_z(\boldsymbol{\rho})|}{4k}\right] + 4k\pi m . \tag{3.56}$$

The sign of q_z is defined by the phase $\arg(Q_z/2k) = -q_z/4k + 4k\pi m$ of the projection $Q_z(\boldsymbol{\rho})$. If the reconstructed modulus $|Q_z(\boldsymbol{\rho})|$ does not reach $4k$, then for a finite potential, $m = 0$. Otherwise, it is necessary to take into account the transitions q_z through the levels $2k\pi m$ or jumps by $4k\pi m$.

Representation of the scattering problem solution as a series like (3.48), (3.49) allows an asymptotically exact solution of the ISP to be obtained. For $\lambda = 2\pi/k \to 0$, the asymptotics of the measured projection Q_z, in accordance with (3.48), (3.49), coincides with the asymptotics of the q_z sought:

$$Q_z(\boldsymbol{\rho}, \lambda = 0) = q_z(\boldsymbol{\rho}, \lambda = 0) \ . \tag{3.57}$$

The derivative of the projection Q_z with respect to λ with (3.49), (3.54) taken into account is expressed through the value of the function q_z and its first derivative with $\lambda \to 0$:

$$\left.\frac{\partial Q_z}{\partial \lambda}\right|_{\lambda=0} = \left.\frac{\partial q_z}{\partial \lambda}\right|_{\lambda=0} - \frac{\mathrm{i}}{8\pi}\, q_z^2(\boldsymbol{\rho}, \lambda = 0) \ . \tag{3.58}$$

If the functional dependence $q(\lambda)$ defined by the type of physical medium is known, the formulas make it possible to reconstruct $q(\boldsymbol{\rho}, k)$. In particular, for a cold ionospheric plasma (3.3),

$$q(\boldsymbol{r}, \lambda) = 4\pi r_{\mathrm{e}} N(\boldsymbol{r}) \left[1 - \frac{\mathrm{i}\lambda\nu(\boldsymbol{r})}{2\pi c}\right] ;$$

hence

$$4\pi r_{\mathrm{e}} \int N(\boldsymbol{r})\ \mathrm{d}z = Q_z(\boldsymbol{\rho}, \lambda = 0) \ ;$$

$$\frac{2r_{\mathrm{e}}}{c} \int \nu(\boldsymbol{r})\, N(\boldsymbol{r})\ \mathrm{d}z = \mathrm{i}Q_z(\boldsymbol{\rho}, \lambda = 0) - \left.\frac{\partial Q_z(\boldsymbol{\rho})}{\partial \lambda}\right|_{\lambda=0} . \tag{3.59}$$

Thus, the projections N and $N\nu$ sought are expressed in terms of the measured projection Q_z and its derivative with respect to the wavelength with $\lambda \to 0$. It should be stressed that these are asymptotically exact solutions of the ISP since subsequent terms of the expansion in asymptotic series (3.48), (3.49) do not affect the result (3.59). However, practical realization of reconstruction according to the scheme (3.59) requires experimental data on the projections $Q_z(\boldsymbol{\rho}, k)$ within a wide frequency range with $k \to \infty$.

It is interesting to note that the first terms of expansions (3.48), (3.49) coincide with the field representation within the scope of the "fifth parameter" method or the Fock–Schwinger method of intrinsic time. This method was elaborated by Fock for the purpose of integrating relativistic wave equations [Fok, 1957]. The additional "fifth parameter" introduced plays the role of intrinsic time in a quasi-classical approximation [Pavlenko, 1976]. Later, the method was developed in Schwinger's works [Schwinger, 1951] and became widely practiced [Pavlenko, 1976; Grib et al., 1980]. Representation of the field within the scope of the "fifth parameter" method makes it possible to allow for the effects of strong scattering as well. The method generalizes most known approximations and in particular cases reduces to the geometric optics approximation and the Born and Rytov approximations [Pavlenko, 1976].

Let us dwell briefly on the interrelation between the above mentioned methods for small-angle scattering by large irregularities. Here, the Lippmann–Schwinger equation (3.22) (for a plane sounding wave) has the form

$$U(\boldsymbol{r}) = 1 - \frac{1}{4\pi}\int \mathrm{d}^3 r_1 \, \frac{q(\boldsymbol{r}_1)\,U(\boldsymbol{r}_1)}{z - z_1}\,\exp\left[\frac{\mathrm{i}k(\boldsymbol{\rho}-\boldsymbol{\rho}_1)^2}{2(z-z_1)}\right] \equiv 1 + \hat{F}qU\,, \tag{3.60}$$

where $\hat{F}$ denotes the Fresnel operator with an approximate Green function. The Born approximation of the field is calculated from the formula $U_{\mathrm{B}} = 1 + \hat{F}q$; the field inside the scatterer is assumed to be equal to the incident one ($U = 1$). Equation (3.60) can also be put in a different way in terms of the complex phase $\Phi - \Phi_0 = \ln U$:

$$\Phi - \Phi_0 = \ln\left[1 + \hat{F}q\exp(\Phi - \Phi_0)\right]\,.$$

Therefore, the Born approximation is $\Phi - \Phi_0 = \ln(1 + \hat{F}q)$ or $U = \exp\left(\hat{F}q\right)$.

It is possible to suggest an approach involving the Rytov method as an approximation for iteration of (3.60):

$$U = 1 + \hat{F}\left[q\exp\left(\hat{F}q\right)\right]\,. \tag{3.61}$$

Calculation of small-angle scattering based on (3.61) is efficiently realized by reducing the Fresnel transform to the fast Fourier transform. The approach (3.61) includes the Born and Rytov approximations and the geometric optics approximation as limiting cases. The first terms of the expansion of the exponential multiplier in (3.61) coincide with those in the intrinsic time method.

In conclusion, let us briefly concentrate on questions concerning the uniqueness of the solutions of ISP obtained. Since all of the above solutions of ISP obtained amount to integral transforms like Fourier or Fresnel, the analysis of the solution uniqueness will be based on analyzing the corresponding properties of the transforms mentioned. Such an analysis has been performed above and conclusions concerning the uniqueness of the solutions for weak scatterers of the types (3.16), (3.21), (3.25) were given in Sect. 3.1. Similar reasoning can be applied also to the solutions discussed in Sect. 3.5. Note that the results on the uniqueness of ISP solutions are known for probing by an infinite set of plane monochromatic waves with differently oriented wave vectors and by a set of plane sounding waves with different frequencies [Berezanskiy, 1955; Hoenders, 1978; Stone, 1987].

3.6 Holographic Approach

As already mentioned in the introduction, nowadays, holographic methods are rapidly developing in applications to the wave fields of different physical natures. In radio techniques, the basic principles of holography (such as reference signal, spectral, spatial transforms, and so on) were used even earlier. However, an intensive development of optical holography and techniques for optical processing made it possible to reveal the fundamental universality of holography as applied to different fields and to obtain effective solutions of a series of problems in applied electrodynamics [Bakhrakh and Kurochkin, 1979]. The problem of three- or even two-dimensional imaging of ionospheric irregularities by holographic methods was stated long ago [Rogers 1956], but good practical results were obtained in this field only in the latest decade. Practical application of radio holography to the investigation of ionospheric irregularities commenced in works [Schmidt, 1972; Schmidt and Taurianen, 1975] where a one-dimensional hologram was synthesized from satellite radio signals measured by a single ground-based receiver and the dimensions and then the position of the irregularity were obtained from the hologram. Later on, other facilities were designed for radio holographic experiments [Stone, 1976b], too. These works, essentially, were based on point source models and made use of optical analogy: the field recorded in the hologram was reemitted "backward" (by numerical simulation) and from the maximum intensity of this field, the position and dimensions of the irregularity were estimated. A similar approach based on the principle of analogy, of course, cannot be used in recovering the true structure of the irregularities; here, solving the inverse scattering problem is required.

In the present section, the theory of the holographic approach will be generalized and its correlation with the foregoing ISP theory will be investigated. The holographic approach is developed in the works [Tereshchenko, 1984, 1987 Tereshchenko and Khudukon, 1981; Kunitsyn et al., 1989] and carried to practical implementation in the Polar Geophysical Institute [Tereshchenko, 1987; Kunitsyn et al., 1989; Tereshchenko et al., 1982]. The term "holography" standing for total record of the field may seem superfluous within the range of radio frequencies, because radio receivers can directly perform such a total recording of the field and holographic reconstruction itself reduces to mere mathematical operations over the data on the measured field. However, in our opinion, the holographic interpretation of this method of solving the ISP is not useless since it has a clear physical sense.

For the sake of simplicity, let us begin by considering the holographic approach in its high-frequency asymptotics, neglecting the effect of the regular ionosphere. Let us introduce the holographically reconstructed field E_{H} that is determined from experimental data on the field scattered from a certain surface S' (not necessarily closed) and is an analogue of the field of a true image known in holography

$$E_{\mathrm{H}}(\boldsymbol{r}) = \int \mathrm{d}\boldsymbol{S}' \, [E(\boldsymbol{r}') \, \nabla' G^*(\boldsymbol{r}-\boldsymbol{r}') - G^*(\boldsymbol{r}-\boldsymbol{r}') \, \nabla' E(\boldsymbol{r}')] \ , \quad (3.62)$$

where ∇' is an operator of differentiation with respect to the $\boldsymbol{r}'$ coordinate of the surface. The intensity I of the hologram of field E with reference wave A is proportional to $A^\star E + E^\star A + |A|^2 + |E|^2$. After the hologram is illuminated by reference wave A, terms are obtained proportional to the fields of imaginary $|A|^2 E$ and real $A^2 E^\star$ images. Remember that integral representation of the form of (3.62) with Green's functions of free space G in the case of a closed surface is equal to the field E (3.31) by virtue of the known Helmholtz–Kirchhoff theorem resulting from (3.9). But also a nonclosed surface of large enough size, as is well known, under certain conditions will give a field quite close to E (3.34), which just takes place in optical holography. Making use of this fact, the field of the real image E_{R} can be found as an integral (3.62) where E^* should enter in the integrand; therefore, $E_{\mathrm{H}} = E^*_{\mathrm{R}}$ will assume the shape of (3.62). In other words, the difference between E_{R} and E_{H} is insignificant, and the latter is introduced only for the convenience of the analysis; if the field of the real image is a field "running" off the surface of the hologram, the introduced field E_{H} is a field "running" to the hologram.

Substituting in (3.62) the expression for E from the Lippman–Schwinger integral equation and changing the order of integration, which is possible owing to the finiteness of $q(\boldsymbol{r},\omega)$ and the integrability of the other integrands, one obtains the following relation for E_{H}:

$$E_{\mathrm{H}}(\boldsymbol{r}) = H(\boldsymbol{r},\boldsymbol{r}_0) + \int \mathrm{d}^3 r_1 \, q(\boldsymbol{r}_1,k) \, E(\boldsymbol{r}_1) \, H(\boldsymbol{r},\boldsymbol{r}_1) \ , \quad (3.63)$$

where the kernel

$$H(\boldsymbol{r},\boldsymbol{r}_1) = \int \mathrm{d}S' \left[\frac{\partial G^*(\boldsymbol{r}-\boldsymbol{r}')}{\partial n'} \, G(\boldsymbol{r}'-\boldsymbol{r}_1) - G^*(\boldsymbol{r}-\boldsymbol{r}') \, \frac{\partial G(\boldsymbol{r}'-\boldsymbol{r}_1)}{\partial n'} \right] \quad (3.64)$$

is defined by the parameters of the recording system.

Equation (3.63), in contrast to the Lippmann–Schwinger equation (3.10), is an ordinary integral equation of the first kind with respect to the function qE, since the other part of the equality (3.63) contains the function $E_{\mathrm{H}}(\boldsymbol{r})$ determined from experimental data and known everywhere, including the interior of the scatterer. Due to the availability of (3.63), the holographic approach makes it possible to reduce the solution of ISP to that of an integral equation of the first kind. Before we proceed with solving the equation, let us consider the form of the kernel in various recording systems.

In a closed surface, H is easily calculated. Passing from integration over the surface to integration in space, according to Green's theorem, we get

$$H(\boldsymbol{r},\boldsymbol{r}_{\mathrm{s}}) = G(\boldsymbol{r}-\boldsymbol{r}_{\mathrm{s}}) - G^*(\boldsymbol{r}-\boldsymbol{r}_{\mathrm{s}}) = \frac{1}{2\pi \mathrm{i}} \frac{\sin k\,|\boldsymbol{r}-\boldsymbol{r}_{\mathrm{s}}|}{|\boldsymbol{r}-\boldsymbol{r}_{\mathrm{s}}|} \ . \quad (3.65)$$

An integral equation with such a kernel dependent on the coordinate difference was obtained in [Porter, 1970; Bojarski, 1981a] and then analyzed in [Porter and Devaney, 1982a,b; Bojarski, 1981b]. The difference between the formulas given in the present section and those in the above mentioned works is that in the latter, the field $E_{\mathrm{R}} = E^*_{\mathrm{H}}$ was as defined not by the total field E as in (3.62) but by the scattered field $E - E_0$. Therefore, in integral equations of the form (3.63) with the kernel (3.65) obtained by Porter and Boyarsky, the term $H(\boldsymbol{r}, \boldsymbol{r}_0)$ is absent. However, practical realization of measurements on a closed aperture is extremely difficult; therefore, the equation with the kernel given by (3.65) is useless for practical applications.

Measurements of the field on the plane $z_{\mathrm{R}} < z$, z_{s} give H in the form [Tereshchenko, 1987]

$$H(\boldsymbol{r}, \boldsymbol{r}_{\mathrm{s}}) = -\frac{\exp[\mathrm{i}k\,|\boldsymbol{r} - \boldsymbol{r}_{\mathrm{s}}|\,\mathrm{sgn}(z - z_{\mathrm{s}})]}{|\boldsymbol{r} - \boldsymbol{r}_{\mathrm{s}}|\,\mathrm{sgn}(z - z_{\mathrm{s}})} ,$$

which can be easily verified if one takes into account the obvious equality $H(\boldsymbol{r}_s, \boldsymbol{r}) = -H^*(\boldsymbol{r}, \boldsymbol{r}_s)$ and symmetry properties. Of primary interest for practical applications is field recording on a limited aperture in a given plane $z = z_{\mathrm{R}}$. Here, we will consider the rectilinear small-angle aperture realized in experiments on ionospheric tomography. Within the scope of paraxial approximation, assuming a distance between the object and the recording plane much higher than the object size and taking into account (3.32) and (3.33), one may use the relation

$$H(\boldsymbol{r}, \boldsymbol{r}_{\mathrm{s}}) \simeq \int \mathrm{d}S'\ [-2\mathrm{i}k\, G^*(\boldsymbol{r} - \boldsymbol{r}')\, G(\boldsymbol{r}' - \boldsymbol{r}_{\mathrm{s}})] .$$

Next, after the usual Fresnel expansion of the kind of (3.18), the expression for H is brought into the form

$$H(\boldsymbol{r}, \boldsymbol{r}_{\mathrm{s}}) = \frac{-2\mathrm{i}k \exp\left[\mathrm{i}k(z_{\mathrm{s}} - z) + \mathrm{i}\frac{k}{2}\frac{(\boldsymbol{\rho} - \boldsymbol{\rho}_{\mathrm{s}})^2}{z_{\mathrm{s}} - z}\right]}{(4\pi)^2(z - z_{\mathrm{R}})(z_{\mathrm{s}} - z_{\mathrm{R}})} \times \int \mathrm{d}^2\rho_{\mathrm{R}} \exp\left[-\mathrm{i}\frac{k}{2d}\,(\boldsymbol{t} - \boldsymbol{\rho}_{\mathrm{R}})^2\right] . \tag{3.66}$$

Integrating over a rectilinear aperture of the size of $2a \times 2b$ in the plane z_{R}, we obtain the formula

$$\begin{aligned} H(\boldsymbol{r}, \boldsymbol{r}_{\mathrm{s}}) \simeq{}& \frac{1}{8\mathrm{i}\pi\,|z - z_{\mathrm{s}}|} \exp\left[\mathrm{i}k\,(z - z_{\mathrm{s}}) + \mathrm{i}\frac{k}{2}\frac{(\boldsymbol{\rho} - \boldsymbol{\rho}_0)^2}{z_{\mathrm{s}} - z}\right] \\ &\times \{C[\gamma(a - t_x)] - \mathrm{i}\sigma S[\gamma(a - t_x)] + C[\gamma(a + t_x)] - \mathrm{i}\sigma S[\gamma(a + t_x)]\} \\ &\times \{C[\gamma(b - t_y)] - \mathrm{i}\sigma S[\gamma(b - t_y)] + C[\gamma(b + t_y)] - \mathrm{i}\sigma S[\gamma(b + t_y)]\} , \end{aligned} \tag{3.67}$$

where H is expressed in terms of the Fresnel integrals, $d = (z - z_{\mathrm{R}})(z_{\mathrm{s}} - z_{\mathrm{R}})/(z_{\mathrm{s}} - z)$, $\gamma = \sqrt{k/\pi\,|d|}$. The two-dimensional vector $\boldsymbol{t} = \boldsymbol{\rho}(z_{\mathrm{s}} - z_{\mathrm{R}})/$

$(z_s - z) - \boldsymbol{\rho}_s(z - z_R)/(z_s - z)$ is located in the recording plane $z = z_R$ and connects the origin with the intersection of the straight line joining the tips of vectors $\boldsymbol{\rho}$ and $\boldsymbol{\rho}_s$ with the recording plane, $\sigma = \operatorname{sgn} d$.

The normalized modulus of the function H at $\boldsymbol{r}_s = (0, 0, z_s)$ as a function of x, z ($y = 0$) is portrayed in Fig. 3.1a for the aperture with dimensions $2a = 10\,\text{km}$, $z_s = 300\,\text{km}$, $\lambda = 2\,\text{m}$. It can be seen that the typical scale of the kernel H modulus abatement along z is of the order of tens of kilometers. The "fine" structure of the kernel in the transverse direction is not resolved at given step sizes along x; it becomes distinguishable when zoomed in (see Fig. 3.1b). The typical oscillations scale is of the order of a few hundred meters. For example, it can be easily drawn that at $z = z_s$,

$$H(\boldsymbol{\rho}, \boldsymbol{\rho}_s = 0, z = z_s) = \frac{-8\mathrm{i}ka\,b}{(4\pi)^2(z_s - z_R)^2} \exp\left(\mathrm{i}\frac{k}{2}\frac{\boldsymbol{\rho}_s^2 - \boldsymbol{\rho}^2}{z_s - z_R}\right)$$
$$\times \frac{\sin(kxa)/(z_s - z_R)}{kxa/(z_s - z_R)} \times \frac{\sin(kyb)/(z_s - z_R)}{kyb/(z_s - z_R)}\,;$$

therefore the scale of oscillations along x is of the order of $\lambda z_s/a \sim 120\,\text{m}$. The phase of the function H minus the component $k(z_s - z)$ is shown in Fig. 3.2. At $z < z_s$, the phase is growing with ρ – the wave front is running off point $\boldsymbol{r}_s$. At $z > z_s$ the opposite is observed – the curvature of the wave front reverses the sign, and the wave is running to the point $\boldsymbol{r}_s$.

At larger distances $|z - z_s|$ in small-angle limits when

$$k\left(\boldsymbol{t} - \boldsymbol{\rho}_R\right)^2/2|d| \gg 1\,,$$

it is possible to pass to the asymptotic approximation in (3.67) or to integrate over the whole plane in (3.66):

$$H(\boldsymbol{r} - \boldsymbol{r}_s) \simeq -\frac{1}{4\pi\,(z_s - z)} \exp\left[\mathrm{i}k\,(z_s - z) + \mathrm{i}\frac{k}{2}\frac{(\boldsymbol{\rho} - \boldsymbol{\rho}_0)^2}{z_s - z}\right]. \tag{3.68}$$

In this case, it can also be seen that the phase of the function is "turning over": above $z = z_s$, the wave is moving to point $\boldsymbol{r}_s$, and lower down, closer to the recording plane, the wave runs off $\boldsymbol{r}_s$. The difference of (3.68) from paraxial approximation of the usual free-space Green function is just that the sign of the phase changes when passing through the plane $z = z_s$; besides, as seen from the more exact relation (3.67), singularity is missing at $\boldsymbol{r} = \boldsymbol{r}_s$.

Making use of (3.68), we obtain, instead of (3.63), an approximate relation for E_H

$$\begin{aligned} E_H(\boldsymbol{r}) = &-\frac{1}{4\pi\,(z_0 - z)} \exp\left[\mathrm{i}k\,(z_0 - z) + \mathrm{i}\frac{k}{2}\frac{(\boldsymbol{\rho} - \boldsymbol{\rho}_0)^2}{z_0 - z}\right] \\ &- \int \frac{q(\boldsymbol{r}_1)\,E(\boldsymbol{r}_1)}{4\pi\,(z_1 - z)} \exp\left[\mathrm{i}k\,(z_1 - z) + \mathrm{i}\frac{k}{2}\frac{(\boldsymbol{\rho} - \boldsymbol{\rho}_1)^2}{z_1 - z}\right] \mathrm{d}^3 r_1\,. \end{aligned} \tag{3.69}$$

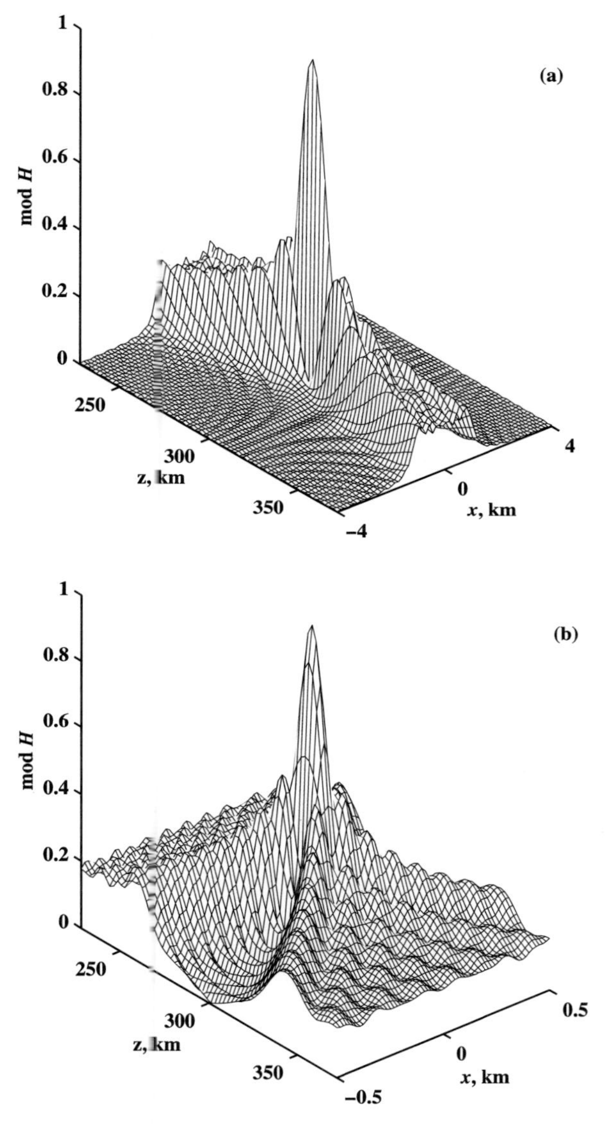

Fig. 3.1. Normalized modulus of the kernel of the holographic reconstruction equation in coordinates x–z for a rectangular aperture (with different zooming)

Comparing (3.69) with a similar paraxial approximation for (3.10), it can be readily seen that in between the scatterer and the hologram (recording plane z_{R}), the field E_{H} is similar to the field E (the scattered wave is running from the scatterer to the hologram), whereas in between the source and the scatterer, the curvature of the phase front reverses the sign (the scat-

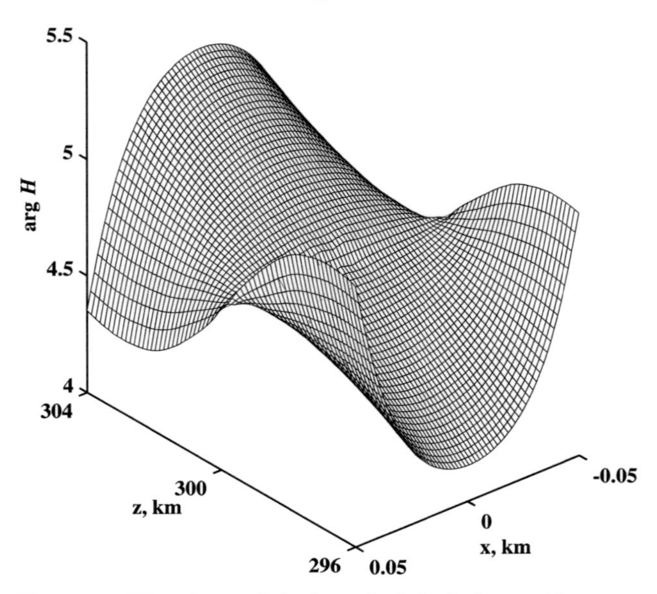

Fig. 3.2. The phase of the kernel of the holographic reconstruction equation in the same coordinates as in Fig. 3.1

tered wave is "focusing" from the source into the scatterer). This feature of a holographically reconstructed field makes it possible to locate the scattering object. By reconstructing the phase of the function E_{H} with a different z, one can determine the coordinates of the scatterer from the change of the phase front curvature. Examples of such location are given in [Tereshchenko, 1987].

Based on the representation (3.67) obtained for the kernel H, it is in principle possible to solve an integral equation of the first kind (3.63) for the function qE from the data on the field E_{H} being reconstructed. The solution of such an equation with the kernel given by (3.65) was analyzed in [Porter and Devaney, 1982a,b; Bojarski, 1981b]. In [Bojarski, 1981b], it is proposed to obtain the field E, inside the scatterer from the found function qE using the Lippmann–Schwinger equation (3.10), after which the solution of ISP reduces to the mere differentiation of $q = (\Delta + k^2)E/(qE)$. It is assumed that this approach to the solution of ISP is as well applicable to strong scatterers. However, the author of [Bojarski, 1981b] does not take into account the following essential fact: the field inside the scatterer varies with the change in the direction of the incident wave; therefore the function qE corresponding to a single direction of the probing wave cannot be reconstructed from a set of projections.

The solution of ISP applicable as well to strong scatterers was considered in Sect. 3.5. Here we will first concentrate on the ISP solution by means of the holographic approach within the scope of the Born and Rytov approximations. Substituting in (3.69) its zero-order approximation for E, i.e., E_0, we obtain

$$
\begin{aligned}
E_{\mathrm{H}} - E_0 = &-\frac{1}{(4\pi)^2 (z_0 - z)} \exp\left[\mathrm{i}k\,(z_0 - z) + \mathrm{i}\frac{k}{2}\frac{(\boldsymbol{\rho} - \boldsymbol{\rho}_0)^2}{z_0 - z}\right] \\
&\times \int \frac{\mathrm{d}^3 r_1}{\zeta_1}\, q(\boldsymbol{r}_1) \exp\left[\mathrm{i}\frac{k}{2}\frac{(\boldsymbol{s}_1 - \boldsymbol{\rho}_1)^2}{\zeta_1}\right] .
\end{aligned}
\tag{3.70}
$$

Here, when passing from (3.69) to (3.70), we used the equality

$$
\begin{aligned}
&\frac{(\boldsymbol{\rho} - \boldsymbol{\rho}_1)^2}{z_1 - z} + \frac{(\boldsymbol{\rho}_0 - \boldsymbol{\rho}_1)^2}{z_0 - z} = \frac{(\boldsymbol{\rho} - \boldsymbol{\rho}_0)^2}{z_0 - z} + \frac{(\boldsymbol{s}_1 - \boldsymbol{\rho}_1)^2}{\zeta_1} , \\
&\zeta_1 = \frac{(z_1 - z)(z_0 - z_1)}{z_0 - z} ,
\end{aligned}
$$

where

$$
\boldsymbol{s}_1 = \boldsymbol{\rho}\frac{z_0 - z_1}{z_0 - z} + \boldsymbol{\rho}_0 \frac{z_1 - z}{z_0 - z}
$$

is a two-dimensional vector in the plane $z = z_1$, connecting the origin with the intersection of the plane with the straight line joining the tips of $\boldsymbol{\rho}$ and $(\boldsymbol{\rho}_0)$. Generally, beyond the scope of the Born approximation, the paraxial approximation of the expression for E_{H} is of the following form:

$$
E_{\mathrm{H}}(\boldsymbol{r}) = E_0(\boldsymbol{r}) - \frac{E_0(\boldsymbol{r})}{4\pi} \int \mathrm{d}^3 r_1 \frac{q(\boldsymbol{r}_1)}{\zeta_1} \frac{E(\boldsymbol{r}_1)}{E_0(\boldsymbol{r})} \exp\left[\frac{\mathrm{i}k}{2\zeta_1}(\boldsymbol{s}_1 - \boldsymbol{\rho}_1)^2\right] .
\tag{3.71}
$$

Defining the complex phase of the holographically reconstructed field as $(E_{\mathrm{H}} - E_0)/E_0)$, we have

$$
\Phi_{\mathrm{H}} \equiv \frac{E_{\mathrm{H}} - E_0}{E_0} \simeq -\frac{1}{4\pi} \int \frac{\mathrm{d}^3 r_1}{\zeta_1} q(\boldsymbol{r}_1) \exp\left[\frac{\mathrm{i}k}{2\zeta_1}(\boldsymbol{s}_1 - \boldsymbol{\rho}_1)^2\right] .
\tag{3.72}
$$

Equation (3.72) is structurally similar to (3.20) but differs significantly from the latter in a physical sense. In (3.20), the complex phase is recorded in the plane of reception and after the distance to the scatterer is determined, it is possible to recover the structure of (3.21). Here, the holographic reconstruction of Φ_{H} is performed consecutively in different planes. When the plane of reconstruction coincides with the position of the scatterer, $1/\zeta_1 \to \infty$, and, if the typical scale of changes in q exceeds the scale of changes in the exponent, the integral (3.72) can be evaluated by the stationary phase method:

$$
\Phi_{\mathrm{H}}(\boldsymbol{\rho}, z_{\mathrm{s}}) \simeq -\frac{\mathrm{i}}{2k} \int q(\boldsymbol{\rho}, z_1)\ \mathrm{d}z_1 ,
\tag{3.73}
$$

i.e., q_z is proportional, within the scope of the Born and Rytov approximations, to the complex phase of a holographically reconstructed field.

The procedure itself of obtaining the complex phase of a holographically reconstructed field within a paraxial approximation also amounts to the Fresnel transform. Indeed, if the distance from the scatterer to the recording plane is sufficiently larger than the size of the scatterer, then $\partial G/\partial n' \simeq ikG$ and instead of (3.72), one can use the formula

$$\begin{aligned} E_{\mathrm{H}}(\boldsymbol{r}) &\simeq -2\mathrm{i}k \int \mathrm{d}^2\rho_{\mathrm{R}}\, E(\boldsymbol{r}_{\mathrm{R}})\, G^*(\boldsymbol{r}-\boldsymbol{r}_{\mathrm{R}}) \\ &\simeq \frac{ik}{2\pi} \int \mathrm{d}^2\rho_{\mathrm{R}}\, \frac{E(\boldsymbol{r}_{\mathrm{R}})}{z-z_{\mathrm{R}}} \exp\left[-\mathrm{i}k(z-z_{\mathrm{R}}) - \mathrm{i}\frac{k}{2}\frac{(\boldsymbol{\rho}-\boldsymbol{\rho}_{\mathrm{R}})^2}{z-z_{\mathrm{R}}}\right]. \end{aligned} \tag{3.74}$$

Expressing E in terms of the first approximation Φ_1 of the complex phase of the measured field $E \simeq E_0 + E_1 \simeq (1+\Phi_1)E_0$ and converting to new variables,

$$\boldsymbol{t}_0 = \frac{(z_0 - z_{\mathrm{R}})\boldsymbol{\rho}}{z_0 - z} - \frac{(z - z_{\mathrm{R}})\boldsymbol{\rho}_0}{z_0 - z}, \quad d_0 = \frac{(z - z_{\mathrm{R}})(z_0 - z_{\mathrm{R}})}{z_0 - z},$$

we obtain

$$\Phi_{\mathrm{H}} \simeq \frac{\mathrm{i}k}{2\pi d_0} \int \mathrm{d}^2\rho_{\mathrm{R}}\, \Phi_1(\boldsymbol{\rho}_{\mathrm{R}}) \exp\left[-\mathrm{i}\frac{k}{2}\frac{(\boldsymbol{t}_0 - \boldsymbol{\rho}_{\mathrm{R}})^2}{d_0}\right]. \tag{3.75}$$

From this, the possibility appears of synthesizing an aperture. It is not necessary to measure E or Φ_1 on a two-dimensional piece of a plane using a spatially fixed emitter. It is possible to measure the field on some line and to move the probing source along another line. Based on the definition of $\boldsymbol{t}_0$, it can be easily shown how variations in $\boldsymbol{\rho}_0$ are equivalent to some variations in $\boldsymbol{\rho}_{\mathrm{R}}$, which makes it possible to convert variations in $\boldsymbol{\rho}_0$ into equivalent ones in $\boldsymbol{\rho}_{\mathrm{R}}$. Next, making use of the identity

$$\frac{(\boldsymbol{t}_0 - \boldsymbol{\rho}_{\mathrm{R}})^2}{d_0} \equiv \frac{(\boldsymbol{s}\boldsymbol{\rho})^2}{\zeta}$$

and passing to integration with respect to $\boldsymbol{s}$ in (3.75), we arrive at

$$\Phi_{\mathrm{H}}(\boldsymbol{r}) \simeq \frac{ik}{2\pi\zeta} \int \mathrm{d}^2 s\, \Phi_1(\boldsymbol{s}) \exp\left[-\mathrm{i}\frac{k}{2}\frac{(\boldsymbol{s}-\boldsymbol{\rho})^2}{\zeta}\right].$$

This equality, with (3.73) taken into account, coincides with (3.21), which displays the equivalence, within the scope of the Born and Rytov approximations, of the holographic approach to the solution of ISP obtained in Sect. 3.2. If the scatterer is strong, the proposed reconstruction procedure within the holographic approach will reduce to reconstruction, using (3.73), of $Q_z(\boldsymbol{\rho})$ and not $q_z(\boldsymbol{\rho})$. Hence, all the results obtained in the previous section relating to strong scatterers remain valid.

The holographic approach to the ISP solution described can be as well generalized to a stratified background ionosphere. Here, when defining E_{H} (3.67) and H (3.68), one should use approximations (3.37) derived in Sect. 3.4 instead of the free-space Green function G. Then, also the expressions for E_{H} and H will change correspondingly. In particular, the kernel H within the paraxial approximation is given by the following formula, instead of (3.71):

$$\begin{aligned} H(\boldsymbol{r},\boldsymbol{r}_{\mathrm{s}}) \simeq & -\frac{2\mathrm{i}k}{(4\pi)^2 g\, g_{\mathrm{s}}} \exp\left[\mathrm{i}k\int_z^{z_{\mathrm{s}}} n\,\mathrm{d}z + \mathrm{i}\frac{k}{2}\frac{(\boldsymbol{\rho}-\boldsymbol{\rho}_{\mathrm{s}})^2}{g_{\mathrm{s}}-g}\right] \\ & \times \int \mathrm{d}^2\rho_{\mathrm{R}} \exp\left[-\mathrm{i}\frac{k}{2d}(\boldsymbol{t}-\boldsymbol{\rho}_{\mathrm{R}})^2\right] . \end{aligned} \tag{3.76}$$

Here, as previously,

$$g_{\mathrm{s}} = \int_{z_{\mathrm{R}}}^{z_{\mathrm{s}}} \frac{\mathrm{d}z}{n}, \quad g = \int_{z_{\mathrm{R}}}^{z} \frac{\mathrm{d}z}{n} \quad \text{are group paths},$$

$$\boldsymbol{t} = \frac{g_{\mathrm{s}}}{g_{\mathrm{s}}-g}\boldsymbol{\rho} - \frac{g}{g_{\mathrm{s}}-g}\boldsymbol{\rho}_{\mathrm{s}}, \quad d = \frac{g_{\mathrm{s}}\, g}{g_{\mathrm{s}}-g} .$$

In a similar way, the formulas for a holographically reconstructed field are transformed, too.

4. Diffraction Radio Tomography of Ionospheric Irregularities

4.1 Fresnel Diffraction Tomography

In forward scattering, the definition of the linear integrals $q_z(\boldsymbol{\rho}, \omega)$ (3.21) from the complex potential $q(\boldsymbol{r}, \omega)$ reduces the ISP to the problem of tomographic reconstruction, i.e., the problem of reconstructing an object from its projections. Achievements in X-ray tomography in recent decades have been responsible for the intense development of tomographic techniques for recovering the structure of nonuniform objects. Reconstruction algorithms applied in practical X-ray tomography are based on a rectilinear approximation of ray trajectories. Mathematically, such problems are reduced to reconstruction of the damping function or the refraction coefficient from the set of linear integrals, i.e., to reconstructing the object from its projections of smaller dimensions. X-ray radiation has been followed for tomographic purposes by practically all of the known kinds of radiation and waves. In tomographic studies using optical, ultrasound, radio, microwaves and other kinds of waves, linear ray approximation does not often lead to good results. Therefore, in recent years, the reconstruction methods making use of refraction and diffraction effects have been developing intensively, and a special term – diffraction tomography – has come into being.

The solutions obtained above of the inverse problem of recovering the scatterer structure in "forward" scattering, with diffraction effects taken into account, can also be considered as part of diffraction tomography. Here, it seems pertinent to use the term Fresnel diffraction tomography, or diffraction tomography in the Fresnel zone [Kunitsyn, 1986b], because in solving that given ISP, Fresnel paraxial approximation was applied. The ISP allowing for diffraction is reduced to a tomographic problem of reconstructing a three-dimensional object from two-dimensional projections $q_z(\boldsymbol{\rho}, \omega)$ (3.21). By turning either the object or the transmitting–receiving system (the z-axis) with respect to the object, it is possible to obtain a series of linear integrals $q_z(\boldsymbol{\rho}, \omega)$ that can be used in tomographic reconstruction of the diffraction coefficient. At the same time, recovering the two-dimensional structure from a set of one-dimensional "sections", i.e., from the functions measured by a single receiver as the satellite-borne transmitter is moving on, is also a tomographic problem. The more so as, from a physical point of view, the single-receiver measurements are integrals over cross sections of an irregularity spectrum, as will be shown below. A set of such integrals makes it possible to reconstruct a two-dimensional cross section of the spectrum, which is equivalent to reconstruction of the two-dimensional integrated projection $q_z(\boldsymbol{\rho}, \omega)$. Similarly, for IP in the framework of the holographic approach (3.72) and the solutions of IP allowing for strong scattering (3.48)–(3.50), (3.44) should also be related to tomographic reconstruction, because as well in these problems, two-dimensional projections are reconstructed from two-dimensional functions measured experimentally by the receiver. A set of two-dimensional projections allows tomographic reconstruction of a three-dimensional structure. The technique for tomographic reconstruction from linear integrals has been worked out quite well, so we shall not consider it here.

The influence of diffraction amounts to the fact that a linear integral (3.21) depends not only on the field "along the ray" but also on the field in the vicinity of intersection of the ray with the plane of measurements ($z = z_{\mathrm{R}}$). If the radiation frequency becomes high, then in the limit $\omega \to \infty$, the integral (3.21) must depend only on the field along the ray (the eikonal approximation with a rectilinear trajectory).

Let a plane sounding wave ($\boldsymbol{\rho}_0 = 0$, $z_0 \to \infty$) impinge on an object; then, by calculating integral (3.21) in the limit of large k ($kr_{\mathrm{m}}^2/2\zeta \gg 1$) using the saddle point method [assuming in accordance with the Rytov approximation that $V(\boldsymbol{s}, \omega) = -4\pi\zeta\Phi_1$], we get

$$q_z(\boldsymbol{\rho}, k) \equiv \int q(\boldsymbol{r}, k)\ \mathrm{d}z \simeq \left(\frac{k}{2\pi\zeta}\right)^2 V(\boldsymbol{\rho}_{\mathrm{R}})\ \frac{2\zeta}{k}\,(-\mathrm{i}\pi) = 2\mathrm{i}k\Phi_1(\boldsymbol{\rho}_{\mathrm{R}})\ . \quad (4.1)$$

In the approximation of geometric optics with a rectilinear ray, the incursion of the complex phase Φ on the object is equal to the integral of the complex refractive index $n(\boldsymbol{r}, \omega) = \sqrt{1 - q/k^2}$:

$$\Phi = \Phi_0 + \Phi_! + \cdots = \mathrm{i}k \int n(\boldsymbol{r}, \omega)\ \mathrm{d}z = \mathrm{i}k \int \sqrt{1 - \frac{q}{k^2}}\ \mathrm{d}z$$
$$\simeq \mathrm{i}k \int \left(1 - \frac{q}{k^2}\right)\ \mathrm{d}z\,.$$

Whence, for the first approximation of the complex phase, we have $\Phi_1 \simeq -\mathrm{i}q_z/(2k)$, which coincides with (4.1), i.e., linear integrals of the kind (3.21) in the limit of high frequencies depend only on the field along the ray. The latter statement means that the approximation applied here also contains, in the high-frequency limit, the approximation of geometric optics with rectilinear trajectories. Limitation in $\boldsymbol{s}$ of the field recording region will result in smoothing the function $q_z(\boldsymbol{\rho}, \omega)$ reconstructed from the limited data $\Phi_1(\boldsymbol{s})$. The equation kernel is defined by the detection system parameters and gives the Rayleigh limit of resolution (Sect. 4.2); then, $q_z(\boldsymbol{\rho}, \omega)$ is the integral taken along a ray of finite "thickness."

With a numerical example, we shall demonstrate the necessity of allowing for diffraction effects in tomographic reconstruction of objects with dimensions r_m comparable to the dimensions of the Fresnel zone $\sqrt{\lambda\zeta}$. As shown in (4.1), the diffraction effects are negligible at $kr_m^2/2\zeta \gg 1$ but can significantly distort the results at $kr_m^2/2\zeta \leq 1$. Figure 4.1a portrays $q_z(\boldsymbol{\rho})$ comprising two irregularities – two Gaussians ($\sim \exp(-\rho^2/r_i^2)$ with the parameters ($r_1/\sqrt{\lambda\zeta} = 5$, $r_2/\sqrt{\lambda\zeta} = 1/3$). If complex phase (3.20) is reconstructed after the radiation has passed through one large irregularity, then $\mathrm{Im}\Phi_1$ is, with high accuracy, proportional to the function $q_z(\boldsymbol{\rho})$ of the large irregularity, according to (4.1). At the same time, the imaginary part of the complex phase Φ_1 (Fig. 4.1b) of the pair of irregularities bears only little

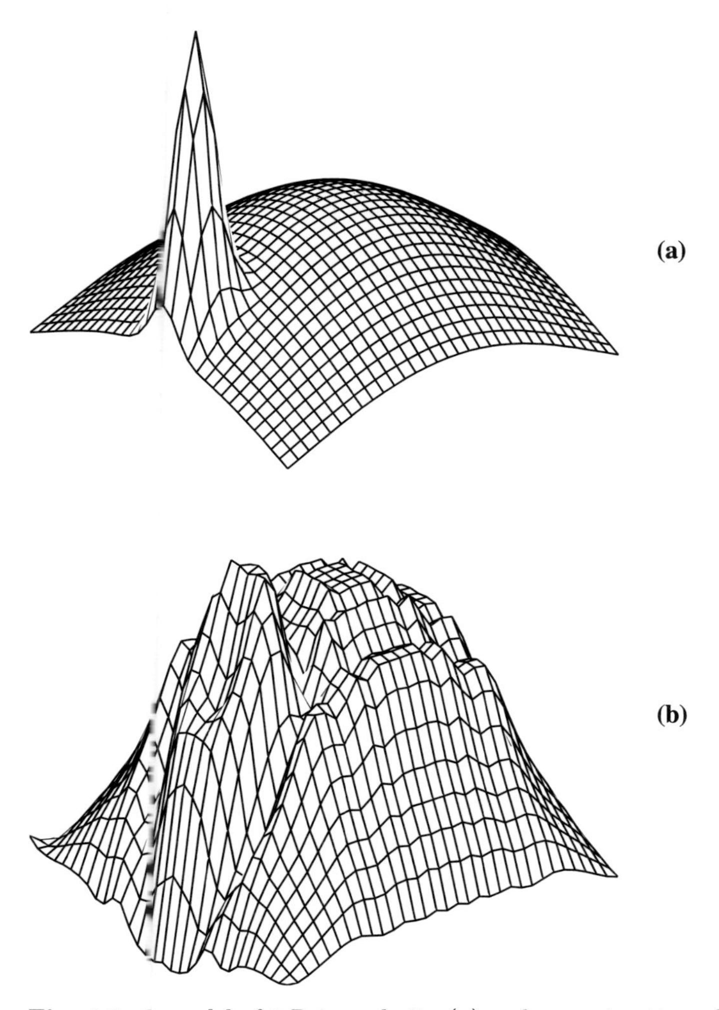

Fig. 4.1. A model of 2-D irregularity (**a**) and reconstruction of the projection of the model (**b**), diffraction effects neglected

resemblance to the projection of the object shown in Fig. 4.1a because of the diffraction effects in the small irregularity. If the Fresnel inversion (3.21) is applied, the pair of irregularities is surely reconstructed quite accurately.

It should be noted that the above solutions of the diffraction tomography problem can also describe diffraction effects in an absorbing medium, even in the presence of absorbing irregularities only, i.e., when variations of the real part of the complex potential can be neglected within a given frequency range. Let us describe in brief the diffraction effects at absorbing irregularities because they are less known. For simplicity, let $\mathrm{Re}\, q(\boldsymbol{r}, \omega) = \mathrm{const}$ and, without loss of generality, assume that $\mathrm{Re}\, q = 0$. One can easily see that a spatially limited absorbing irregularity will produce, in accordance with (3.9)–(3.14), diffraction effects when the irregularity dimensions are comparable to the Fresnel zone $\sqrt{\lambda\zeta}$. For example, the field of a thin absorbing disk of radius a in the far zone (the Born approximation)

$$E_1 \simeq E_0 \frac{-\gamma}{2k} \frac{ika^2}{\zeta} \exp\left(\frac{iks^2}{2\zeta}\right) \frac{J_1(ksa/\zeta)}{ksa/\zeta}$$

has an angular distribution given by the Bessel function. Here,

$$\gamma \equiv -\int \mathrm{Im}\, q(r, \omega)\, \mathrm{d}z > 0$$

is the integral coefficient of absorption. It should be recalled that $\mathrm{Im}\, q < 0$. The complex phase of the aperture in the absorbing screen (the Rytov approximation) is expressed in the following way:

$$\Phi_1 \simeq \frac{-\gamma}{2k} \left[1 + \frac{ika^2}{\zeta} \exp\left(\frac{iks^2}{2\zeta}\right) \frac{J_1(ksa/\zeta)}{ksa/\zeta}\right] .$$

The solutions of the problem of propagation through an unlimited absorbing layer within the Born and Rytov approximations have the form (the summary field $E = E_0 + E_1$, $E = \exp \Phi$)

$$E \simeq E_0 \left(1 - \frac{\gamma}{2k}\right) \text{ (Born approximation)};$$

$$\Phi_1 \simeq -\frac{\gamma}{2k},$$

$$E = E_0 \exp\left(-\frac{\gamma}{2k}\right) \text{ (the Rytov approximation)}.$$

Hence, it is clear that formulas (3.13), (3.14) do correctly describe weak absorption when the following expansion is valid:

$$\mathrm{i}kn = \mathrm{i}k\sqrt{1 - \frac{q}{k^2}} \simeq \mathrm{i}k\left(1 - \frac{\mathrm{Re}\, q}{2k^2}\right) + \frac{\mathrm{Im}\, q}{2k} .$$

Diffraction at irregularities of the absorption coefficient can be noticeable even when the influence and variations of the real coefficient of refraction can be neglected. Therefore, the diffraction effects mentioned are significant as well for tomography in the case of absorption if the dimensions of the irregularities studied are comparable with the Fresnel radius.

In the last decades, a lot of works on diffraction tomography have been published; for a bibliography, the reader is referred to the reviews [Langeberg, 1987; Tabbara et al., 1988; Bolomey, 1989; Burov et al., 1986; Bukhshtaber and Maslov, 1990]. Here, it would be pertinent, without pretending to completeness of the review, to give a brief description of the well-known studies and to compare their results with ours. Most studies are based on various modifications of the expression likely to have been obtained first in [Wolf, 1969] relating a three-dimensional Fourier-image $\hat{q}(\boldsymbol{\varkappa})$ of the function $q(\boldsymbol{r}, k)$ to the angular spectrum of the field $\hat{E}$ or its complex phase:

$$\begin{aligned}\hat{q}(\boldsymbol{\varkappa}) &\equiv \frac{1}{(2\pi)^3}\int q(\boldsymbol{r})\exp(\mathrm{i}\boldsymbol{\varkappa r})\,\mathrm{d}^3 r \\ &= \frac{\mathrm{i}}{\pi}(\varkappa_z + k)\exp[-\mathrm{i}\,(\varkappa_z + k)\,z]\,\hat{E}(\boldsymbol{\varkappa}_\rho, z)\;.\end{aligned} \tag{4.2}$$

Here, the incident sounding plane wave propagates along the z-axis; $E_0 = \exp(\mathrm{i}kz)$, $\hat{E}(\boldsymbol{\varkappa}_\rho, z) = (2\pi)^{-2}\int \mathrm{d}^2\rho\; E(\boldsymbol{\rho}, z)\exp(-\mathrm{i}\boldsymbol{\varkappa}_\rho\boldsymbol{\rho})$ is the angular spectrum. Each measurement of the scattered field on a plane gives a two-dimensional slice of the spectrum of the object cut by the surface of a sphere (the Ewald sphere): $(\boldsymbol{\varkappa} + \boldsymbol{k}_0)^2 = \boldsymbol{k}'^2 = k^2$, where $\boldsymbol{k}_0 = (0, 0, k)$ is the wave vector of the sounding wave and $\boldsymbol{k}'$ is that of the scattered wave. After the Fourier space is filled by turning the probing wave (z-axis) or the object, the function $q(\boldsymbol{r}, \omega)$ can be obtained using interpolation in the frequency or space domain by means of an inverse Fourier transform. Field measurement by a single satellite-borne receiver (along the x-axis) is equivalent to line reception. In this case, the Fourier transform (with respect to x) of the measured field is the Fourier transform of the spectrum $\hat{E}(\varkappa_x, \varkappa_y)$ by $\varkappa_y$. In the works on diffraction tomography of the last decades [Devaney, 1982, 1984; Mueller et al., 1979; Slaney and Kak, 1983; Dickens and Winbow, 1997; Johnson et al., 1984; Duchene et al., 1985; Langeberg, 1987; Tabbara et al., 1988; Bolomey, 1989; Leone et al., 1999; Rao and Carin, 2000, and others] different approximations, including the Born and the Rytov approximations, were studied as well as iterative schemes for stronger scatterers [Prosser, 1969, 1976; Johnson and Tracy, 1983; Ney et al., 1984; Lesselier et al., 1985; Datta and Bandyopadhyay, 1986; Caorsi et al., 1988; Burov et al., 1989; Park et al., 1996; Tsihrintzis and Devaney, 2000], which has been already mentioned in Sect. 3.5. Generalizations of diffraction tomography have been obtained from the algorithms known in ordinary tomography, in particular, the filtered back-propagation algorithm (FBP) [Devaney, 1982] generalizing the back-projection algorithm.

In recent years, a series of important modifications of the algorithm was proposed, in particular, the FBP algorithm version based on the expansion of the propagation operator in terms of eigenfunctions [Mast et al., 1997]. In [Xiaochuan et al., 1999], it was shown that for normal operation of the algorithm, it is enough to have a limited range of scanning aspects $(0-3\pi/2)$. The corresponding versions of algorithms with complete and limited sets of aspects are reported in [Chen et al., 2000; Anastasio and Pan, 2001]. In [Testorf and Fiddy, 1999], by means of the McCutchen theorem, the FBP algorithm is extended to more general experimental geometries. In [Mast, 1999], a multifrequency FBP algorithm is described that makes it possible to carry out 4-D tomographic reconstruction of nonstationary objects in the space-time domain.

On the one hand, as regards the formulation of the problem, the ISP being solved here has a more general sense since the complex potential $q(\boldsymbol{r},\omega)$ is reconstructed by means of probing point sources, including the stratified background medium (Sect. 3.4). On the other hand, the specificity of this particular physical problem where only small-angle approximation is used, makes it possible to apply the Fresnel paraxial approximation and to reduce the problem to Fresnel transforms (3.21), (3.24). Therefore, the results obtained within the small-angle approximation, in weak scattering, follow from (4.2). Indeed, making use of the paraxial approximation $\varkappa_\rho^2 \ll k^2$, the Ewald sphere surface can be substituted by a paraboloid:

$$\varkappa_z + k = -k + \sqrt{k^2 - \varkappa_\rho^2} + k \simeq k - \frac{\varkappa_\rho^2}{2k} \, .$$

Then, the angular spectrum of the field defines the plane section of the Fourier image of the irregularity

$$\hat{q}(\varkappa) \simeq \frac{\mathrm{i}k}{\pi} \exp\left[-\mathrm{i}\left(k - \frac{\varkappa_\rho^2}{2k}\right) z\right] \hat{E}_1(\varkappa_\rho, z) \simeq \hat{q}(\varkappa_\rho, \varkappa_z = 0) \, . \tag{4.3}$$

Substituting $\hat{q}(\varkappa_\rho, 0)$ in the formula for $q_z(\boldsymbol{\rho})$, we can write

$$\begin{aligned} q_z(\boldsymbol{\rho}, k) &\equiv \int q(\boldsymbol{r}, k)\,\mathrm{d}z = \int \mathrm{d}z \int q(\hat{\varkappa}) \exp(\mathrm{i}\varkappa \boldsymbol{r})\, \mathrm{d}^2\varkappa \\ &= 2\pi \int \hat{q}(\varkappa_\rho, 0) \exp(\mathrm{i}\varkappa_\rho \boldsymbol{\rho})\, \mathrm{d}^2\varkappa_\rho \, ; \end{aligned}$$

having integrated over $\mathrm{d}^2\varkappa_\rho$, we get a particular case of the Fresnel transform (3.21) with a plane sounding wave $V = -4\pi z E_1/E_0 = -4\pi z E_1 \exp(-\mathrm{i}kz)$.

Noteworthy are the advantages of the approach employing the Fresnel transform in many practically important cases of small receiving apertures. This approach to solving the ISP based on (3.21) makes it possible to apply and generalize to diffraction tomography a lot of methods developed in ordinary linear tomography, including three-dimensional tomography, and in some cases, it helps also to avoid using the interpolation procedure. The Fresnel transform itself inherently allows efficient computational realization by reducing to the fast Fourier transform. And, last and most important, such an approach permits solving the ISP from small-angle scattering data in the case of large scatterers, both weak and strong. It could be hardly expected that other iterative methods (different from those using asymptotic field approximations) would not be developed for solving such ISR for scatterers thousands and tens of thousands as large as the wavelength. We have started using this approach beginning from the works [Kunitsyn, 1986a; Tereshchenko and Khudukon, 1981] and made it applicable in practice (Sect. 4.6).

Measurements at one receiving chain perpendicular to the satellite path give data for reconstructing a two-dimensional image of the irregularity. Several receiving chains spaced about several hundred kilometers apart make it possible to recover the three-dimensional structure of the object from a set of its two-dimensional cross sections. The methods and algorithms of three-dimensional tomography were developing intensively in recent decades, including algorithms for small-angle tomography [Pikalov and Preobrazhenskiy, 1987a; Orlov, 1975a, 1975b; Ra and Cho, 1981; Minerbo et al., 1980]. In [Orlov, 1975a, 1975b], the conditions for the completeness of the set of two-dimensional projections were revealed. It was found that for the procedure of three-dimensional reconstruction to be proper, it is necessary and sufficient that a closed region on the sphere of aspect angles should have common points with any circumference of a large circle. Integrating over the whole semisphere contains excessive information. The case of limited angles lumped in a certain part of the semi-sphere was considered in [Ra and Cho, 1981]. Several schemes are known for measuring the projection data, in particular, the planar and equatorial ones. For the problem of three-dimensional radio tomography of ionospheric scattering irregularities, algorithms for small-angle tomography are required. Most algorithms developed for small-angle tomography belong to the class of iterative algorithms involving various techniques of allowing for a priori information. One of the commonest algorithms of this type is that of Herschberg–Papulis, to be exact, a series of algorithms of this kind.

4.2 Reconstructing Irregularities from a Limited Data Set

Let us introduce the data functions (characteristic functions) $B(\boldsymbol{p})$, $B(\boldsymbol{s})$, $B_\omega(\omega)$ defining the domain where the field is specified and equal to zero outside this region limited with respect to the variables $\boldsymbol{p}$, $\boldsymbol{s}$, ω. Inside the region, where the field is known, the functions $B(\boldsymbol{p})$, $B(\boldsymbol{s})$, $B_\omega(\omega)$ may be equal to some known function (with preliminary data processing) or to unity (without preliminary data processing). Let the functions $\mathcal{Q}(\boldsymbol{r})$ and $\mathcal{Q}_z(\boldsymbol{\rho},\omega)$

be defined as transformations of the experimental data on the field by formulas (3.16), (3.25), (3.21), respectively; the experimental data are known only within limited regions given by the introduced data functions, i.e.,

$$\mathcal{Q}(\boldsymbol{r}) = \int B(\boldsymbol{p})\, W(\boldsymbol{p}) \exp(\mathrm{i}\boldsymbol{p}\boldsymbol{r})\ \mathrm{d}^3 p\,; \tag{4.4}$$

$$\mathcal{Q}_z(\boldsymbol{\rho},\omega) = \left(\frac{k}{2\pi\zeta}\right)^2 \int B(\boldsymbol{s})\, V(\boldsymbol{s},\omega) \exp\left[-\frac{\mathrm{i}k\,(\boldsymbol{s}-\boldsymbol{\rho})^2}{2\zeta}\right] \mathrm{d}^2 s\,; \tag{4.5}$$

$$\begin{aligned}\mathcal{Q}(\boldsymbol{r}) = \int \frac{B_\omega(\omega)\exp(2\mathrm{i}kz)}{\pi c}\,\mathrm{d}\omega \\ \times \left(\frac{k}{2\pi\zeta}\right)^2 \int B(\boldsymbol{s})\, V(\boldsymbol{s},\omega) \exp\left[-\frac{\mathrm{i}k\,(\boldsymbol{s}-\boldsymbol{\rho})^2}{2\zeta}\right] \mathrm{d}^2 s\,.\end{aligned} \tag{4.6}$$

Similarly, it is possible to introduce $\tilde{N}(\boldsymbol{r})$, $\tilde{\nu}(\boldsymbol{r})$ via transformations (3.17) or (3.26). Then the functions $\mathcal{Q}(\boldsymbol{r})$ and $\mathcal{Q}_z(\boldsymbol{\rho},\omega)$ known from experiment are related to the corresponding solutions of the inverse problem $q(\boldsymbol{r})$ and $q_z(\boldsymbol{\rho},\omega)$ by the following integral equations.

While reconstructing from the field in the Fraunhofer zone,

$$\mathcal{Q}(\boldsymbol{r}) = \int q(\boldsymbol{r}')\, B(\boldsymbol{r}-\boldsymbol{r}')\,\mathrm{d}^3 r' \tag{4.7}$$

with the kernel

$$b(\boldsymbol{r}) = \frac{1}{(2\pi)^3}\int B(\boldsymbol{p}) \exp(\mathrm{i}\boldsymbol{p}\boldsymbol{r})\ \mathrm{d}^3 p\,.$$

While reconstructing from the field in the Fresnel zone (forward scattering),

$$\mathcal{Q}_z(\boldsymbol{\rho},\omega) = \int q_z(\boldsymbol{\rho},\omega)\, b(\boldsymbol{\rho},\boldsymbol{\rho}',\omega)\ \mathrm{d}^2\rho' \tag{4.8}$$

with the kernel

$$b(\boldsymbol{\rho},\boldsymbol{\rho}',\omega) = \left(\frac{k}{2\pi\zeta}\right)^2 \int B(\boldsymbol{s}) \exp\left\{\frac{\mathrm{i}k}{2\zeta}\left[(\boldsymbol{s}-\boldsymbol{\rho}')^2 - (\boldsymbol{s}-\boldsymbol{\rho})^2\right]\right\} \mathrm{d}^2 s\,.$$

While reconstructing from the field in the Fresnel zone (backward scattering),

$$\mathcal{Q}(\boldsymbol{r}) = \int q(\boldsymbol{r}')\, b(\boldsymbol{\rho},\boldsymbol{\rho}',z-z')\ \mathrm{d}^2\rho'\,\mathrm{d}z' \tag{4.9}$$

with the kernel

$$\begin{aligned}b(\boldsymbol{\rho},\boldsymbol{\rho}',z-z') = \int \frac{B_\omega(\omega)}{\pi} \exp[2\mathrm{i}k\,(z-z')]\ \mathrm{d}k \\ \times \left(\frac{k}{2\pi z}\right)^2 \int B(\boldsymbol{s}) \exp\left\{\frac{\mathrm{i}k}{2\zeta}\left[(\boldsymbol{s}-\boldsymbol{\rho}')^2 - (\boldsymbol{s}-\boldsymbol{\rho})^2\right]\right\} \mathrm{d}^2 s\,.\end{aligned}$$

By direct substitution of kernels b and integration, one can make sure that (4.7)–(4.9) reduce to determining $\mathcal{Q}(\boldsymbol{r})$ and $\mathcal{Q}_z(\boldsymbol{\rho},\omega)$ (4.4)–(4.6). Relations (4.7)–(4.9) are integral equations of the first kind with respect to the functions $\mathcal{Q}(\boldsymbol{r})$ and $\mathcal{Q}_z(\boldsymbol{\rho},\omega)$, since $\mathcal{Q}(\boldsymbol{r})$ and $\mathcal{Q}_z(\boldsymbol{\rho},\omega)$ become known after appropriate transformation of experimental data. The methods for solving such equations are known [Tikhonov and Arsenin, 1977; Khurgin and Yakovlev, 1971]. For exact measurements of the field, $q(\boldsymbol{r})$ and $q_z(\boldsymbol{\rho},\omega)$ can be reconstructed with an unlimited resolution. The accuracy of the reconstruction of the irregularity depends on the character of errors in the data $\mathcal{Q}(\boldsymbol{r})$ and $\mathcal{Q}_z(\boldsymbol{\rho},\omega)$. On the basis of integral equations (4.7)–(4.9), it is possible to create an algorithm for reconstruction of $q(\boldsymbol{r})$ and $q_z(\boldsymbol{\rho},\omega)$ tolerant of disturbances in $\mathcal{Q}(\boldsymbol{r})$ and $\mathcal{Q}_z(\boldsymbol{\rho},\omega)$. A particular realization of the algorithm depends on the noise level in the detection system [Khurgin and Yakovlev, 1971].

It should be noted that the functions $\mathcal{Q}(\boldsymbol{r})$ and $\mathcal{Q}_z(\boldsymbol{\rho},\omega)$ themselves are, as seen from (4.7)–(4.9), "smoothed" solutions of the ISP, i.e., are solutions of the ISP with certain resolution (the Rayleigh limit) [Khurgin and Yakovlev, 1971] defined by the width of the "peak" of the equation kernel, depending on the parameters of the receiving system. By solving (4.7)–(4.9), it is in principle possible to obtain a resolution higher than the Rayleigh limit. However, in ionospheric measurements, as well as in any other geophysical observations, the data are usually significantly noised, therefore a resolution only slightly exceeding the Rayleigh limit is possible.

Thus, the functions $\mathcal{Q}(\boldsymbol{r})$ and $\mathcal{Q}_z(\boldsymbol{\rho},\omega)$ found from experimental data measured within a limited domain where the field is given are the solutions of the inverse problem of reconstructing the scattering irregularities with the Rayleigh limit resolution, which is sufficient in most practical applications. Besides, such solutions of inverse problems with a finite resolution limit are tolerant of errors in measurements of the field in uniform metrics l^∞ (the more so in l^2 metrics), which is evident since $\mathcal{Q}(\boldsymbol{r})$ and $\mathcal{Q}_z(\boldsymbol{\rho},\omega)$ are combinations of (4.4)–(4.6) Fourier and Fresnel transforms of finite limited functions. Calculation of the system resolution amounts to estimating the width of the kernel peak of (4.7)–(4.9) from the known parameters of the recording transmitter–receiver system.

To take examples, let the domain of definition of $V(\boldsymbol{s},\omega)$ along the x- and y-axes have the limits $-s_\mathrm{m} \le s_x,\ s_y \le s_\mathrm{m}$, which can be achieved by varying the coordinates of both the transmitter $\boldsymbol{\rho}_0$ and the receivers $\boldsymbol{\rho}_\mathrm{R}$ along the corresponding axes, in accordance with the definition of $\boldsymbol{s}$ included in (3.20) or (3.23). Then, the estimate of the kernel modulus (4.8)–(4.9) ($B = 1$ in the measurement region) with $y = y'$, $z = z'$ has the form

$$|b(\varkappa - \varkappa', 0, 0)| \sim \left| \left[\frac{k(\varkappa - \varkappa')\, s_\mathrm{m}}{\zeta} \right]^{-1} \sin \left[\frac{k(\varkappa - \varkappa')\, s_\mathrm{m}}{\zeta} \right] \right| .$$

Hence, the resolution of such a recording system of detection along the x-axis (and similarly along the y-axis) is defined by the expression

$$\delta_y = \delta_x = \frac{\pi\zeta}{ks_{\mathrm{m}}} = \frac{\lambda}{2s_{\mathrm{m}}} \frac{(z_0 - z_{\mathrm{s}})(z_{\mathrm{s}} - z_{\mathrm{R}})}{(z_0 - z_{\mathrm{R}})} . \tag{4.10}$$

If the function $V(\boldsymbol{s}, \omega)$ is given over a narrow frequency range $\omega_0 - \Delta\omega \leq \omega \leq \omega_0 + \Delta\omega$, then for the kernel of (4.9), we can obtain (with $x = x'$, $y = y'$) the following:

$$|b(0, 0, z - z')| \sim \left| \frac{2\Delta\omega(z - z')}{c} \right|^{-1} \sin\left[\frac{2\Delta\omega(z - z')}{c} \right] .$$

Note that the resolution along z is inversely proportional to the spatial dimension of the pulse,

$$\delta_z = \frac{\pi c}{2\Delta\omega} = \frac{\pi}{2\Delta k} . \tag{4.11}$$

Estimates for a square receiving system located in the Fraunhofer zone orthogonal to the direction of the sounding wave propagation (the z-axis) and seen at the aperture angle $\Theta \ll 1$ from the irregularity lead to the known results:

$$\delta_x = \delta_y \sim \frac{\lambda}{\Theta} , \ \delta_z \sim \frac{\lambda}{\Theta^2} .$$

In a similar way, it is possible to determine the resolution of reconstructing the interior of the irregularities also in other, more complicated cases of field measurements, including the methods employing preliminary processing of the data. In particular, if multifrequency sounding ($\omega_0/c - \Delta \leq \omega/c \leq \omega_0/c + \Delta$) is applied and only one chain of receivers (one side of the square along the y-axis) is used, the resolution is

$$\delta_x \sim \frac{1}{\Theta\Delta} , \ \delta_y \sim \frac{\lambda_0}{\Theta} , \ \delta_z \sim \frac{1}{\Theta^2\Delta} ,$$

which is consistent with the results [Gusev and Kunitsyn, 1982].

4.3 Linear-Array and Single-Point Measurements

Particular cases of the foregoing results are linear-array measurements or reception of the signals from a satellite-borne transmitter by a single receiver. This is the simplest scheme to be realized in experiment, and therefore it is worthwhile discussing it in detail. As before, the satellite is assumed moving along the x-axis. One can easily see that in this case of forward scattering, the following integral of $q(\boldsymbol{r}, \omega)$ [Kunitsyn, 1986a] can be reconstructed:

$$\begin{aligned} \mathcal{Q}_{zy}(x) &\equiv \int q_z(\boldsymbol{\rho}, \omega) \exp\left(\frac{\mathrm{i}ky^2}{2\zeta} \right) \mathrm{d}y \\ &= \frac{k}{2\pi\zeta} \int B(s_x)\, V(s_x, 0) \exp\left[-\frac{\mathrm{i}k(s_x - x)^2}{2\zeta} \right] \mathrm{d}^2 s_x . \end{aligned} \tag{4.12}$$

In VHF radio probing with $\lambda = 2$ m and $\zeta \sim 200\,\mathrm{km}$, typical in the ionosphere, the value of $\sqrt{\zeta/k}$ is about 0.25 km; therefore, when integrating with respect to y in (4.12), a layer of the order of several hundred meters is being "cut out" within which the irregularities are integrated along z. In other words, relation (4.12) allows reconstructing a one-dimensional structure of irregularities $\mathcal{Q}_{zy}(x)$ integrated along z within a layer "cut out" by the plane containing the satellite path. The functional dependence $\mathcal{Q}_{zy}(x)$ gives an estimate of the scale of irregularities. Such dependence is obtained, as mentioned above, in a single-receiver experiment. The moving satellite emits a signal at a frequency ω. Having passed through the ionosphere, this signal is measured [either the complex phase $\Phi(s_x)$ or the field] by one ground-based receiver $\boldsymbol{\rho}_{\mathrm{R}} = 0$. In this case, the coordinate x_0 of the transmitter changes and $s_x = x_0(z_{\mathrm{s}} - z_{\mathrm{R}})/(z_0 - z_{\mathrm{R}})$. After transformation (4.12) of the detected phase, we get the dependence:

$$\mathcal{Q}_{zy}(x) = -\frac{2k(s_{\mathrm{s}} - z_{\mathrm{R}})}{z_0 - z_{\mathrm{R}}} \int \Phi(x_0) \exp\left[\frac{\mathrm{i}k}{2\zeta}\left(\frac{z_{\mathrm{s}} - z_{\mathrm{R}}}{z_0 - z_{\mathrm{R}}} x_0 - x\right)^2\right] \mathrm{d}x_0 \,.$$

It should be noted that to achieve the transverse resolution $\delta \sim 200\,\mathrm{m}$, (4.10) is necessary, i.e., a continuous phase measurement is required during an interval when the satellite covers the distance $2x_{\mathrm{m}} = \lambda(z_0 - z_{\mathrm{s}})/\delta \approx 7\,\mathrm{km}$ (with $z_0 - z_{\mathrm{s}} \sim 700\,\mathrm{km}$) which takes about one second.

Linear-array measurements of the field in backscatter allow recovering the two-dimensional structure of irregularities using multifrequency probing. This scheme can be realized by one stationary transmitter ($\boldsymbol{\rho}_0 = 0$) and a chain of the receivers arranged along the x-axis. Then, it follows from (4.5) that in narrowband ($|\Delta| \ll k_0$) multifrequency ($k = k_0 + \Delta$) sounding, it is possible to reconstruct the following integral of $q(\boldsymbol{r})$ with respect to y:

$$\begin{aligned}\mathcal{Q}_y(x,z) \equiv \int q(r) \exp\left(\frac{\mathrm{i}k_0 y^2}{2\zeta}\right) \mathrm{d}y &= \int \frac{B_\omega \exp(2\mathrm{i}kz)}{\pi c} \mathrm{d}\omega \\ &\times \frac{k}{2\pi\zeta} \int B(s_x)\, V(s_x, 0, \omega) \exp\left[-\frac{\mathrm{i}k(s_x - x)^2}{2\zeta}\right] \mathrm{d}s \,.\end{aligned} \tag{4.13}$$

At frequencies typical in VHF radio sounding, the value of $\sqrt{\zeta/x_0}$ is of the order of hundreds of meters; therefore, the functional $\mathcal{Q}_y(x,z)$ is a two-dimensional section of the ionosphere by the plane $y = 0$.

Similarly obtained are the relations for a holographically reconstructed field measured in a linear array. Let us introduce the function $\tilde{\Phi}_{\mathrm{H}}(x,z)$:

$$\tilde{\Phi}_{\mathrm{H}}(x,z) = \int \Phi_{\mathrm{H}}(\boldsymbol{r})\ \mathrm{d}y\ \exp\left(\frac{\mathrm{i}k}{2\zeta} y^2\right) .$$

Let us substitute Φ_{H} given by (3.75) in the formula and integrate the expression with respect to y and the resulting delta function $\delta(s_y)$ with respect to s_y; then,

$$\tilde{\Phi}_{\mathrm{H}}(x,z) = \mathrm{i} \int \mathrm{d}s_x \, \Phi_1(s_x, 0) \exp\left[-\frac{\mathrm{i}k}{2\zeta}(s_x - x)^2 \right] . \tag{4.14}$$

One can see that $\tilde{\Phi}_{\mathrm{H}}$ is a one-dimensional Fresnel transform of the data in the single-point linear-array measurements. On the other hand, substituting the expression for $\Phi_1(s_x, 0)$ from (3.20) in (4.14) and integrating it with respect to s_x, we obtain the relation between $\tilde{\Phi}_{\mathrm{H}}$ and the scattering potential:

$$\tilde{\Phi}_{\mathrm{H}}(x,z) = -\frac{\mathrm{i}}{2k} \int \mathrm{d}y' \, \mathrm{d}z' \exp\left(-\frac{\mathrm{i}ky'^2}{2\zeta} \right) q(x, y', z') . \tag{4.15}$$

The relation given is simplified in limiting cases. If the dimension of the Fresnel zone is much smaller than the typical scale of the potential q, then

$$\tilde{\Phi}_{\mathrm{H}}(x,z) \simeq -\frac{\mathrm{i}}{2k} \sqrt{\frac{2\pi\zeta}{k}} \exp\left(-\mathrm{i}\frac{\pi}{4} \right) \int \mathrm{d}z' \, q(x, 0, z') . \tag{4.16}$$

Otherwise, when the Fresnel zone is much greater than the scale of the potential q,

$$\tilde{\Phi}_{\mathrm{H}}(x,z) \simeq -\frac{\mathrm{i}}{2k} \int \mathrm{d}z' \, \mathrm{d}y' \, q(x, y', z') . \tag{4.17}$$

Thus, the relations obtained make it possible to reconstruct some transformations of the scattering potential in single-point linear-array measurements owing to the movement of the satellite.

4.4 Diffraction Radio Tomography Based on Discrete Data

In reconstructing ionospheric irregularities in radio tomography experiments, questions concerning data discretization and reconstruction procedures are essential. Sounding radio signals are always recorded by point-like receivers. The signal of each receiver recording radio waves from a moving satellite-borne transmitter is also sampled by the electronic circuit at a given sampling frequency. And, finally, integral transformations used in reconstruction of the scattering irregularities can be realized as numerical computations only after being appropriately discretized. Thus, generally, both the type of the data recorded and the nature itself of numerical processing and reconstructing the objects necessitate performance of purely digital operations. However, in most cases when considering theoretically solutions of the ISP, it does not seem advisable to turn immediately to discrete formulas. Discrete analogues of reconstruction formulas are, as a rule, rather cumbersome; besides, a digital form complicates analyzing the results obtained. On the other hand, transition from continuous relations to their discrete analogues in most cases does not cause much difficulty and is performed uniquely. Therefore, in view of further practical reconstruction, the continual formulas may be considered a compact form of their digital analogues.

Here, the couple of transformations (3.20), (3.21) are considered to illustrate the transition to discrete analysis of the reconstruction procedure and related questions. Methods for digital processing of signals and fields have been adequately developed and covered in the literature [Prett, 1982; Dadzhion and Mersero, 1988; Yaroslavskiy, 1987]. So the task of this section is to describe in brief the application of the known results to particular integral transformations.

To begin with, in (3.20), (3.21), we shall change to the dimensionless variables $\boldsymbol{P} = \boldsymbol{\rho}/\sqrt{\lambda\zeta}$, $\boldsymbol{S} = \boldsymbol{s}/\sqrt{\lambda\zeta}$ normalized to the Fresnel zone radius. Introducing the quantities

$$F_q(\boldsymbol{P}) \equiv \frac{q_z(\boldsymbol{\rho})\,\lambda}{4\pi}\exp\left(\mathrm{i}\pi P^2\right) \text{ and } F_v(\boldsymbol{S}) \equiv \frac{V(\boldsymbol{s})}{4\pi\zeta}\exp\left(-\mathrm{i}\pi S^2\right) \tag{4.18}$$

from (3.20), (3.21), we obtain for them a pair of integral Fourier transforms

$$\begin{aligned} F_q(\boldsymbol{P}) &= \int F_v(\boldsymbol{S})\exp(-2\mathrm{i}\pi\boldsymbol{SP})\,\mathrm{d}^2S\,,\\ F_v(\boldsymbol{S}) &= \int F_q(\boldsymbol{P})\exp(+2\mathrm{i}\pi\boldsymbol{SP})\,\mathrm{d}^2P\,. \end{aligned} \tag{4.19}$$

By definition, $F_q(\boldsymbol{P})$ is a finite function due to $q_z(\rho)$ being finite. Let the object dimensions along the x-axis not be outside the segment $[-P_{0x}, P_{0x}]$, along the y-axis – $[-P_{0y}, P_{0y}]$, respectively, i.e., the support of the function $F_q(\boldsymbol{P})$ is included in the given rectangle. Then $F_v(\boldsymbol{S})$ will be an infinite analytical function with finite spectrum $F_q(\boldsymbol{P})$. According to the Kotelnikov–Shannon theorem, $F_v(\boldsymbol{S})$ is represented in the form of an infinite series in functions of counts with sampling intervals $\Delta S_x = 2/P_{0x}$, $\Delta S_y = 2/P_{0y}$:

$$\begin{aligned} F_v(S_x, S_y) = \sum_{m_x}\sum_{m_y} F_v\left(\frac{x^m}{2P_{0x}}, \frac{y^m}{2P_{0y}}\right) \sinh(2\pi P_{0x}S_x - \pi m_x)\\ \times \sinh(2\pi P_{0y}S_y - \pi m_y)\,. \end{aligned}$$

By substituting the above representation in (4.19) and integrating with respect to S_x, S_y, we arrive at the relation

$$F_q(\boldsymbol{P}) = \sum_{m_x}\sum_{m_y} F_v\left(\frac{m_x}{2P_{0x}}, \frac{m_y}{2P_{0y}}\right)\frac{\exp[-2\mathrm{i}\pi\,(P_x\Delta S_x m_x + P_y\Delta S_y m_y)]}{2P_{0x}2P_{0y}}\,. \tag{4.20}$$

Making use of the notion of a signal really indistinguishable at the level ε [Slepyan, 1976], it is advisable to introduce a finite function that is really indistinguishable from $F_v(\boldsymbol{S})$ at the given level. Let the support of this function be included in the rectangle $[-S_{0x}, S_{0x}] \times [-S_{0y}, S_{0y}]$. Then, according to the Landau–Pollack theorem [Slepyan, 1976; Landau and Pollak, 1961], the approximate dimension of the set of all functions $F_v(\boldsymbol{S})$ that are finite at the given level with support in the rectangle $[-S_{0x}, S_{0x}] \times [-S_{0y}, S_{0y}]$ and the finite spectrum $\sup F_q(\boldsymbol{P}) \in [-P_{0x}, P_{0x}] \times [-P_{0y}, P_{0y}]$, is close to $N_x N_y = (2S_{0x} \times 2P_{0x})(2S_{0y} \times 2P_{0y})$. To put it differently, because of the finite accuracy in measuring signals, both the signal and its spectrum can be considered finite. Limitation of the spectrum support leads to a finite sampling frequency and $F_q(\boldsymbol{P})$, $\Delta P_x = 1/2S_{0x}$, $\Delta P_y = 1/2S_{0y}$, with the number of discretization intervals along the Cartesian axes being equal to $N_x = 2S_{0x}/\Delta S_x = 2P_{0x}/\Delta P_x$, $N_y = 2S_{0y}/\Delta S_y = 2P_{0y}/\Delta P_y$. Hence, assuming $P_x = n_x \Delta P_x$, $P_y = n_y \Delta P_y$, one can easily proceed from (4.18), (4.19) to a pair of discrete transformations for $\tilde{F}_v = F_v/\Delta P_x \Delta P_y$ and $\tilde{F}_q = F_q$:

$$
\begin{aligned}
\hat{F}_q(n_x, n_y) &= \frac{1}{N_x N_y} \sum_{m_x=-N_x/2}^{N_x/2-1} \sum_{m_y=-N_y/2}^{N_y/2-1} \hat{F}_v(m_x, m_y) \\
&\quad \times \exp\left[-2\mathrm{i}\pi\left(\frac{m_x n_x}{N_x} + \frac{m_y n_y}{N_y}\right)\right], \\
\hat{F}_v(m_x, m_y) &= \sum_{n_x=-N_x/2}^{N_x/2-1} \sum_{n_y=-N_y/2}^{N_y/2-1} \hat{F}_q(n_x, n_y) \\
&\quad \times \exp\left[-2\mathrm{i}\pi\left(\frac{m_x n_x}{N_x} + \frac{m_y n_y}{N_y}\right)\right].
\end{aligned}
\tag{4.21}
$$

Hereinafter in numerical modeling, the algorithms of the fast Fourier transform are applied and the numbering of sums m_x, n_x is shifted from 0 to $N_x - 1$; the same for m_y, n_y.

It is the dimensions of the reception region S_{0x}, S_{0y} and, consequently, the sampling intervals ΔP_x, ΔP_y that determine the resolutions introduced in Sect. 4.2. The minimal resolvable interval δ_x over x is equal to $\delta_x = \sqrt{\lambda\zeta}$, $\Delta P_x = \sqrt{\lambda\zeta}/2S_{0x} = \lambda\zeta/2S_{\mathrm{m}}$, which is perfectly consistent with (4.10). Limiting the reception region always leads to a limiting resolution or discretization interval as well as finite spectrum functions. Therefore, generally speaking, in a limited reception region, it is possible to proceed immediately from (4.20) to the finite discrete Fourier transform (4.21).

The transition from other continual reconstruction formulas containing Fourier and Fresnel transforms to their discrete forms can be performed in accordance with the above scheme.

4.5 Effects of Noise and Distortion on the Reconstruction of Irregularities

In practice, recovering the structure of irregularities depends, to a considerable extent, on the insensitivity (stability) of the reconstruction procedure to various noises and distortions that inevitably arise in measurement and experimental data processing. The influence of distortions and noises can be estimated by numerical simulation of the reconstruction process. Let us exemplify the modeling procedure by a brief description of two-dimensional Fresnel reconstruction. Here, the function $q_z(\boldsymbol{\rho})$ is given that characterizes a two-dimensional structure of an irregularity of electron density and an efficient collision frequency. Then, using formula (4.18), the value of $F_q(\boldsymbol{P})$ is calculated. Having performed the discrete Fourier transform (4.21) corresponding to (4.19), we obtain $F_v(\boldsymbol{S})$ and, hence, the field $V(\boldsymbol{S})$ on a discrete grid. This step of modeling relates to the forward problem of scattering, with the distance thought to be easily calculated within appropriate approximations. Next, a noise can be added to the data obtained on the field, then the inverse discrete Fourier transform (4.21) is performed corresponding to (4.19), and finally we obtain the result of the reconstruction of $\tilde{q}_z(\boldsymbol{\rho})$. The influence of distortions is simulated in the same way: the transform parameters are changed or "distorted" at the step of the inverse transform.

Numerical modeling of the influence of noises showed sufficient stability in the reconstruction procedure. This was not unexpected, as mentioned above, since the solution of the ISP of the type (4.4)–(4.6) with a Rayleigh resolution limit, obtained from the data measured in a limited region, is resistant to perturbations in the data recorded. Figure 4.2a portrays three real irregularities $q_z(\boldsymbol{\rho})$ – three Gaussians of different sizes on a 32×32 grid. The actual size of the image frame in units of the Fresnel radius scale size $\sqrt{\lambda\zeta}$ is 5×5. After the complex phase of the field scattered by these irregularities was calculated, the data V were distorted by noise. Then, the reconstruction was carried out; the quality of that proved quite high even with comparatively big errors in the data on the field [Andreeva and Kunitsyn, 1989]. For example, Fig. 4.2b shows the result of reconstruction of the given irregularities from the data on Φ_1 with an additive Gaussian complex noise; the rms of the noise was 0.05 times the maximum amplitude of the changes in the real and imaginary parts of the field complex phase. Figure 4.2c displays reconstruction with twice as much noise in the data (rms being 0.1). If preliminary data processing is done, even higher noise levels will not appreciably affect the reconstruction results.

Problems of optimizing the parameters of the measuring system will not be considered here. Suffice it to say that the theory for linear transformation of the data has been elaborated quite well. In radio tomography, the situation is somewhat more complicated. Generally, the problems of radio probing of ionospheric irregularities are not reduced to a linear setup scheme, i.e., the "scatterer → field" transformation is not linear, since the kernel of inte-

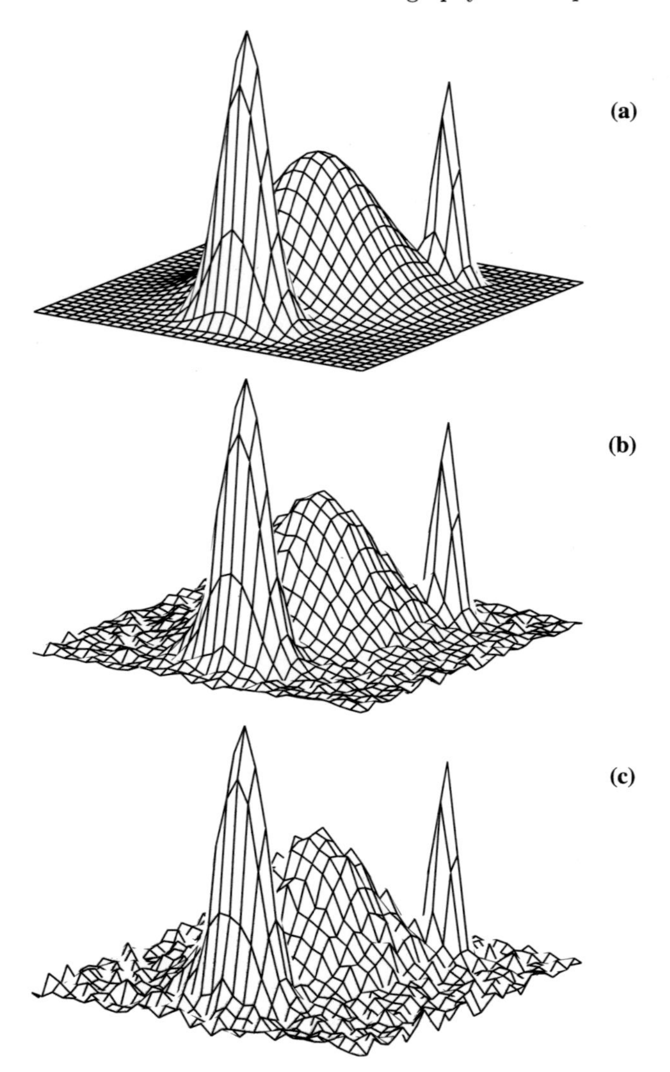

Fig. 4.2. A model of a 2-D irregularity (**a**) and reconstructions of the irregularity projection with 5% (**b**) and 10% (**c**) Gaussian noise in the data

gral equation (3.10) depends on the scatterer in an unknown way. However, even with approximate approaches, for example, the Born approximation where the kernel is determined explicitly, the problem of optimizing radio sounding is to be stated in a different way. The setup parameters (geometry and sensitivity) define the kernel of the integral transform; therefore, the optimization problem should be stated generally, including the kernel parameters, i.e., the "scatterer $\rightarrow$ field" transformation operator itself should be selected by the optimization algorithm from a certain given class. Such a general statement of the problem has been poorly studied; only some particular results in spectroscopy are reported [Preobrazhenskiy and Sedel'nikov, 1984]. If the parameters of the recording system are fixed, then, within linear approximations where the reconstruction procedure reduces to Fourier and/or Fresnel transforms, the analysis of optimization problems is simplified. Owing to the known properties of these transformations, the problem of optimized data processing and suppressing the noises amounts to the well-elaborated procedure of data filtering and subsequent inverse transformation [Prett, 1982; Dadzhion and Mersero, 1988; Yaroslavskiy, 1987].

For practical applications, numerical estimation of noise effects is of interest. It is suitable to describe the level of noise effects in the l^∞ and l^2 metrics. In a discrete reconstruction procedure, it is more convenient to use their discrete, normalized analogues, i.e., the difference between the function reconstructed from noised data $\tilde{f}$ and the true function f will be estimated by the quantities

$$\delta(l^\infty) = \frac{\max\limits_i \left| f_i - \tilde{f}_i \right|}{\max\limits_i |f_i|} ; \quad \delta(l^2) = \left[\frac{\sum\limits_i \left(f_i - \tilde{f}_i \right)^2}{\sum\limits_i f_i^2} \right]^{1/2} , \tag{4.22}$$

where summation and the choice of the maximum are performed over all discretes of the function to be reconstructed.

Figure 4.3 displays, on a bilogarithmic scale, the dependence of a normalized disturbance of reconstructed two-dimensional structures [i.e., the differ-

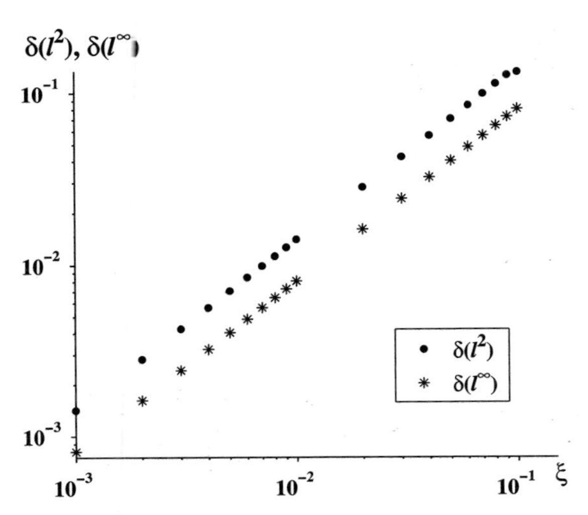

Fig. 4.3. Normalized disturbance of the reconstruction of the model in Fig. 4.2a as a function of normalized noise

ence between the true real function $q_z(\boldsymbol{\rho})$ and the reconstructed function $\tilde{q}_z$ in the $\delta(l^\infty)$ and $\delta(l^2)$ metrics] on the amplitude Δ of normalized (to the maximum amplitude of the field variation) complex Gaussian noise. In brief, Fig. 4.3 shows the dependence of a normalized disturbance on a normalized noise. Simulation was performed for the irregularities portrayed in Fig. 4.2a. One can see that normalized deviations of the recovered two-dimensional structures are comparable with the noise level. The absolute disturbances in the imaginary part of the reconstructed function are close to those in the real part, although the true function is real.

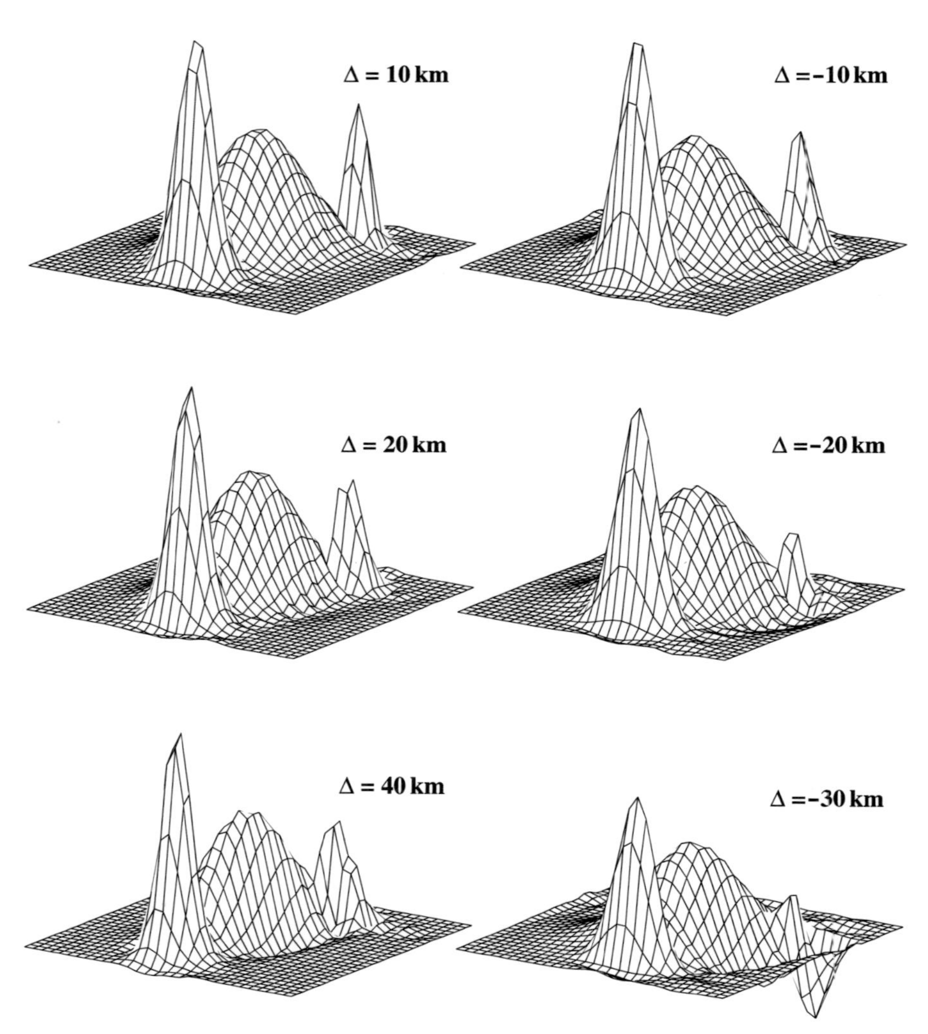

Fig. 4.4. Reconstruction of the model in Fig. 4.2a with different errors in determining the coordinates of irregularities

Distortions of irregularities reconstructed by radio tomographic methods arise because of inaccurately determined distances to the irregularities and, as a consequence, incorrectly given parameters of integral transformations. The effect of distortions was modeled according to the scheme described above [Andreeva and Kunitsyn, 1989]. For the sake of clearness, the same model (Fig. 4.2a) was chosen as an object to be reconstructed. The inaccurately determined coordinate of the scatterer $z'_{\mathrm{s}} = z_{\mathrm{s}} - \Delta$ with an error Δ was substituted in the parameters of the Fresnel transform (3.21). In calculations, numerical values of $z_0 - z_{\mathrm{s}} = 700\,\mathrm{km}$, $z_{\mathrm{s}} - z_{\mathrm{R}} = 300\,\mathrm{km}$, $\lambda = 2\,\mathrm{m}$ typical in satellite sounding were used, with the Fresnel radius scale size $\sqrt{\lambda\zeta}$ being 0.65 km, the frame size being $3.2 \times 3.2\,\mathrm{km}$, and the transverse resolution δ being of the order of a hundred meters. Figure 4.4 portrays the reconstructed two-dimensional structures with different errors in determining the coordinates Δ measured in kilometers; Δ is successively equal to $+10, -10, +20, -30, +40\,\mathrm{km}$. One can see that errors of tens of kilometers have a small effect on reconstruction. Errors Δ of the order of 20 kilometers do distort the object, but its prominent details are still reconstructed quite well. Only errors exceeding 30–40 km noticeably distort the reconstructed irregularity and can therefore lead to incorrect conclusions about the structure of the object. The values of reconstruction errors in normalized metrics (4.22) are shown in Fig. 4.5. Reconstruction errors have a well-pronounced minimum and within the range of $\pm(5 - 10)\,\mathrm{km}$ the errors are not big ($\leq 0.1 \div 0.2$). Note that this minimum coincides in order of magnitude with the longitudinal δ_z resolution of the system; here, the

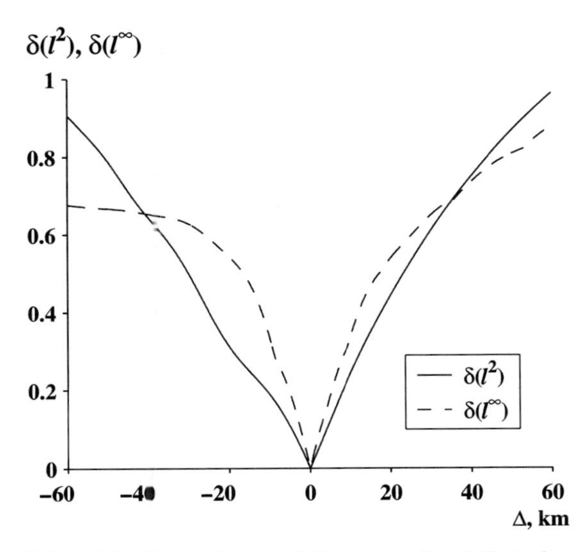

Fig. 4.5. Dependence of the normalized disturbance of the reconstruction of the model in Fig. 4.2a on the errors in determining the coordinates of the irregularities

transversal resolution $\delta \sim \lambda/\Theta \sim 100\,\mathrm{m}$, the aperture angle $\Theta \sim 0.02$, and $\delta_z \sim \lambda/\Theta^2 \sim 5\,\mathrm{km}$. Therefore, for a high-quality reconstruction, it is sufficient to determine the distance to the scatterer with an accuracy of longitudinal resolution of the recording system. The method for determining the distance from variations in the "phase" curvature of the reconstructed function (see Sect. 3.3) makes it possible to achieve this accuracy, even with considerable noise.

4.6 Reconstruction of Isolated Irregularities

The reconstruction of a 3-D structure of isolated irregularities with scale sizes significantly exceeding the wavelength but at the same time comparable with the Fresnel radius can be divided into two steps. In the first step, a two-dimensional structure of irregularities integrated in the direction of the wave propagation is recovered; next, by performing measurements at different angles, one can obtain the three-dimensional structure of the scatterer by tomographic reconstruction of the two-dimensional data. In the further consideration, it is more convenient to analyze reconstructing two-dimensional structures in terms of the holographic approach (Sect. 3.6). As shown in Chap. 3, an equivalent alternative is also possible when two-dimensional reconstruction is carried out by integral transform of the field data after the distance to the scatterer has been obtained. Experiments on 3-D reconstruction of irregularities that require the field to be measured at a few spaced receiving chains have never been carried out up to now. In this section, results will be presented of the 2-D reconstruction of data obtained at a single chain of receivers. Note that the methods used here for recovering 2-D structures from a set of one-dimensional data measured by each receiver also should be attributed to the scope of tomography since here reconstruction is done from a set of "projections".

If the amplitude and phase of satellite signals are measured at a chain of receivers perpendicular to the ground projection of the satellite path, one can synthesize a 2-D hologram, then carry out 3-D reconstruction of the holographic field and find the 2-D structure as a set of two-dimensional distributions of its amplitude and phase (real and imaginary parts) for a series of consecutive height levels in between the satellite and the ground level. From the behavior of the phase of the holographically reconstructed field, the location of the scatterer can be determined (see Sect. 3.6). And a 2-D structure of irregularities integrated in the direction of the sounding wave propagation is directly related to the holographically reconstructed field (3.73).

Input data for holographic reconstruction are the quantities (2.43) measured in the experiment:

$$I = (A_0 + \Delta A)\exp(\mathrm{i}\phi_\mathrm{d} + \mathrm{i}\Delta\phi) .$$

The fact that the changes in the field along the x and y axes are measured at different levels (at ground level and at satellite altitude) results in the necessity to pass from a real experiment to an equivalent hologram. Calculation is carried out in accordance with (3.75). The discrete analogue of formula (3.75) for reconstructing the phase of a holographically recovered field is

$$\Phi_{\mathrm{H}}(l\Delta x, m\Delta y) = \sum_{l'=-L/2}^{L/2-1} \sum_{m'=-M/2}^{M/2-1} \Phi_1(al', bm') \times \exp\left[\frac{-\mathrm{i}k}{2\zeta}\left(a^2 l'^2 + b^2 m'^2\right) + 2\pi\mathrm{i}\left(\frac{ll'}{L} + \frac{mm'}{M}\right)\right] \times \frac{\mathrm{i}k}{2\pi\zeta} \exp\left[-\mathrm{i}\frac{2\pi^2\zeta}{k}\left(\frac{l^2}{L^2a^2} + \frac{m^2}{M^2b^2}\right)\right], \quad (4.23)$$

$$s_x = \frac{z_0 - z}{z_0 - z_{\mathrm{R}}}(\Delta x_0)l \equiv al\,, \quad s_y = \frac{z - z_{\mathrm{R}}}{z_0 - z_{\mathrm{R}}}(\Delta y_{\mathrm{R}})m \equiv bm\,,$$

$$\zeta = \frac{(z - z_{\mathrm{R}})(z_0 - z)}{z_0 - z_{\mathrm{R}}}\,,$$

where L is the number of points from which radiation is received, M is the number of receivers, and $x = \dfrac{\lambda\zeta}{a\,L}$, $y = \dfrac{\lambda\zeta}{bM}$ are elementary intervals along the x and y axes.

A holographically reconstructed field can be calculated in (4.23) at different heights by varying parameter z. The resulting information can be displayed as plots of the amplitude and phase of the reconstructed field at a series of altitudes.

Most difficult in processing the data is locating the scatterer. In optics, this problem is solved by means of a priori information about the holographed object using a subjective criterion for the sharpness of the screen edge. It was shown earlier by analyzing models [Tereshchenko, 1987] that such a criterion does not always provide true information when considering the scatter from large-scale irregularities. Besides, difficulties arise for sure, caused by the necessity to discern the image in the background traces of defocused images of irregularities located at other altitudes. Therefore phase changes of a reconstructed image should be deemed key data in determining the position of the scatterer. In optics, interference patterns are measured and phase changes by π rad should be fixed for resolving the irregularities; in radio physics, it is possible to measure phase values less than π and thus to achieve [2 signal-to-noise]$^{0.5}$ times higher resolution if the signal-to-noise ratio exceeds unity. In [Datta and Bandyopadhyay, 1986], the method for locating irregularities from data on intensity changes has been subjected to criticism based on the assertion that the reconstructed intensity changes are an image of amplitude

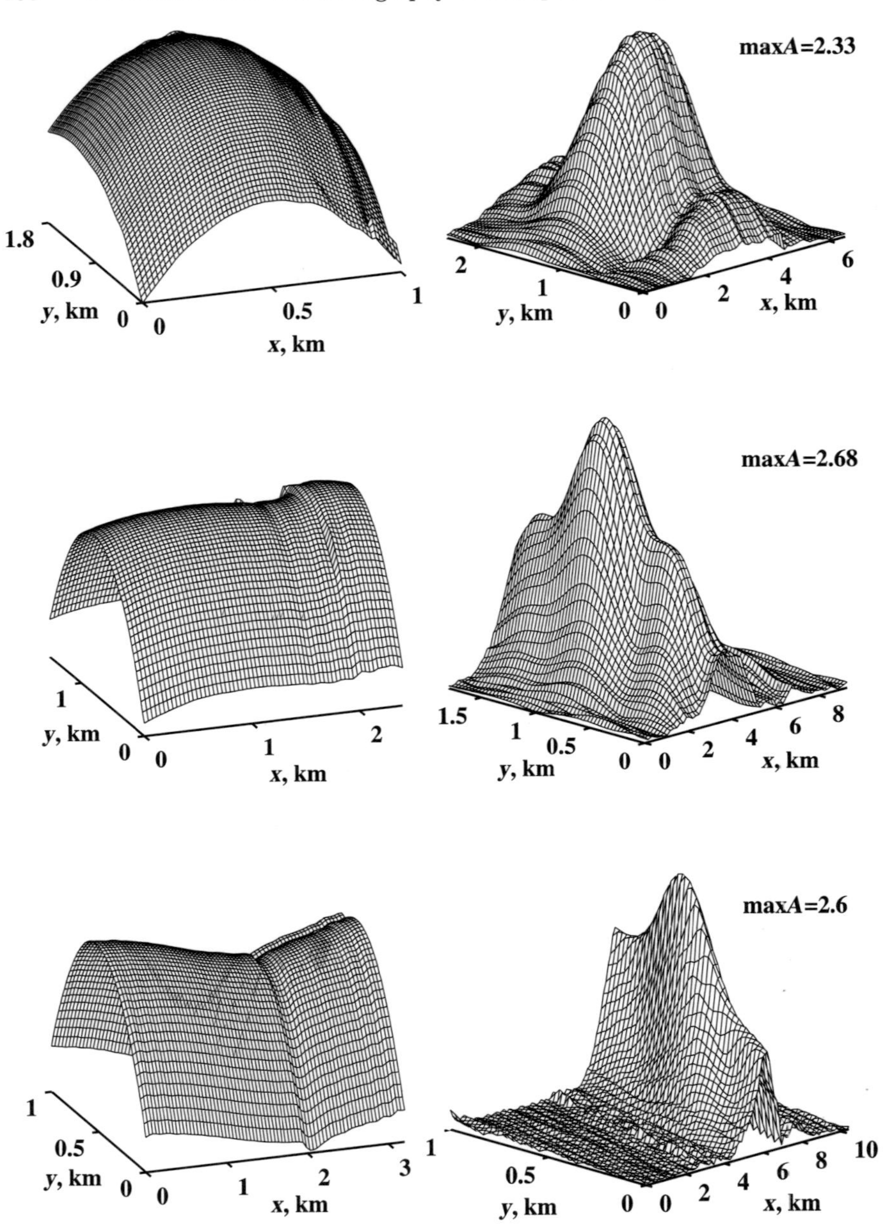

Fig. 4.6. Reconstructions of the complex phase of holographically reconstructed field Φ_{H} (phases – left column, amplitudes – right column) at various altitudes (500, 300, and 200 km) for the experiment on November 17, 1984

fluctuations that develop with the wave propagation due to initially acquired phase perturbations in the incident field. Therefore, the altitudes obtained from intensity changes will be closer to the ground level than the true heights of the irregularities that produced the field perturbations. Note that such an assertion is valid for the results of an E_{H} reconstruction. However, if Φ_{H} is reconstructed, then in weak scattering, as has been already mentioned, the intensity maximum will define the lower edge of the scattering irregularity.

Equation (4.23) was used for the holographic reconstruction of experimental data. Calculation of the Fourier transform has been carried out using a FFT (fast Fourier transform) algorithm. Figure 4.6 portrays the results of reconstruction (at a series of altitudes) of the phase and amplitude of $\Phi_{\mathrm{H}}(x, y)$ (3.72) from the data obtained on November 17, 1984 at 09:02:14 UT near Murmansk. Phases of $\Phi_{\mathrm{H}}(x, y)$ reconstructed at altitudes (from top to bottom) 500, 300, and 200 km are shown on the left. The spans of phase changes are, respectively, 54.4, 47.6, and 57.8 rad. On the right are shown 2-D functions $|\Phi_{\mathrm{H}}(x, y)|$. The satellite flew at about 1025 km in altitude above the ground level, and its path lay close to the zenith. The length of the hologram in the x direction was 10 km, and in the direction perpendicular to the satellite path projection, it was 1.55 km. The region of the focused image is located at about 350 km in height; the resolution along x is 0.25 km, that in the y direction is 1 km, and the lengthwise resolution is of the order of 30 km. The system of Cartesian coordinates corresponds to real scale sizes along the x and y axes. The modulus of the reconstructed function $|\Phi_{\mathrm{H}}|$ is connected (3.73) with distribution of density fluctuations integrated along the lengthwise coordinate for this particular case by the formula $\int N\,\mathrm{d}z = 10^{10}$ A electron/cm^2. In the figures, corresponding maximum amplitude A values are indicated. In the vicinity of the true position of the scatterer, the reconstructed 2-D function $|\Phi_{\mathrm{H}}(x, y)|$ represents a 2-D structure of the scattering irregularity in 10^{10} electron/cm^2 units. The curvature of the "phase front" of reconstructed $\Phi_{\mathrm{H}}(x, y)$ according to (3.27) undergoes a reversal on passing through the region of the actual location of the scatterer. Such a change in curvature of the phase front is clearly seen along the x-axis. No curvature change is observed along the y-axis because of the large size of the irregularity and insufficient transverse resolution of the recording system. Therefore, the analysis of the phase of the holographically reconstructed field makes it possible to determine the height of the irregularities with high accuracy.

Often a hologram contains information about a few scatterers. In this case, the reconstructed pattern has a more complicated structure, but, as for an isolated irregularity, it is possible to localize a group of irregularities in the ionosphere from the data on phase changes of $\Phi_{\mathrm{H}}(x, y)$ [Kunitsyn and Tereshchenko, 1990]. Figure 4.7 corresponds to May 5, 1987 and exhibits those rare situations when the irregularity is "resolved" perpendicularly to the satellite path direction. Reconstruction heights are (from top to bottom) 400, 300, and 200 km. Phase change spans are 5, 2.3, and 2.8 rad. The scatterer

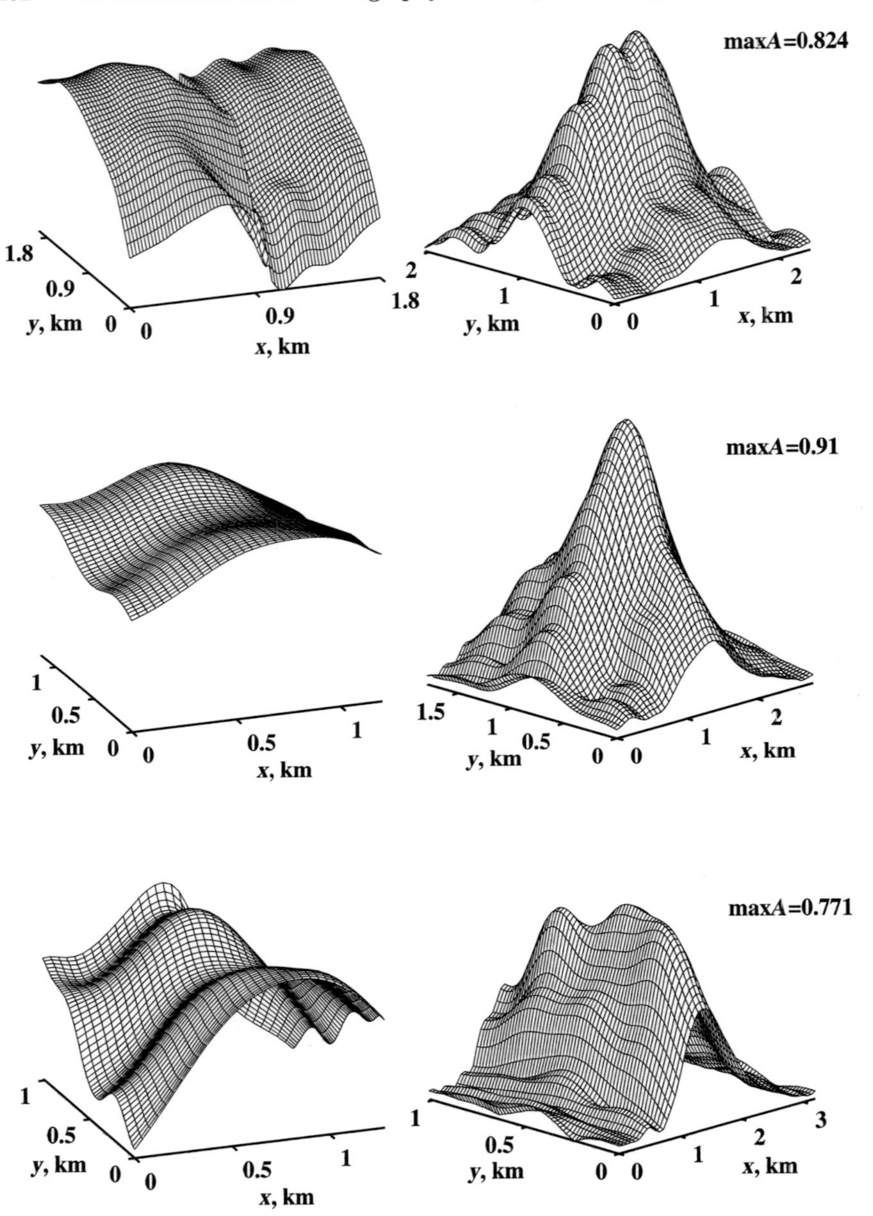

Fig. 4.7. Reconstructions of the complex phase of holographically reconstructed field Φ_H (phases – left column, amplitudes – right column) at various altitudes (400, 300, and 200 km) for the experiment on May 5, 1987

in Fig. 4.6 in accordance with criterion (3.51) is strong but condition (3.52) is satisfied, and therefore, reconstruction of the true distribution of the electron density should be based on transform (3.56). Figure 4.8 portrays the reconstruction of a 2-D structure of such a scatterer. 2-D structures of two isolated irregularities in 10^{10} electron/cm^2 units are shown in conventional colors in Plates 14a and 14b.

Practical implementation of a complete holographic experiment, as has been shown above, is rather difficult technically. At the same time, the reception of radio signals from orbital satellites at a single ground-based antenna allows synthesizing a one-dimensional hologram. In this case, since the irregularities have certain horizontal dimensions, the results obtained from a linear hologram will depend on the one-dimensional parameters of the irregularities.

The data on the wave field that are sort of "gained" by satellite will give a 1-D hologram. In accordance with (4.14), function $\tilde{\Phi}(x, z)$ can be reconstructed from $\Phi_1(s_x, 0, 0)$.

In experimental work employing linear holograms, the question about the relationship between the parameters of medium and the results of holographic reconstruction is important. It follows from the above formulas that, depending on the transverse dimensions of the scatterer, different relationships are possible between $\Phi_1(s_x, z)$ and $\tilde{\Phi}_{\mathrm{H}}(x, z)$ and, therefore, different ways of passing from intensity $|\tilde{\Phi}_{\mathrm{H}}(x, z_{\mathrm{s}})|$ to the distribution of density fluctuations. Therefore, in experiments where use is made of a linear hologram, a priori information is desirable on the dimensions of the scattering body in the direction transverse to the hologram. This requirement can be canceled

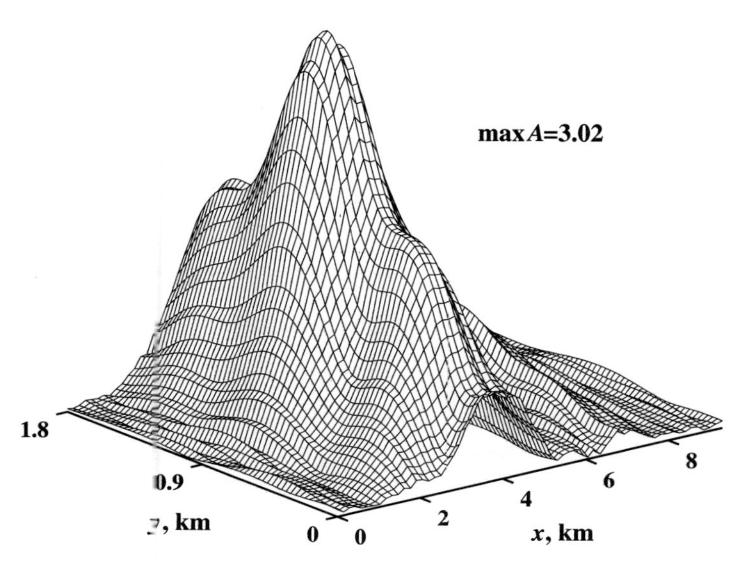

Fig. 4.8. Restoration of the 2-D structure of a strong scatterer (experiment on November 17, 1984)

if the field is measured at a receiving chain arranged perpendicularly to the satellite path. If typical scales of the field in the direction perpendicular to the projection of the satellite path are much greater than the Fresnel radius, for the sake of saving time and simplicity of processing, one can carry out a 1-D reconstruction of the hologram instead of reconstructing the 2-D one. Similarly, this approach can be used if the size of the diffraction pattern is

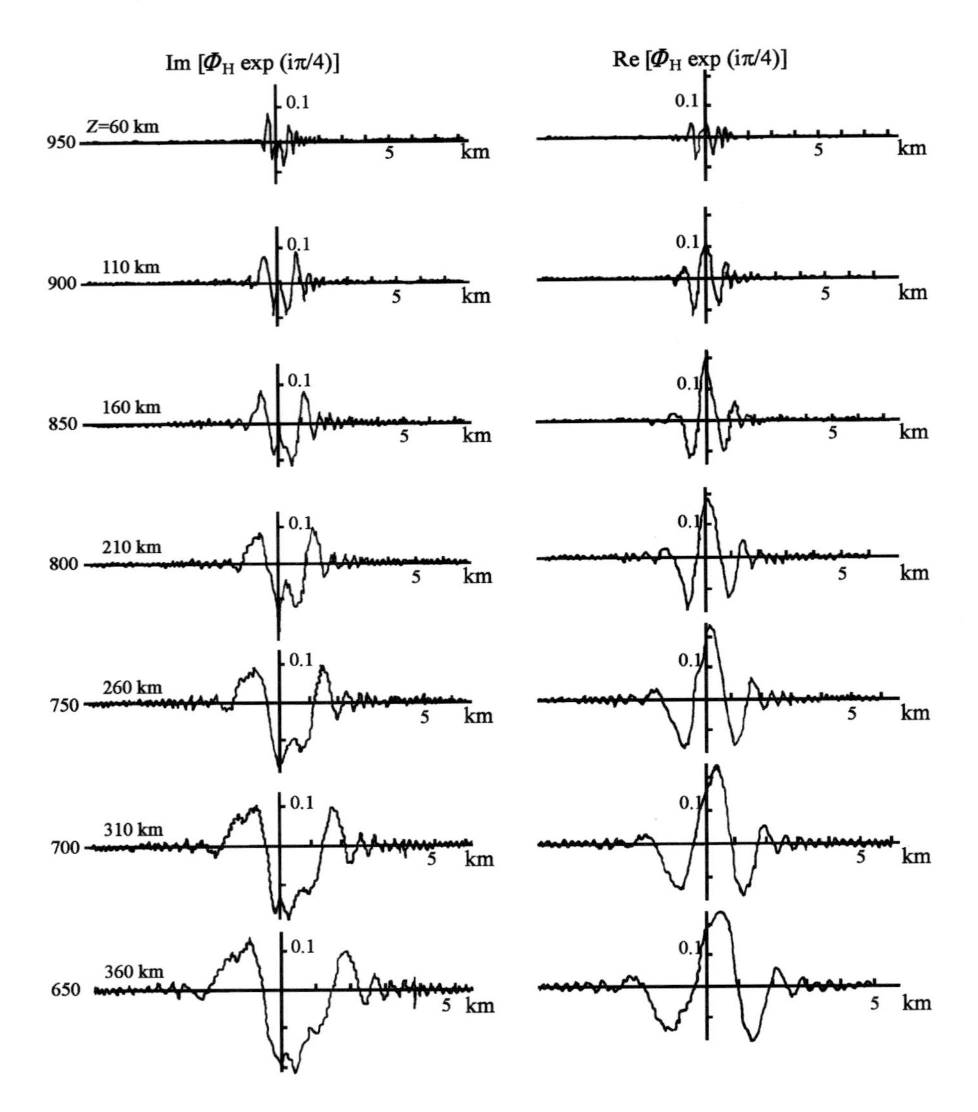

Fig. 4.9. Reconstructions of the function $\Phi_H \exp(i\pi/4)$ (real part – left column, imaginary – right column) at various altitudes for the experiment on May 5, 1987

much less than the Fresnel radius. Here, for weak scattering in the first case, relationship (4.16) follows from the reconstruction results, and (4.17) is valid in the second case.

Thus, if one carries out the reconstruction of $\Phi_{\mathrm{H}} \exp(\mathrm{i}\pi/4)$ (to be exact, its real and imaginary parts), then for irregularities with dimensions greater than the Fresnel radius in the y direction and less than or comparable to the Fresnel radius in the x direction, the following type of change in its imaginary part will take place: reversal of the change of its curvature direction when passing the location of the irregularity. Such behavior is caused by the character of phase changes of the holographically reconstructed field discussed in model calculations and it is a good tool for locating the irregularities.

Similarly, vanishing of the imaginary part of $\Phi_{\mathrm{H}} \exp(-\mathrm{i}\pi/4)$ can be used for determining the height of ionospheric irregularities having a transversal scale size less than the Fresnel radius.

Analysis of Fig. 4.9 portraying the result of the reconstruction of $\Phi_{\mathrm{H}} \exp(-\mathrm{i}\pi/4)$ from 1-D data on May 5, 1987 (that were used in their full for 3-D reconstruction in Fig. 4.7) shows the possibility of resolving, from a 1-D hologram, an irregularity with dimensions less than the Fresnel radius along x and greater than the Fresnel radius along y in the presence of a larger size background irregularity.

5. Statistical Radio Tomography of a Randomly Inhomogeneous Ionosphere

5.1 Reconstruction of a Randomly Inhomogeneous Ionosphere

Very often the ionosphere contains large volumes filled with a great number of electron density irregularities of different sizes. This is typical of equatorial and polar regions, especially in the nighttime. Of course, in such cases, reconstruction of separate realizations of the disturbed dynamic ionosphere makes little sense, and of more interest is reconstruction of the statistical parameters of a turbulent randomly inhomogeneous ionosphere, such as the correlation function or the spectrum of electron density fluctuations. The statistical approach is explicitly informative of the structure of a disturbed ionosphere. It allows studying generation mechanisms of the irregularities and their effect on propagating radio signals. A randomly inhomogeneous ionospheric plasma causes fluctuations or scintillations in the radio waves coming through the ionosphere. Getting rid of such scintillations is a problem of greatest importance in radio location and communication systems. Separation in time and space of the receiving systems and signal coding schemes helps to weaken the influence of scintillations, but design of such schemes requires information about the statistical parameters of electron density irregularities.

Fluctuations of radio signals caused by ionospheric irregularities have been investigated for a long time. Scintillation of stellar radio waves, of the signals from satellite-borne transmitters, UHV radio stations, etc. have been studied. The basic features and the probability of scintillation appearance were determined from data obtained in numerous experiments. The results of these investigations are reported in a number of reviews [Getmansev and Eroukhimov, 1967; Booker, 1958; Yen and Swenson, 1964; Aarons et al., 1971; Aarons, 1982; Crane, 1977]. The problem of the determination of the parameters of a randomly inhomogeneous ionosphere from scintillation measurements has a wide bibliography starting from the work [Booker et al., 1950] where it was first proposed to account for the ionospheric effect on propagating radio signals using the model of a thin screen that caused phase changes in the waves. This approach proved to be promising, and in the subsequent investigations, a correlation was established between the scale of the diffraction pattern observed on the ground level and the parameters of the phase screen in the ionosphere. In this way, a set of techniques was developed for describing the ionosphere as a phase screen, such as the similarity method, correlation analysis, dispersion analysis, and their modifications [Briggs et al., 1950; Philips and Spenser, 1955; Golley and Rossiter, 1970; Baker, 1981]. Using the above methods, spectrum parameters of irregularities were also obtained, such as the power index, in the case of the power-law spectrum, and the characteristic scale size, in the case of the Gaussian spectrum.

All of the described approaches to the tomographic reconstruction of a randomly inhomogeneous ionosphere are confined to arbitrariness in the choice of the model, which results in ambiguity of the solution. This is a common defect of all model approaches most significant in the lack of a priori or empirical data available for choosing a model. In spite of numerous experimental and theoretical investigations, not one specific spectral model can be deemed preferable, and the information obtained by now about which spectral shape is more or less suitable bears just an estimating sense. In this connection, of particular interest is the nonmodel (direct) inverse problem of reconstructing the spectrum of ionospheric irregularities, which will be discussed below.

5.2 Stochastic Inverse Problems

Tomographic reconstruction of the statistical structure of the medium requires obtaining projections or cross sections of the volume of interest from measurements of the scattered field. The most suitable mathematical techniques for tomography are integral equations relating the recovered structure of the media to the measured field reflecting the projections of this structure, i.e., the integrals over lower order variety. The purpose of the present subsection is to derive such integral equations. In most practical ionospheric applications, it is usually enough to reconstruct the correlation function of electron density fluctuations. Therefore, in the further consideration when speaking about reconstruction of the statistical structure of the medium, we will mean reconstructing the correlation function of fluctuations in the medium or its spectrum. That very reconstruction of the statistical structure of a randomly inhomogeneous medium from the measured statistics of the scattered field we will call the statistical inverse scattering problem.

The theory describing propagation and scattering of waves in randomly inhomogeneous media is developed quite well. There are a lot of publications (papers and monographs) on the subject [Tatarskii, 1971; Rytov et al., 1989; Ishimaru, 1978; Fante, 1975; Klyatzkin, 1980; Zuev et al., 1988]. A series of results for weak fluctuations was obtained within the scope of the well-known approximations by means of the smooth disturbances method or using the Born approximation. Strong fluctuations were analyzed using integral equations of the Dyson and Bete–Solpeter type, Feynman's diagrams, and also the Huygens–Fresnel generalized principle, and the parabolic equation method. It is known that, under some assumptions, these methods are equivalent [Rytov et al., 1989; Ishimaru, 1978; Strohben, 1968; Yura, 1972]. Therefore, we will use below the expressions derived within the scope of the parabolic equation techniques.

In the description of a field scattered by a randomly inhomogeneous ionosphere, it is pertinent to use statistical parameters such as the first and second coherence functions of the second order. The use of these functions has the advantage (over second moments of amplitude and phase) that in the Markov approximation, a simple formula relates the first coherence function to the structural function of density fluctuations. Furthermore, the second coherence function can be used for location of the scattering volume and determination of the density fluctuation correlation function.

The distribution of electron density in the ionosphere can be written in the form $N = N_0 + \delta N$ where $N_0(z) = \langle N \rangle$ is the average density and δN are zero-mean fluctuations of the electron density. Angular brackets denote an ensemble average. Suppose that the average density and statistical parameters of the fluctuations remain constant during the period of observations. Let us also assume unity mean permittivity. The possibility of generalization allowing for the effect of a regular background ionosphere will be considered later. Besides, we will neglect also the influence of the geomagnetic field on radio waves as well as wave depolarization. The assumptions made are valid for VHF waves used in ionospheric tomography experiments. Field $\boldsymbol{E}$ obeys Helmholtz equation (3.9) with complex potential

$$q(\boldsymbol{r}, \omega) = \frac{4\pi r_{\mathrm{e}}\, \delta N(\boldsymbol{r})}{1 + \mathrm{i}\, \nu_{\mathrm{eff}}(\boldsymbol{r})/\omega} .$$

The imaginary part of q usually can be neglected.

Mathematically, the statistical inverse scattering problem reduces to a determination of the statistical parameters of q from the statistical parameters of measured field $\boldsymbol{E}$. Here, to shorten computations, it is convenient to use two different coordinate systems. Let us call the first, "laboratory" and "fixed" reference frame $\boldsymbol{R} = (X, Y, Z)$, a "global" coordinate system. Its origin is at the receiving site and the Z-axis points vertically upward; the plane $Y = 0$ contains the satellite trajectory. In the further consideration, the spherically symmetric regular ionosphere could be easily taken into account by introducing appropriate curvilinear coordinates (h, τ), as was done in Chap. 2.

It is expedient to match the origin of the second Cartesian coordinate system $\boldsymbol{r} = (x, y, z) = (\boldsymbol{\rho}, z)$ to one of the satellite positions. The z-axis points to the center of the receiving system. We shall use a few of such local coordinate systems with different orientations defined by satellite motion. These coordinate systems are introduced for convenience in the problem consideration so that in each local coordinate system, the probing wave is scattered nearly "forward" into a narrow cone, which allows reduced calculations.

Let us assume that in a "global" coordinate system R spherical probing wave,

$$E = -\frac{1}{4\pi R}\exp(\mathrm{i}\, kR)$$

is incident on the irregular layer with Z_{d} and Z_{u} being its lower and upper boundaries. The altitude of the satellite is Z_0. Let us introduce a function $v(\boldsymbol{r}) = E(\boldsymbol{r})/E_0$ – normalized field. In narrow-cone scattering after "Fresnel expansion" (3.18), (3.19) of the exponents of Green's functions, from Lippman–Schwinger's equation (3.10), let us turn to the integral equation for v:

$$v(\boldsymbol{r}) = 1 - \frac{1}{4\pi}\int \mathrm{d}^3 r'\, q(\boldsymbol{r}')\,\frac{v(\boldsymbol{r}')}{\zeta}\exp\left[\mathrm{i}\,\frac{k}{2\zeta}(\boldsymbol{s}-\boldsymbol{\rho}')^2\right]. \tag{5.1}$$

This equation is similar to (3.22). Equation (5.1) is written in local coordinates; as earlier $\boldsymbol{r_0} = (\boldsymbol{\rho}_0, z_0)$ are the coordinates of the satellite,

$$\boldsymbol{s} = \frac{z' - z_0}{z - z_0}\boldsymbol{\rho} + \frac{z - z'}{z - z_0}\boldsymbol{\rho_0}, \qquad \zeta = \frac{(z' - z_0)(z - z')}{z - z_0}.$$

To simplify the formulas, let us assume that the satellite is located exactly at the origin of the local coordinates $\boldsymbol{r_0} = (0, 0)$. If necessary, it is easy to go over to a more general case.

Equations for the averaged field and second moments of function v can be derived directly from (5.1) which is a paraxial approximation of the Lippmann–Schwinger equation. To reduce the derivation procedure and to make use of the known results, let us pass from (5.1) to a parabolic equation using new variables

$$\xi = \frac{1}{z}, \qquad \boldsymbol{\sigma} = \frac{\boldsymbol{\rho}}{z}, \qquad \xi' = \frac{1}{z'}.$$

Then

$$\frac{k(\boldsymbol{s}-\boldsymbol{\rho}')^2}{\zeta - \zeta'} = \frac{k(\boldsymbol{\sigma} - \xi'\boldsymbol{\rho}')^2}{2(\xi - \xi')},$$

so that v satisfies the differential equation

$$\left(-2\mathrm{i}k\frac{\partial}{\partial\xi} + \Delta_\sigma - \xi^{-2}q\right) v(\xi, \boldsymbol{\sigma}) = 0. \tag{5.2}$$

Equation (5.2) stems from (5.1) after differentiating it and with the relationship taken into account for the fundamental solution of the Schrödinger operator [Hörmander, 1986] :

$$\left(-2\mathrm{i}k\frac{\partial}{\partial\xi} + \Delta_\sigma\right)\frac{\exp\left[-\frac{\mathrm{i}k}{2}(\boldsymbol{\sigma} - \xi'\boldsymbol{\rho}')^2/(\xi' - \xi)\right]}{-4\pi(\xi' - \xi)}\,\theta(\xi' - \xi) \tag{5.3}$$
$$= \delta(\xi' - \xi)\,\delta(\boldsymbol{\sigma} - \xi'\boldsymbol{\rho})\,;$$

the existence of the unit step $\theta(z'-z) = \theta(\xi'-\xi)$ in the integrand (5.1) as well as in (3.22) is implied by the terms of the derivation of (5.1). After passing to the parabolic equation, one can use the known results and the techniques developed for deducing equations for the first and second moments of v [Tatarskii, 1971; Rytov et al., 1989; Ishimaru, 1978; Fante, 1975; Klyatzkin, 1980; Zuev et al., 1988]. Under the conventional assumption on the Gaussian distribution of the stochastic field q and its δ-correlation along z

$$B(\boldsymbol{r}_1, \boldsymbol{r}_2) = \langle q(\boldsymbol{\rho}_1, z_1)\; q(\boldsymbol{\rho}_2, z_2)\rangle = \delta(z_2 - z_1)\, A_q(\boldsymbol{\rho}_2 - \boldsymbol{\rho}_1)\;; \tag{5.4}$$

after averaging of the parabolic equation (5.2) and making use of the Furutsu–Novikov equation [Rytov et al., 1989; Ishimaru, 1978], we arrive at

$$\left(-2\mathrm{i}k\frac{\partial}{\partial\xi} + \Delta_\sigma\right)\langle v\rangle + \frac{\mathrm{i}}{4k}\xi^{-2}\, A_q(0)\,\langle v\rangle = 0\,. \tag{5.5}$$

In common Cartesian variables, (5.5) is

$$\left[-2\mathrm{i}k\left(\frac{\partial}{\partial z} + \frac{\boldsymbol{\rho}}{z}\frac{\partial}{\partial\boldsymbol{\rho}}\right) + \Delta_\rho + \frac{\mathrm{i}}{4k}\, A_q(0)\right]\langle v\rangle = 0\,. \tag{5.6}$$

The boundary condition at the upper edge Z_u of the layer of irregularities (and corresponding z_u) is $\langle v(\boldsymbol{\rho}, z_\mathrm{u})\rangle = 1$.

Similarly, in $(\xi, \boldsymbol{\sigma})$ coordinates, one can reiterate the derivation, from (5.2), of the equations for second moments $V_{1,1}(\boldsymbol{\rho}_1, \boldsymbol{\rho}_2, z) = \langle v(\boldsymbol{\rho}_1, z)\;\; v^*(\boldsymbol{\rho}_2, z)\rangle$ – the second-order first coherence function and the second-order second coherence function $V_{2,0}(\boldsymbol{\rho}_1, \boldsymbol{\rho}_2, z) = \langle v(\boldsymbol{\rho}_1, z)\;\; v(\boldsymbol{\rho}_2, z)\rangle$. Since the derivation procedure is standard and described in the papers referred to above, here we will give only the resulting formula obtained using new variables $(\boldsymbol{\rho}, z)$ and $\boldsymbol{\rho}_+ = (\boldsymbol{\rho}_1 + \boldsymbol{\rho}_2)/2$, $\boldsymbol{\rho}_- = \boldsymbol{\rho}_2 - \boldsymbol{\rho}_1$:

$$\begin{aligned}
&\left\{2\mathrm{i}k\left(\frac{\partial}{\partial z} + \frac{\boldsymbol{\rho}_-}{z}\frac{\partial}{\partial\boldsymbol{\rho}_-}\right) + 2\mathrm{i}k\left(\frac{\boldsymbol{\rho}_+}{z}\frac{\partial}{\partial\boldsymbol{\rho}_+}\right) + 2\frac{\partial^2}{\partial\boldsymbol{\rho}_+\partial\boldsymbol{\rho}_-}\right.\\
&\left.+\frac{\mathrm{i}}{2k}\left[A_q(0) - A_q(\boldsymbol{\rho}_-)\right]\right\} V_{1,1}(\boldsymbol{\rho}_+, \boldsymbol{\rho}_-, z) = 0\,,\\
&V_{1,1}(\boldsymbol{\rho}_+, \boldsymbol{\rho}_-, z) = 1\,;
\end{aligned} \tag{5.7}$$

$$\begin{aligned}
&\left\{2\mathrm{i}k\left(\frac{\partial}{\partial z} + \frac{\boldsymbol{\rho}_-}{z}\frac{\partial}{\partial\boldsymbol{\rho}_-}\right) + 2\frac{\partial^2}{\partial\boldsymbol{\rho}_-^2} + \frac{1}{2}\frac{\partial^2}{\partial\boldsymbol{\rho}_+^2} + 2\mathrm{i}k\frac{\boldsymbol{\rho}_+}{z}\frac{\partial}{\partial\boldsymbol{\rho}_+}\right.\\
&\left.+\frac{\mathrm{i}}{2k}\left[A_q(0) - A_q(\boldsymbol{\rho}_-)\right]\right\} V_{2,0}(\boldsymbol{\rho}_+, \boldsymbol{\rho}_-, z) = 0\,,\\
&V_{2,0}(\boldsymbol{\rho}_+, \boldsymbol{\rho}_-, z) = 1\,.
\end{aligned} \tag{5.8}$$

Equations (5.6)–(5.8) are written in a local coordinate system where the layer with irregularities (between Z_d and Z_u) is inclined. However, the fluctuation parameters of the probing wave are affected mostly by the paraxial region of the order of the Fresnel radius ($\sim \sqrt{\lambda\xi}$). Therefore, neglecting side effects, let us replace the inclined layer by that orthogonal to the z-axis with boundary altitudes z_{u}, z_{d}, based on the relation $z\cos\theta \to Z_0 - Z$ where θ is the angle between the vertical line in the global coordinate system R and the line of sight, i.e., the ray from the receiver to the satellite. The appearance of additional terms like $\frac{\boldsymbol{\rho}}{z}\frac{\partial}{\partial z}$ in (5.6)–(5.8), in contrast to traditional parabolic equations for moments, is caused by extracting the spherical probing wave E_0 from the scattered field.

The solution of equations like (5.6)–(5.8) for a statistically uniform scattering layer will be independent of the summary variable $\boldsymbol{\rho}_+$. In the future consideration, we will use the symbol $\boldsymbol{\rho}$ instead of $\boldsymbol{\rho}_-$ implying that the arguments for the desired functions contain only the difference coordinate. Hence, instead of (5.6)–(5.8), one can write simpler equations where $\Delta_\perp \equiv \partial^2/\partial\boldsymbol{\rho}^2$:

$$\left\{2\mathrm{i}k\left(\frac{\partial}{\partial z} + \frac{\boldsymbol{\rho}}{z}\frac{\partial}{\partial\boldsymbol{\rho}}\right) + \frac{\mathrm{i}}{2k}\left[A_q(0) - A_q(\boldsymbol{\rho})\right]\right\} V_{1,1}(\boldsymbol{\rho}, z) = 0\,,$$
$$V_{1,1}(\boldsymbol{\rho}, z_{\mathrm{u}}) = 1\,; \tag{5.9}$$

$$\left\{\mathrm{i}k\left(\frac{\partial}{\partial z} + \frac{\boldsymbol{\rho}}{z}\frac{\partial}{\partial\boldsymbol{\rho}}\right) + \Delta_\perp + \frac{\mathrm{i}}{4k}\left[A_q(0) - A_q(\boldsymbol{\rho})\right]\right\} V_{2,0}(\boldsymbol{\rho}, z) = 0\,,$$
$$V_{2,0}(\boldsymbol{\rho}, z_{\mathrm{u}}) = 1\,. \tag{5.10}$$

The function

$$\langle v(\boldsymbol{r})\rangle = \exp\left[-\frac{1}{8k^2}A_q(0)(z - z_{\mathrm{u}})\right]$$

is the solution of (5.6) wherefrom the relation for the average field is obtained:

$$\langle E(\boldsymbol{r})\rangle = E_0(\boldsymbol{r})\exp\left[-\frac{1}{8k^2}A_q(0)(z - z_{\mathrm{u}})\right]\,. \tag{5.11}$$

Therefore, if we introduce a function γ normalized by averaged field $\gamma(\boldsymbol{r}') = E(\boldsymbol{r})/\langle E(\boldsymbol{r})\rangle$, for its second order moments,

$$\begin{aligned} \Gamma_{1,1}(\boldsymbol{\rho}, z) &= \langle \gamma(\boldsymbol{\rho}_2, z)\ \gamma^*(\boldsymbol{\rho}_2, z)\rangle\,, \\ \Gamma_{2,0}(\boldsymbol{\rho}, z) &= \langle \gamma(\boldsymbol{\rho}_2, z)\ \gamma(\boldsymbol{\rho}_2, z)\rangle\,, \end{aligned} \tag{5.12}$$

the following equations can be obtained from (5.9)–(5.11):

$$\begin{aligned} &\left[\frac{\partial}{\partial z} + \frac{\boldsymbol{\rho}}{z}\frac{\partial}{\partial\boldsymbol{\rho}} - \frac{1}{4k^2}A_q(\boldsymbol{\rho})\right]\Gamma_{1,1}(\boldsymbol{\rho}, z) = 0\,, \\ &\Gamma_{1,1}(\boldsymbol{\rho}, z_{\mathrm{u}}) = 1\,; \end{aligned} \tag{5.13}$$

$$\begin{aligned} &\left[\mathrm{i}k\left(\frac{\partial}{\partial z} + \frac{\boldsymbol{\rho}}{z}\frac{\partial}{\partial\boldsymbol{\rho}}\right) + \Delta_\perp + \frac{\mathrm{i}}{4k}A_q(\boldsymbol{\rho})\right]\Gamma_{2,0}(\boldsymbol{\rho}, z) = 0\,, \\ &\Gamma_{2,0}(\boldsymbol{\rho}, z_{\mathrm{u}}) = 1\,. \end{aligned} \tag{5.14}$$

The solution of (5.13) is a function

$$\Gamma_{1,1}(\boldsymbol{\rho}, z) = \exp\left[\frac{1}{4k^2}\int_{z_u}^{z} \mathrm{d}z' A_q\left(\boldsymbol{\rho}\frac{z'}{z}\right)\right] . \tag{5.15}$$

The integral equation corresponding to (5.14) is

$$\begin{aligned}\Gamma_{2,0}(\boldsymbol{\rho}, z) = 1 + \frac{\mathrm{i}}{16\pi k}\int \mathrm{d}^3 r' \frac{A_q(\boldsymbol{\rho}')}{\zeta}\,\Gamma_{2,0}(\boldsymbol{\rho}', z') \\ \times \exp\left[\frac{\mathrm{i}k}{4\zeta}(\boldsymbol{s}-\boldsymbol{\rho}')^2\right] .\end{aligned} \tag{5.16}$$

Equation (5.14) follows from (5.16) and is analogous to (5.1) or (3.22). As done in conversion from (5.1) to (5.2), one can pass to the variables $(\boldsymbol{\sigma}, \xi)$ and after differentiation of (5.16) with (5.3) taken into account, one arrives at (5.14). The integral equation obtained is valid even for $\rho_0 \neq 0$, $z_0 \neq 0$ in accordance with the definitions of $\boldsymbol{s}$ and ζ given after formula (5.1). All of the parabolic equations obtained above can be easily rewritten for the probing source off the coordinate origin by mere substitution of $z - z_0$ for z and $\boldsymbol{\rho} - \boldsymbol{\rho}_0$ for $\boldsymbol{\rho}$.

Integral equation (5.16) together with the formula (5.15) can serve as basic relations for tomographic reconstruction methods that will be expounded in the next sections. Note that the essential difficulty in solving (5.16) for $A_q(\boldsymbol{\rho})$ is that the second coherence function is unknown within the layer of irregularities, i.e., the kernel of (5.16) is not defined. Otherwise, the inverse problem would not exist at all: the data on $\Gamma_{2,0}(\boldsymbol{\rho}, z)$ within the scattering irregular structure define $A_q(\boldsymbol{\rho})$ straight off and completely through differential equation (5.14). Therefore, it is expedient to use here also a holographic approach to the inverse problem that makes it possible to obtain an equation of the first kind. By analogy with holographically reconstructed field (3.70), let us introduce a function Γ_{H}:

$$\Gamma_{\mathrm{H}}(\boldsymbol{\rho}, z) = \frac{\mathrm{i}k}{4\pi}\int \mathrm{d}^2\rho_R\, \Gamma_{2,0}(\boldsymbol{\rho}_{\mathrm{R}}, z_{\mathrm{R}}) \exp\left[-\frac{\mathrm{i}k}{4d}(\boldsymbol{\rho}_{\mathrm{R}} - \boldsymbol{t})^2\right] , \tag{5.17}$$

where

$$\boldsymbol{t} = \boldsymbol{\rho}(z_{\mathrm{R}} - z_0)/(z - z_0) - \boldsymbol{\rho}_0(z_{\mathrm{R}} - z)/(z - z_0) ,$$
$$d = (z_{\mathrm{R}} - z)(z_{\mathrm{R}} - z_0)/(z - z_0) .$$

Function Γ_H results from integral transformation of the data on the second coherence function in the receiving plane $(\boldsymbol{\rho}_{\mathrm{R}}, z_{\mathrm{R}})$. It is possible to reconstruct $\Gamma_{\mathrm{H}}(\boldsymbol{\rho}, z)$ in any plane including the interior of the layer of irregularities. Substituting (5.16) in (5.17) and integrating the equation for $\mathrm{d}^2\boldsymbol{\rho}_{\mathrm{R}}$, taking into account identity $(\boldsymbol{t} - \boldsymbol{\rho}_{\mathrm{R}})^2/d \equiv (\boldsymbol{s} - \boldsymbol{\rho})^2/\zeta$, one obtains to the following integral equation containing the function $\Gamma_{\mathrm{H}}(\boldsymbol{\rho}, z)$:

$$1 - \frac{\mathrm{i}}{16\pi k} \int \mathrm{d}^3 r' \frac{A_q(\boldsymbol{\rho}')}{\zeta} \Gamma_{2,0}(\boldsymbol{\rho}', z') \exp\left[\frac{ik}{4\zeta}(\boldsymbol{s} - \boldsymbol{\rho}')^2\right] \\ = \Gamma_{\mathrm{H}}(\boldsymbol{\rho}, z) . \tag{5.18}$$

Here $\boldsymbol{s}$ and ζ are defined in the same way as in (5.1). The basic equation in the holographic approach is similar in structure to integral equation (3.71) in the deterministic case.

5.3 Reconstructing Projections of Correlation Functions

Experimental measurements of the coherence functions of the waves passed through a randomly inhomogeneous medium make it possible to determine the projections of the correlation functions of complex potential q fluctuations. In accordance with (5.4), $A_q(\boldsymbol{\rho})$ is a projection of the correlation function, i.e.,

$$\int B(\boldsymbol{\rho}, z)\,\mathrm{d}z = A_q(\boldsymbol{\rho}) .$$

Function $A_q(\boldsymbol{\rho})$ can be obtained as well from the data on the first coherence function (5.15) as from integral equations (5.16) or (5.18). Let us first consider weak scattering. Assume that the irregularities are concentrated in a rather extended layer that can, generally, be inclined in the direction of an incident probing wave. The most general case of the arbitrary shape of an irregular region will be analyzed in the next section. If the thickness of the layer exceeds the Fresnel scale size $\sqrt{\lambda\zeta}$ of the paraxial zone that mostly contributes to the fluctuations of the field, one can neglect side effects and consider a layer orthogonal to the ray instead of an inclined layer, with appropriate changes in the thickness of the layer. In general, the layer is not necessarily thick. It can also be thin, but in this case the along-ray Rayleigh resolution of the recording system will act as effective thickness of the layer, since a shift of the medium by the distance of the order of the ray-aligned resolution has less effect on the field.

In the weak scatter approximation $|\Gamma_{2,0} - 1| \ll 1$, hence, the deviation of $\Gamma_{2,0}$ from unity in the integrand (5.16) can be neglected. One can invert the Fresnel transform with respect to the variable $\boldsymbol{\rho}'$. The inversion is feasible if the ζ dependence on z' is neglected. This is admissible for the thickness of a layer less than the along-ray Rayleigh resolution, which is of the order of tens–hundreds of kilometers in ionospheric experiments. Applying an inverse Fresnel transformation to (5.16) in the same way as we did previously in

Chaps. 3 and 4, we arrive at

$$\begin{aligned}\int A_q(\boldsymbol{\rho})\,\mathrm{d}z &= A_q(\boldsymbol{\rho})\,L\\ &= \frac{k^3}{\mathrm{i}\pi\zeta}\int \mathrm{d}^2 s\,[\Gamma_{2,0}(\boldsymbol{s})-1]\,\exp\left[-\mathrm{i}\frac{k}{4\zeta}(\boldsymbol{s}\boldsymbol{\rho})^2\right].\end{aligned} \tag{5.19}$$

Here, L is the thickness of the layer with irregularities, and ζ contains the fixed coordinate of the layer z_s, $\zeta = (z - z_\mathrm{s})(z_\mathrm{s} - z_0)/(z - z_0)$. As follows from (5.16), $\Gamma_{2,0}$ is a function of $\boldsymbol{s}$. Relation (5.19) defines the projection $A_q(\boldsymbol{\rho})$ of the correlation function from the data on $\Gamma_{2,0}$, given the coordinates of receiver $\boldsymbol{r}$, transmitter $\boldsymbol{r}_0$, and scattering region z_s. A two-dimensional projection $A_q(\boldsymbol{\rho})$ can be obtained from measurements at a receiving chain oriented perpendicularly to the satellite path along the y-axis and synthesizing the aperture (along the x-axis) of moving satellites, since variable $\boldsymbol{s}$ contains independent coordinates of the transmitter and the receivers [Tereshchenko and Khudukon, 1981].

Coordinate z_s is obtained by the same method as described in Sect. 3.3. So long as the projection of correlation function is a symmetrical smooth function with its maximum at a central point ($\boldsymbol{\rho} = 0$), a simpler approach is more advantageous that makes it possible to obtain z_s from single receiver data. In this case of reception by a linear synthesized aperture, the following integral is reconstructed:

$$\begin{aligned}\tilde{A}_q(x) &\equiv \int A_q(\boldsymbol{\rho})\,\exp\left(\mathrm{i}\frac{k}{4\zeta}y^2\right)\mathrm{d}y\\ &= \frac{4k^2}{\mathrm{i}L}\int \mathrm{d}s_x\,[\Gamma_{2\,0}-1]\,\exp\left[-\mathrm{i}\frac{k}{4\zeta}(s_x - x)^2\right].\end{aligned} \tag{5.20}$$

This follows from (5.19) similarly to relation (4.12). Performing one-dimensional reconstruction of $A_q(x)$ with different ζ' or z_s', one obtains a set of one-dimensional functions of the difference between the given and true coordinates of the layer $\Delta' = z_\mathrm{s}' - z_\mathrm{s}$. It can be easily shown that the reconstructed one-dimensional projection $A_q(x, \Delta)$ is a convolution of true projection $\tilde{A}_q(x)$ with a kernel function of the form

$$(\Delta)^{-1/2}\exp\left\{\left[-\mathrm{i}\frac{k}{4\Delta}(x - x')^2\right]\right\}$$

and can be developed as a series in powers of Δ including the true $\tilde{A}_q(x)$ and its derivatives:

$$\begin{aligned}\tilde{A}_q(x,\Delta) \simeq \tilde{A}_q(x) &+ \frac{\mathrm{i}\Delta}{2\pi k}\left(\frac{\partial^2\tilde{A}_q}{\partial x^2} + \frac{\mathrm{i}k}{2\zeta}\tilde{A}_q\right)\\ &- \frac{\Delta^2}{4\pi k^2}\left(\frac{\partial^4\tilde{A}_q}{\partial x^2} - \mathrm{i}\frac{3k}{\zeta}\frac{\partial^2\tilde{A}_q}{\partial x^2} - \frac{3k^2}{4\zeta^2}\tilde{A}_q\right) + 0\left(\frac{\Delta^3}{k^3}\right),\end{aligned} \tag{5.21}$$

where Δ is proportional to Δ' and $\Delta = \Delta' \left(2z_{\mathrm{s}} - z_0 - z\right) / \left(z - z_0\right)$. Usually, the imaginary part of A_q is negligible in VHF radio probing, that is, $A_q = \left(4\pi r_{\mathrm{e}}\right)^2 A_{\mathrm{N}}$ is proportional to the projection of the correlation function of electron density. Then, after separating the real and imaginary parts of the reconstructed function, one arrives at

$$\begin{aligned} \operatorname{Re} \tilde{A}_q(x, \Delta) &\simeq \left(1 - \frac{\Delta}{4\pi\zeta} + \frac{3}{16\pi}\frac{\Delta^2}{\zeta^2}\right) \tilde{A}_q(x) - \frac{1}{4\pi}\frac{\Delta^2}{k^2}\frac{\partial^4}{\partial x^4} \tilde{A}_q(x) + \dots , \\ \operatorname{Im} \tilde{A}_q(x, \Delta) &\simeq \frac{\Delta}{2\pi k}\left(1 + \frac{3}{2}\frac{\Delta}{\zeta}\right)\frac{\partial^2}{\partial x^2} \tilde{A}_q(x) + \dots . \end{aligned} \tag{5.22}$$

From this, the following qualitative conclusions can be drawn. If the transform coordinate $z_{\mathrm{s}}{}'$ coincides with the true position of the layer z_{s} ($\Delta = 0$), the real part of $\tilde{A}_q(x, \Delta)$ coincides with the true one-dimensional projection $\tilde{A}_q(x)$, and the imaginary part of $\tilde{A}_q(x, \Delta)$ becomes zero. Conventionally, the reconstructions of $\tilde{A}_q(x, \Delta)$ with different Δ can be displayed as corresponding to different heights $z_{\mathrm{s}}{}'$; however, it should not be forgotten that each $\tilde{A}_q(x, \Delta)$ is a ground projection of the correlation function with corresponding transformation parameter $z_{\mathrm{s}}{}'$ (or ζ). As the plane of reconstruction "moves away" from the true position of the layer, Fresnel oscillation will appear due to the kernel of the convolution integral; the scale of the oscillations are increased with the distance to the layer. When passing through $\Delta = 0$, the oscillations of the imaginary part of $\tilde{A}_q(x, \Delta)$ change the sign (turn over). A minimum will take place in the center of the image in the lower half-plane ($\Delta > 0$), and a maximum will be observed in the upper half-plane since the second derivative of $\tilde{A}_q(x)$ in (5.22) is negative. Smooth symmetrical projection $\tilde{A}_q(x)$ has a maximum in the center and reduces to zero at the edges; hence its second derivative in the center at $x = 0$ is always negative, and its fourth derivative is positive. The real part of $\tilde{A}_q(x)$ is nearly symmetrical with respect to Δ (up to Δ/ζ) in the vicinity of $\Delta = 0$. Here, as ($\pm\Delta$) moves away from the true position of the layer, a crevasse appears in the center of the main maximum due to the term in (5.22) containing the fourth derivative of $\tilde{A}_q(x)$. At $\Delta = 0$, the Fresnel oscillations vanish, and the real part of $\tilde{A}_q(x)$ is "focusing".

Figure 5.1 portrays the results of reconstruction of $\tilde{A}_q(x, \Delta)$ with different Δ or $z_{\mathrm{s}}{}'$ for Gaussian function $\tilde{A}_q(x) \sim \exp\left(-x^2/a^2\right)$ and the scale a of the order of the Fresnel radius. It is clearly seen in the figure how the imaginary part of $q(x)$ comes through zero and the maximum changes into the minimum at the height of 600 km. The appearance of a trough is also noticeable in the center of the reconstructions of the real parts of $\tilde{A}_q(x)$ accompanied by increasing Fresnel oscillations.

The situation becomes more complicated if there are several layers at different altitudes in the ionosphere. Reconstruction of the projections of spectra $\tilde{A}_q(x)$ corresponding to different ionospheric layers is hampered by the superposition of "defocused images" of other layers; however, determination of the

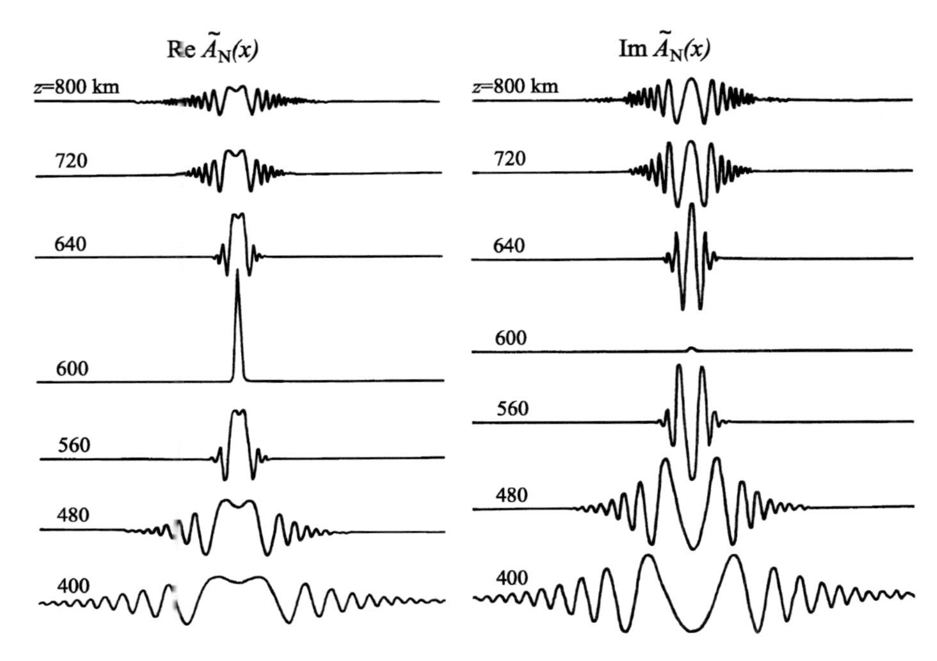

Fig. 5.1. Result of the holographic reconstruction of a Gaussian function

heights of the layers is not difficult, as previously, because $\tilde{A}_q(x)$ is a linear superposition of a series of summands. Figure 5.2 shows the results of simulation. Gaussian correlation functions with $a = 0.1$ km were defined at 600 and 840 km heights. With the given position of the layers, it is possible to identify certainly the above mentioned features such as the central "focusing" of the real part and the appearance of dips on both sides of the true position of the layer as well as the change of the central maximum into a minimum in the imaginary part.

The above stated peculiarities seen in the simulated reconstructions are also confirmed by the analysis of the experimental data. Let us consider scintillations of satellite radio signals measured on Feb. 4, 1987 at 10:01 UT (Fig. 5.3) and Oct. 29, 1979 at 20:35 UT. In both cases, the satellite trajectories lay very close to the geomagnetic zeniths of the receiving sites; the distances from the receiver to the satellites were $Z_0 = 1014$ and $Z_0 = 1050$ km, respectively. The first case presents weak scintillations: the variance of the logarithmic relative amplitude was $\sigma_\chi^2 = 0.008$, and the phase variance was $\sigma_{\Delta\Phi}^2 = 0.007$; the second case describes strong scintillations with $\sigma_\chi^2 = 0.36$, $\sigma_{\Delta\Phi}^2 = 5.7\ \text{rad}^2$.

Consider the case of weak scintillations. To separate the regular trend, the original data were filtered at 0.3 Hz. The regular components of the amplitude and the phase thus obtained are shown in Fig. 5.3 by smooth lines; the phase

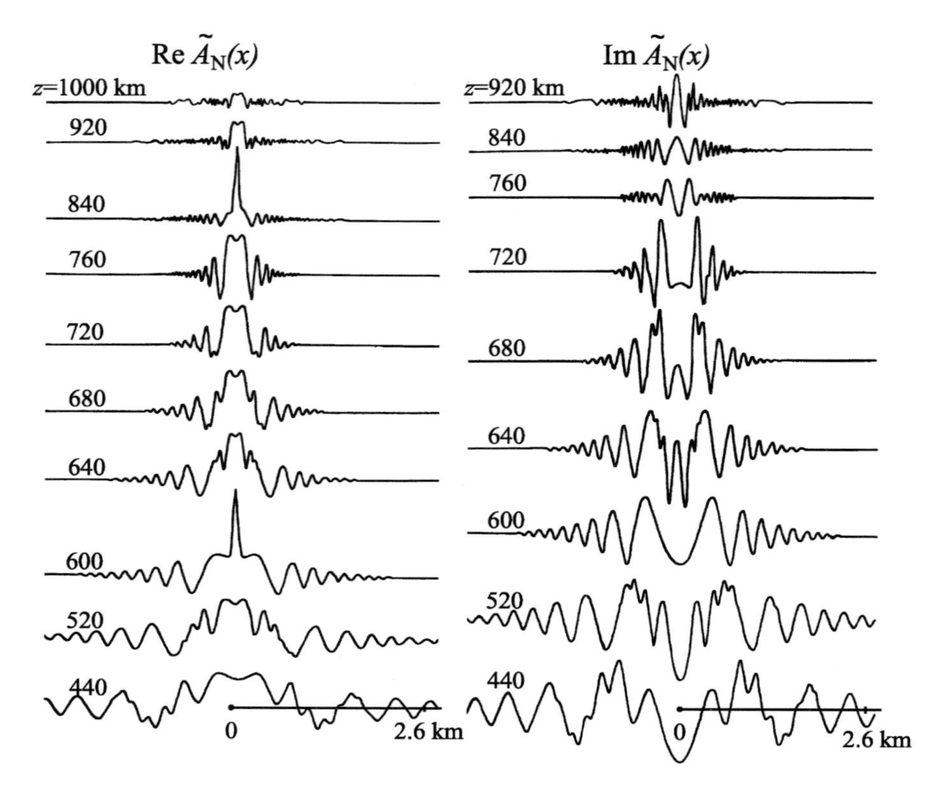

Fig. 5.2. Holographic reconstruction of two Gaussians

fluctuations and variations of logarithmic relative amplitude are portrayed in Fig. 5.4. Input data for the reconstruction (the calculated deviation from unity of the second-order first coherence function $\Delta\Gamma_{1,1}$ and $\mathrm{Re}\,\Delta\Gamma_{2,0}$ and $\Im\,\Delta\Gamma_{2,0}$ are presented in Figs. 5.5 and 5.6, respectively.

In the presence of a single thin layer of irregularities in the ionosphere, (5.15) yields the following equation for $\Delta\Gamma_{1,1} \equiv \Gamma_{1,1} - 1$:

$$\Delta\Gamma_{1,1} \approx \frac{4\pi^2 r_\mathrm{e}^2}{2} L A_\mathrm{N} \frac{x\,(Z_0 - Z_\mathrm{s})}{Z_0} ,$$

i.e., $\Delta\Gamma_{1,1}$ is proportional to the correlation function of the electron density fluctuations integrated in the propagation direction of the probing wave.

On the other hand, nearly the same quantity can be independently obtained from the reconstruction of one-dimensional $\Delta\Gamma_{2,0}$:

$$\begin{aligned} \frac{\tilde{A}_q(x)\,L}{k^2} &= \frac{4\pi^2 r_\mathrm{e}^2}{k^2}\,\tilde{A}_\mathrm{N}(x)\,L \\ &= \frac{4\pi^2 r_\mathrm{e}^2}{k^2} L \int \mathrm{d}y\; A_\mathrm{N}(\boldsymbol{\rho})\; \exp\left[\mathrm{i}k \frac{Z_0}{4Z_\mathrm{s}\,(Z_0 - Z_\mathrm{s})} y^2\right] , \end{aligned}$$

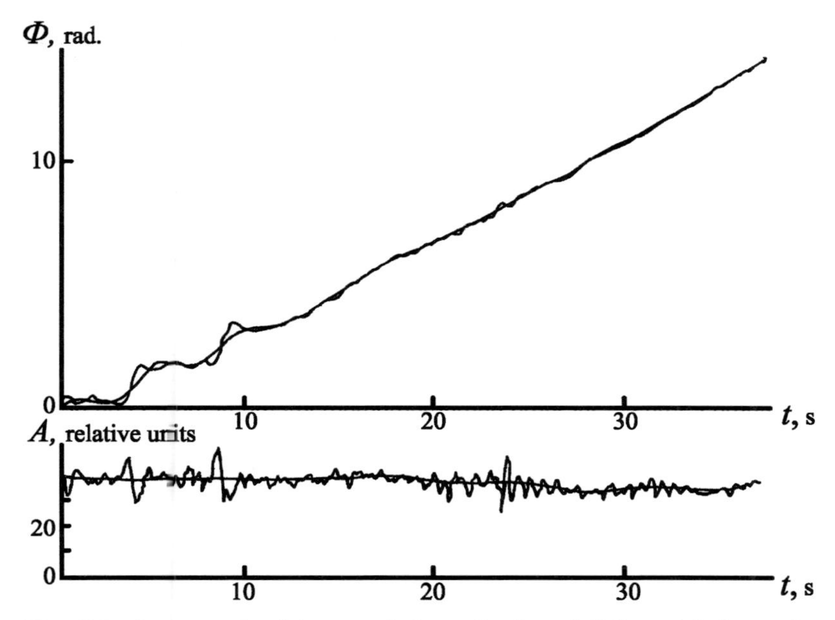

Fig. 5.3. An example of the recorded amplitude and differential phase of a satellite radio signal

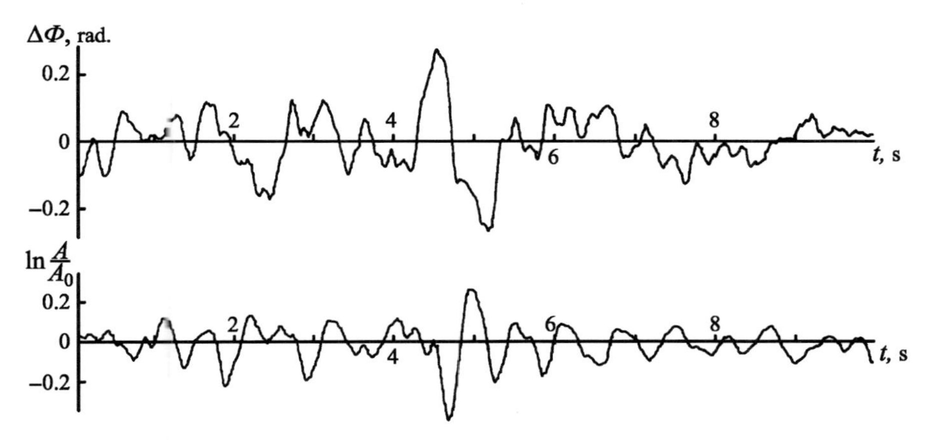

Fig. 5.4. Variations in the phase and logarithmic relative amplitude caused by ionospheric irregularities

which allows, in weak scattering, comparing the results of the determination of the correlation function of electron density fluctuations obtained by two independent methods. Notice that from the reconstruction of the second-order second correlation function, it is possible to determine the height of the scattering layer and to evaluate the scaling factor $(Z_0 - Z_s)/Z_0$ contained in the expression for the second-order first coherence function.

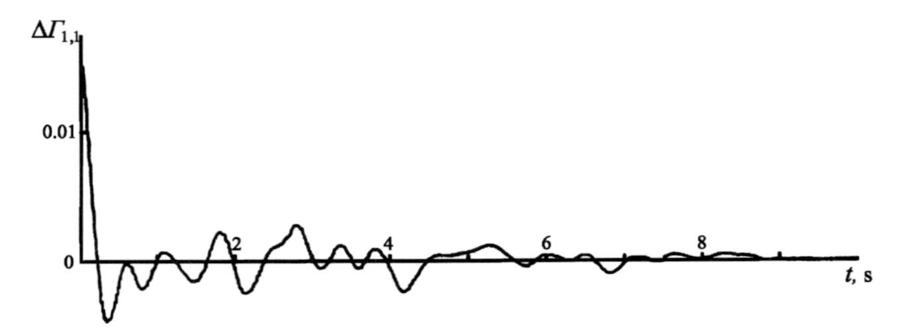

Fig. 5.5. Deviation from unity of the first coherence function of the second order

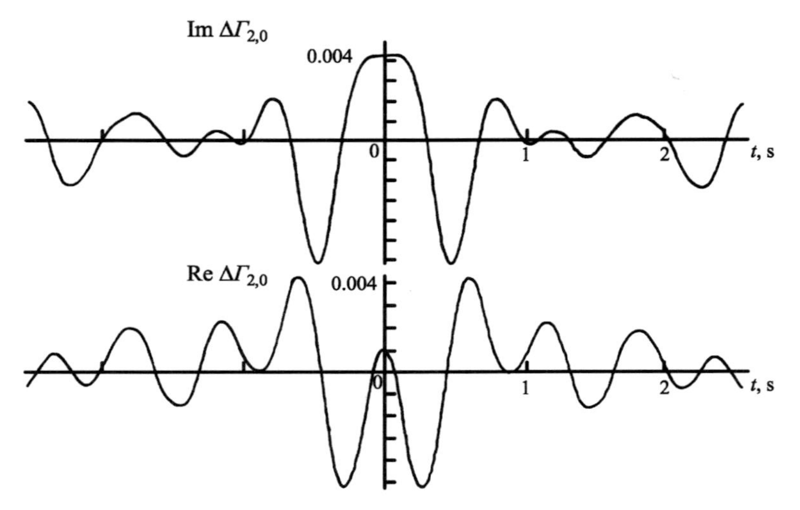

Fig. 5.6. Input data for reconstruction

The results of the reconstruction of $\Delta\Gamma_{2,0}$ are shown in Fig. 5.7. Comparison with the simulated results presented above displays good agreement with the reconstruction of the scattering layer with the typical scale of the correlation function perpendicular to the satellite path direction less than the Fresnel radius. Therefore, it is possible in this case to write an approximate formula: $\tilde{A}_{\mathrm{N}}(x) \approx \int A_{\mathrm{N}}(\boldsymbol{\rho})\,\mathrm{d}y$. Comparison of the structure of $\Delta\Gamma_{1,1}(x)$, the scaling factor taken into account, with the reconstructed data for $Z_0 - Z = 800\,\mathrm{km}$, that is, for the height above the ground level $Z = 240\,\mathrm{km}$, shows good agreement.

Let us now proceed with strong scintillations. Figure 5.8 shows the results of the reconstruction of the function $\tilde{A}_{\mathrm{N}}(x)$ in its real and imaginary parts. Figure 5.8a portrays the behavior of $\mathrm{Re}\,\tilde{A}_{\mathrm{N}}(x)$ with Z. To avoid overdetailing the picture, the whole set of calculated curves (at nearly 160 height

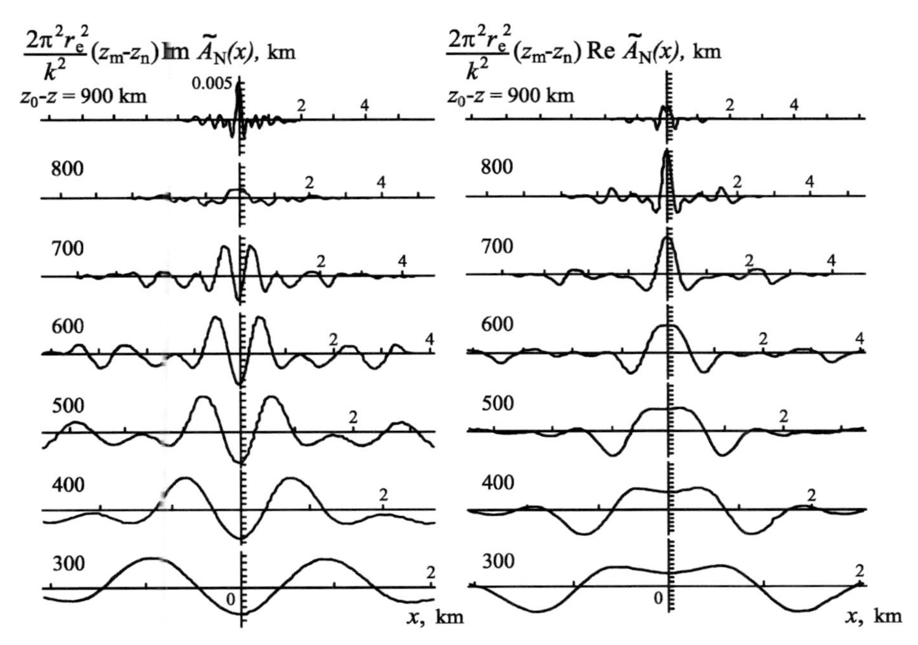

Fig. 5.7. Result of reconstruction for weak scintillation

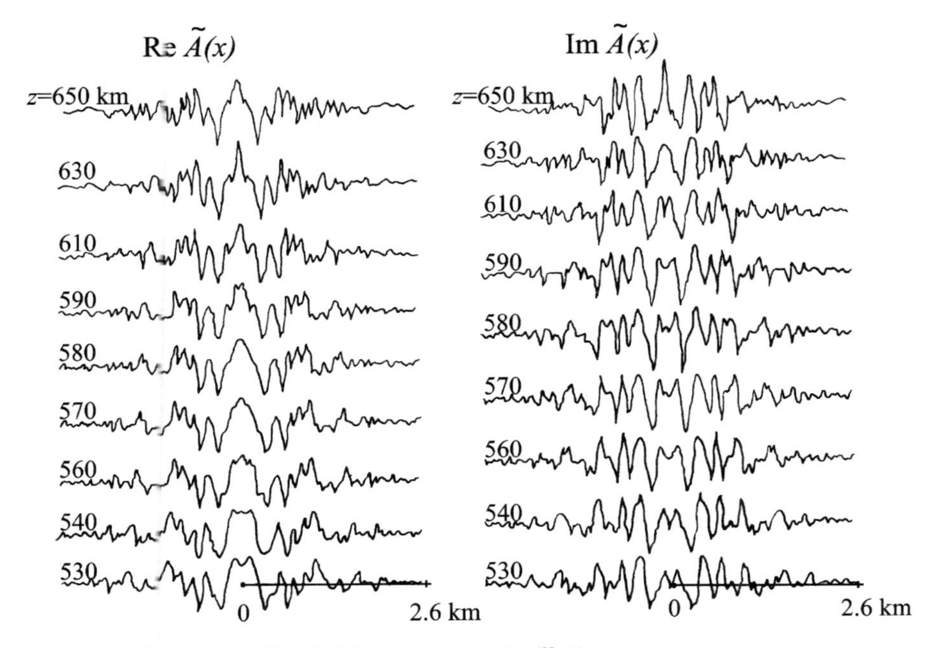

Fig. 5.8. Same as in Fig. 5.7 for a strong scintillation

levels) is not shown but only those for heights close to the layers' altitudes. At $Z = 520\,\text{km}$, $\operatorname{Re}\tilde{A}_{\text{N}}(x)$ has a minimum at $x = 0$ that disappears with increasing Z. The amplitude grows with Z, reaching its maximum at $Z = 560\,\text{km}$. A small peak in the vicinity of $x = 0$ lies on a pedestal produced by another irregular layer whose image is "focused" at $Z = 310\,\text{km}$. At $Z > 560\,\text{km}$, the amplitude of the maximum near $x = 0$ decreases up to the level $Z = 580\,\text{km}$ and starts growing again at higher altitudes. A local minimum in the vicinity of $x = 0$ vanishes at $Z = 610\,\text{km}$, and a maximum appears increasing with Z and reaching its highest at $Z = 630\,\text{km}$. At $Z > 630\,\text{km}$, the amplitude diminishes, a minimum arises near $x = 0$, and the image becomes "defocused." Consider now the imaginary part of the reconstructed function. At $Z = 530\,\text{km}$, $\operatorname{Im}\tilde{A}_{\text{N}}(x)$ has a minimum in a negative domain. The depth of the minimum reduces with Z, and at $Z = 560\,\text{km}$, a maximum appears instead, that is, the changes seen in the reconstructed experimental data are similar to the simulated results for $\operatorname{Im}\Gamma_{\text{N}}(x)$ near the scattering layer. The maximum weakens and disappears at $Z = 590\,\text{km}$, and a new minimum arises near $x = 0$. From $Z = 600\,\text{km}$ to $Z = 620\,\text{km}$, the depth of the minimum decreases a little, and the minimum degenerates into a maximum. At this very height, as we saw previously, the behavior of the imaginary part as well indicates the presence of another layer of irregularities at $Z = 630\,\text{km}$.

The information obtained about the z_{s} coordinate of the scattering layer makes it possible to reconstruct the projection of the correlation function from (5.15) for the first coherence function. If the thickness of the layer is much less than the distance to that ($L \ll z_{\text{s}}$), one can factor the projection of the correlation function outside the integral sign in (5.15), and

$$A_q\left(\boldsymbol{\rho}\,\frac{z_{\text{s}}}{z}\right) \simeq \frac{4k^2}{L} \ln \Gamma_{1,1}(\boldsymbol{\rho}, z) \; . \tag{5.23}$$

Note that in this approach, we imposed no limitation on the strength of fluctuations, and, to be exact, the only limitations are those implied by the terms of applicability of a parabolic equation. Constraints on the thickness of the layer are inessential. If the layer is thick, one will obtain fan-beam projections that, in a statistically homogeneous layer, by exhaustion of the rays, reduce to parallel projections. For weak scintillations when $\Gamma_{1,1}$ only faintly departs from unity,

$$1 + \frac{A_q\left(\boldsymbol{\rho}\,\frac{z_{\text{s}}}{z}\right)}{4k^2} \simeq \Gamma_{1,1}(\boldsymbol{\rho}, z) \; . \tag{5.24}$$

The projection of the correlation function can also be reconstructed using a holographic approach based on (5.18). As the plane of reconstruction approaches the scattering layer, $\zeta \to 0$ and $\boldsymbol{s} \to \boldsymbol{\rho}$ that makes it possible to calculate asymptotically the inner integral with respect to $\boldsymbol{\rho}$ in the left part of (5.18). This gives a relation between Γ_{H} and the projection of the correlation function:

$$1 + \frac{A_q(\boldsymbol{\rho})}{4k^2} \int \Gamma_{2,0}(\boldsymbol{\rho}, z')\,\mathrm{d}z' \simeq \Gamma_{\mathrm{H}}(\boldsymbol{\rho}, z) \;. \tag{5.25}$$

For weak scattering, $\Gamma_{2,0}(\boldsymbol{\rho}, z') \simeq 1$, and the expression changes into a formula similar to (5.24); the only differences are the scales of the reconstructed correlation function:

$$1 + \frac{A_q(\boldsymbol{\rho})\, L}{4k^2} \simeq \Gamma_{\mathrm{H}}(\boldsymbol{\rho}, z) \;. \tag{5.26}$$

The integral equations obtained allow reconstructing the projections of the correlation functions also for strong scintillations within the scope of the parabolic equation on which the above mentioned integral relations are based. First, it is possible to reconstruct projections $A_q(\boldsymbol{\rho})$ of the correlation function of a strongly fluctuating medium directly from the first coherence function (5.23), once the scattering layer is located. Reconstruction of the projections in strong fluctuations is possible as well from the second coherence function by the holographic approach. Let us describe in brief the essence of the method considering the reconstruction of projections $A_q(\boldsymbol{\rho})$ from the second coherence function as an example. Reconstruction of the projections of the correlation function for strong fluctuations in the medium is similar to the approach applied in reconstructing strongly scattering objects (Sect. 3.5). Integral equation (5.16), after taking the inverse Fresnel transform of it, yields the reconstruction of the projections of the product $A_q(\boldsymbol{\rho})\, \Gamma_{2,0}(\boldsymbol{\rho}, z)$, similarly to (5.19):

$$\begin{aligned} \int A_q(\boldsymbol{\rho})\; \Gamma_{2,0}(\boldsymbol{\rho},\, z')\; \mathrm{d}z' = \frac{k^3}{\mathrm{i}\pi\zeta} \int \mathrm{d}^2 s\; [\Gamma_{2,0} - 1] \\ \times \exp\left[-\mathrm{i}\frac{k}{4\zeta}\,(\boldsymbol{s} - \boldsymbol{\rho})^2\right] \;. \end{aligned} \tag{5.27}$$

Next, by "rotating" the z-axis (i.e., by varying the directions of the probing rays), one can reconstruct $A_q(\boldsymbol{\rho})$ from the set of projections thus obtained by means of iterative procedure (3.50). Note that for strong fluctuations, due to rescattering, the projection $A_q(\boldsymbol{\rho})$ of the correlation function in some specified direction depends not only on the data on the second coherence function in that single direction. In this case, contributions from all probing directions need to be taken into account in reconstructing the projections.

Similarly, iterative reconstruction is also possible within the holographic approach. Here, once (5.16) is reconstructed, it is possible to reconstruct $A_q(\boldsymbol{\rho})$ from multidirectional probing.

The approaches developed for the statistical inverse scattering problem can be generalized to a stratified regular background ionosphere. Similarly to the deterministic case considered, one can pass to a parabolic equation of the type (3.38) with the group path variable. After this, the solutions of the inverse problem – projections of correlation functions (5.19), (5.20), (5.27) will take the form of the Fresnel transform with respect to new variables $\boldsymbol{\rho}$ and ζ containing corresponding group paths, similarly to (3.39). The holographic approach may also be generalized to a stratified background ionosphere. In this case, the appropriate kernel $H(\boldsymbol{r}, \boldsymbol{r}')$ should be used (3.76).

5.4 Tomography of Randomly Inhomogeneous Isotropic Media

Irregularities of electron density may aggregate into rather extensive ionospheric volumes of arbitrary shape. For various physical reasons, the structures of such turbulent regions can be quite different; for example, extended layers are encountered in the ionosphere as well as arbitrary solid forms. Reconstruction of projections of correlation functions for statistically homogeneous layers was considered above. In the present subsection, we will describe a more general approach to the reconstruction of the correlation function within arbitrary domains filled up with irregularities. Reconstruction of an arbitrary correlation function from measured integral data seems unreal since the function depends on six variables. But in many cases, randomly inhomogeneous media may be treated as statistically quasi-homogeneous with slowly changing statistical properties, so that the field correlation length along the difference argument is significantly less than the scale size of changes in the variance and correlation coefficient. In other words, the correlation function of complex potential q can be written as

$$B(\boldsymbol{r}_1, \boldsymbol{r}_2) = \sigma^2(\boldsymbol{R})\, \mathcal{K}(\Delta \boldsymbol{r}, \boldsymbol{R}) \,. \tag{5.28}$$

In a statistically homogeneous isotropic medium, $\mathcal{K} = \mathcal{K}(\Delta \boldsymbol{r})$. It is possible to reconstruct the correlation coefficient $\mathcal{K}$ and the variance of fluctuations σ^2 of a quasi-uniform field from the measured statistical parameters of the probing a wave.

Let us first consider the case when the correlation coefficient depends only on the absolute value of the difference coordinate and does not change in space. Such randomly inhomogeneous fields with constant correlation coefficients and varying intensities of fluctuations only, can be called "additive" in the sense that density variations in the medium (for example, electron density) affect the variance of fluctuations, but they do not change the shape of the functional dependence of the correlation coefficient. Fluctuations of the medium at different points are similar; the only change is in their intensity. Assuming, as previously, delta-correlation of the field along z, the direction of propagation of the probing wave,

$$B(\boldsymbol{R},\, \Delta\boldsymbol{r}) = \sigma^2(\boldsymbol{R})\; K(\boldsymbol{\rho})\; \delta(\Delta z) \quad \text{or}$$
$$\int B(\boldsymbol{R},\, \Delta\boldsymbol{r})\; \mathrm{d}(\Delta z) = \sigma^2(\boldsymbol{R})\; K(\boldsymbol{\rho})\;, \tag{5.29}$$

one can reiterate the derivation of equations obtained in previous subsections, replacing the projection of correlation function $A_q(\boldsymbol{\rho})$ in the integrands by $\sigma^2(\boldsymbol{R})\; K(\boldsymbol{\rho})$ or, in local coordinates, by $\sigma^2(z)\; K(\boldsymbol{\rho})$. Recall that $\boldsymbol{\rho}$ in the argument of the correlation coefficient is a transverse (cross-ray) component of the difference coordinate $\Delta\boldsymbol{r} = (\boldsymbol{\rho},\, \Delta z)$.

Expression (5.11) for an average field contains an integral of the variance of fluctuations in the direction of the propagation of $K(0)$:

$$\langle E(\boldsymbol{r})\rangle = E_0(\boldsymbol{r})\; \exp\left[-\frac{K(0)}{8k^2}\int_{z_\mathrm{u}}^{z} \sigma^2(z)\; \mathrm{d}z\right]\,. \tag{5.30}$$

The first coherence function changes in the same way:

$$\Gamma_{1,1}(\boldsymbol{\rho},\, z) = \exp\left[\frac{1}{4k^2}\int_{z_\mathrm{u}}^{z} \sigma^2(z')\; K\left(\boldsymbol{\rho}\,\frac{z'}{z}\right)\, \mathrm{d}z'\right]\,. \tag{5.31}$$

The factors $\sigma^2(z)$ also appear in integral equations (5.16) and (5.18) for Γ_H and the second coherence function:

$$\Gamma_{2,0}(\boldsymbol{\rho},\, z) = 1 + \frac{\mathrm{i}}{16\pi k}\int \mathrm{d}^3r'\, \frac{\sigma^2(z')\; K(\boldsymbol{\rho}')}{\zeta}\, \Gamma_{2,0}(\boldsymbol{r}')$$
$$\times \exp\left[\mathrm{i}\frac{k}{4\zeta}\,(\boldsymbol{s}-\boldsymbol{\rho}')^2\right]\;; \tag{5.32}$$

$$1 - \frac{\mathrm{i}}{16\pi k}\int \mathrm{d}^3r'\, \frac{\sigma^2(z')\; K(\boldsymbol{\rho}')\; \Gamma_{2,0}(\boldsymbol{r}')}{\zeta}\, \exp\left[\mathrm{i}\frac{k}{4\zeta}\,(\boldsymbol{s}-\boldsymbol{\rho}')^2\right]$$
$$= \Gamma_\mathrm{H}(\boldsymbol{\rho},\, z)\;. \tag{5.33}$$

For this once, it is no longer possible to obtain the formula for the projections of the correlation coefficient of type (5.19), (5.23)–(5.26) directly from the above equations, since (5.30)–(5.33) contain an unknown function $\sigma^2(\boldsymbol{R})$. However, reconstruction of the correlation coefficient becomes feasible once the spatial distribution of $\sigma^2(\boldsymbol{R})$ is recovered. Thus, the problem of reconstruction of the correlation function of the medium divides into two separate tasks: (i) recovering $\sigma^2(\boldsymbol{R})$ and (ii) reconstructing $\mathcal{K}(\Delta r)$ from the measured projections of $K(\boldsymbol{\rho})$.

Formulas (5.30), (5.31) for the average field and first coherence function allow determining the integrals of σ^2 in the directions of propagation of the probing wave (the z-axis). Using a set of integrals over the region with irregularities, one comes to the classical tomographic problem of determination of the function from linear integrals. However, in the ionosphere as well as in other geophysical media, the number of projections and the angular range of rays probing an irregular volume are usually limited. Therefore, in this case, one should consider an aspect-limited problem. Such a problem has been already discussed in Chap. 2. Here, in a statistical consideration, the situation is even simpler since the measurements of an average field give the very linear integrals but not their differences, as do phase measurements in global radio tomography. Thus, illumination of the turbulent volume by a large enough set of rays over an appropriate angle range allows reconstructing the fluctuation intensity field $\sigma^2(\boldsymbol{R})$. Multireceiver measurements in a plane containing the satellite path make it possible to reconstruct a two-dimensional section of $\sigma^2(\boldsymbol{R})$. A set of cross-path receiving chains will give a data set for reconstructing the three-dimensional structure of the volume with irregularities. Simulated results described in the previous section and the experiments on global radio tomography (Chap. 2) prove the possibility of reconstruction of rather complex structures by means of a few receivers.

Information about the two-dimensional distribution of fluctuation intensity $\sigma^2(\boldsymbol{R})$ within an irregular volume is sufficient for reconstruction of appropriate sections (projections) of the correlation function. The techniques for reconstructing the projections were discussed in the previous subsections of the monograph. Hence, we will give here only slightly modified [due to variations in $\sigma^2(\boldsymbol{R})$] formulas:

$$\int \sigma^2(z') \, K\left(\boldsymbol{\rho}\,\frac{z'}{z}\right) \mathrm{d}z' = 4k^2 \, \ln \Gamma_{1,1}(\boldsymbol{\rho},\, z) \;. \tag{5.34}$$

The first coherence function defines the set of fan projections with known weights. If the size of an irregular volume is much less than the distance s_z to it, the equation reduces to

$$K\left(\boldsymbol{\rho}\,\frac{z_{\mathrm{s}}}{z}\right) \int \sigma^2(z') \, \mathrm{d}z = 4k^2 \, \ln \Gamma_{1,1}(\boldsymbol{\rho},\, z) \;. \tag{5.35}$$

Within the scattering volume, holographically reconstructed function Γ_{H} is asymptotically ($\zeta \to 0$) associated with the projections of the product $K(\boldsymbol{\rho})\, \sigma^2(z)\, \Gamma_{2,0}(\boldsymbol{\rho},\, \boldsymbol{z})$ which follows from (5.33) like (5.25):

$$1 + \frac{1}{4k^2} \int \mathrm{d}z' \, K(\boldsymbol{\rho}) \, \sigma^2(z') \, \Gamma_{2,0}(\boldsymbol{\rho},\, z') \simeq \Gamma_{\mathrm{H}}(\boldsymbol{\rho},\, \boldsymbol{z}) \;. \tag{5.36}$$

Similarly, the projection of the above product can be obtained from (5.32) after taking an inverse Fresnel transform:

$$\begin{aligned} \int \mathrm{d}z' \, K(\boldsymbol{\rho}) \, \sigma^2(z') \, \Gamma_{2,0}(\boldsymbol{\rho},\, z') = \frac{k^3}{\mathrm{i}\pi\zeta} \int \mathrm{d}^2 s \, \left[\Gamma_{2,0}(\boldsymbol{s}) - 1\right] \\ \times \exp\left[-\frac{\mathrm{i}k}{4\zeta}\,(\boldsymbol{s}-\boldsymbol{\rho})^2\right] . \end{aligned} \tag{5.37}$$

Obtaining a set of projections of the product $\left[\sigma^2(z') \, K(\boldsymbol{\rho}) \, \Gamma_{2,0}(\boldsymbol{\rho},\, z')\right]$ by illuminating the scattering volume in different directions, we can reconstruct the three-dimensional structure of the product $F(\boldsymbol{r}) \equiv \left[\sigma^2(z') \; K(\boldsymbol{\rho})\right]$ using an iterative procedure of type (3.50). Recovered $F(\boldsymbol{r})$ allows reconstructing $K(\boldsymbol{\rho})$:

$$K(\boldsymbol{\rho}) = \sigma^{-2}(z) \; F(\boldsymbol{r}) \left\{ 1 + \frac{\mathrm{i}}{16\pi k} \int \mathrm{d}^3 r' \, \frac{F(\boldsymbol{r}')}{\zeta} \times \exp\left[\frac{\mathrm{i}k}{4\zeta} (\boldsymbol{s} - \boldsymbol{\rho})^2 \right] \right\}^{-1} . \tag{5.38}$$

In weak scattering, formulas (5.36)–(5.38), naturally, become simpler. The projection of the correlation coefficient is determined directly by Γ_{H} from (5.33):

$$K(\boldsymbol{\rho}) = 4k \left(\Gamma_{\mathrm{H}} - 1 \right) \left[\int \mathrm{d}z \, \sigma^2(z') \right]^{-1} . \tag{5.39}$$

In contradistinction to strong fluctuations (5.38), the projection of the correlation coefficient is now completely determined only by the data on the second correlation function only in a given direction:

$$K(\boldsymbol{\rho}) = \frac{k^3}{\mathrm{i}\pi\zeta} \int \mathrm{d}^2 s \, \left[\Gamma_{2,0}(\boldsymbol{s}) - 1 \right] \exp\left[-\mathrm{i} \frac{k}{4\zeta} (\boldsymbol{s} - \boldsymbol{\rho})^2 \right] \times \left[\int \mathrm{d}z' \, \sigma^2(z') \right]^{-1} . \tag{5.40}$$

Equations (5.34)–(5.40) define the projections of correlation functions of the complex potential of the medium. A set of projections allows tomographic reconstruction of the correlation function or its spectrum. Let us first consider reconstruction of a two-dimensional cross section from a set of one-dimensional projections. Owing to (5.28), (5.29), the projections $K(\boldsymbol{\rho})$ are integrals of three-dimensional correlation coefficient $\mathcal{K}(\Delta \boldsymbol{r}) \equiv \mathcal{K}(\boldsymbol{\rho}, \Delta z)$:

$$\int \mathcal{K}(\Delta \boldsymbol{r}) \, \mathrm{d}(\Delta z') = K(\boldsymbol{\rho}) . \tag{5.41}$$

Next, we shall use global ("laboratory") coordinates $\boldsymbol{R} = (X, Y, Z)$ associated with the medium and a series of local coordinate systems $\boldsymbol{r} = (\boldsymbol{\rho}, z)$ with their z-axes oriented in the direction of the probing wave propagation. Hereinafter, for the sake of uniformity, we will denote the argument Δz of the correlation coefficient by z, bearing in mind, however, that z is a difference coordinate of the correlation function and correlation coefficient, like $\boldsymbol{\rho}$ is a difference transverse coordinate. In the above described reference systems, the one-dimensional x-projections of $K(x, y)$ (at $y =$ const) can be written as follows:

$$K(x, y) = \int \mathcal{K}(x, y, z) \, \mathrm{d}z = \int \mathrm{d}^2 R \, \mathcal{K}(\boldsymbol{\mathfrak{R}}, y) \, \delta(x - \boldsymbol{\mathfrak{R}} \boldsymbol{n}) . \tag{5.42}$$

Here, a two-dimensional vector $\boldsymbol{\mathfrak{R}} = (X, Y)$ is introduced in the global coordinate system; $\boldsymbol{n} = (\cos\theta, \sin\theta)$ is a unit vector determining the direction of the probing ray in $\boldsymbol{\mathfrak{R}}$ that is a straight line making angle θ with the

X-axis. The delta function in the integrand means that integration in the plane (X, Y) $(Y = \text{const})$ comes over a set of straight lines parallel to the probing direction.

Let us introduce a two-dimensional Fourier image $\hat{\mathcal{K}}(\varkappa)$, $\varkappa = (\varkappa_x, \varkappa_z)$ of two-dimensional function $\mathcal{K}(x, y = \text{const}, z)$, or a spectrum of cross section

$$\hat{\mathcal{K}}(\varkappa, y) = \int \mathcal{K}(x, z) \exp(\mathrm{i}\boldsymbol{k}\mathfrak{R}) \, \mathrm{d}^2\mathfrak{R} \,. \tag{5.43}$$

The one-dimensional Fourier image $\hat{K}_\theta(\chi)$ of projection $K(x, y = \text{const})$ (at angle θ) of two-dimensional function $\mathcal{K}(x, z)$ is a section of two-dimensional Fourier image $\hat{\mathcal{K}}$ (5.43). This is the so-called central section theorem [Pikalov and Preobrazhenskiy, 1987b]. One can easily prove its validity by taking the one-dimensional Fourier transform of (5.42):

$$\begin{aligned} \hat{K}_\theta(\chi) \equiv \int K_\theta(x) \exp(\mathrm{i}\chi x) \, \mathrm{d}x = \int \mathrm{d}^2 R \, \exp(\mathrm{i}\chi \mathfrak{R}\boldsymbol{n}) \, \mathcal{K}(\mathfrak{R}) \\ = \hat{\mathcal{K}}(\chi \boldsymbol{n}) \end{aligned} \tag{5.44}$$

In other words, Fourier transform $\hat{K}_\theta(\chi)$ of one-dimensional projection $K_\theta(x)$ defines two-dimensional Fourier image $\hat{\mathcal{K}}(\boldsymbol{x})$ of the section of the correlation coefficient at the straight line $\boldsymbol{x} = \chi \boldsymbol{n}$. From the set of projections obtained at different angles, it is possible to find values of $\hat{K}(\boldsymbol{x})$ at a series of straight lines from (5.44) and then reconstruct the spectrum of the section $\hat{K}(\varkappa)$. The entire three-dimensional $\boldsymbol{p} = (\varkappa_x, p_y, \varkappa_z)$ spectrum $\hat{\mathcal{K}}(\boldsymbol{p})$ of the three-dimensional correlation coefficient can be reconstructed, for example, from the sets of two-dimensional spectra obtained of the section $(y = \text{const})$ (5.43):

$$\hat{\mathcal{K}}(\boldsymbol{p}) \equiv \int \mathcal{K}(\boldsymbol{r}) \exp(\mathrm{i}\boldsymbol{p}\boldsymbol{r}) \, \mathrm{d}^3 r = \int \hat{\mathcal{K}}(\varkappa, y) \exp(\mathrm{i}p_y y) \, \mathrm{d}y \,. \tag{5.45}$$

A lot of tomographic methods are known for reconstructing functions from their projections [Natterer, 1986; Nolet, 1987], including those based on inversion of a ray transform over the set of incomplete angular data [Palamodov, 1990]. In experiments in ionospheric radio tomography, it is possible to implement the scheme of measuring comparatively few projections of a limited angular range, similar to the situation arising in aspect-limited tomography with a restricted range of probing angles, as well as in other applications. In aspect-limited tomography, methods of inverse projection and algebraic reconstruction, like those used in Chap. 2, work well. However, the statistical problem has an important peculiarity, that is, the correlation function is rather smooth and "simple" and centrally symmetrical. Therefore, in this case it is convenient to expand the functions into appropriate series.

Let us expand the two-dimensional spectrum $\hat{\mathcal{K}}(\varkappa)$ in a series of orthogonal polynomials. By virtue of the central symmetry of the spectrum, it is more convenient to use polar coordinates $\chi = \sqrt{\varkappa_x^2 + \varkappa_y^2}$, $\theta = \arctan(\varkappa_y / \varkappa_x)$:

$$\hat{\mathcal{K}}(\varkappa) \equiv \hat{\mathcal{K}}(\chi, \theta) = \sum_n a_n(\chi) \; P_n(\theta) \,. \tag{5.46}$$

Naturally, polynomials $P_n(\theta)$ satisfy the condition of central symmetry $P_n(\theta) = P_n(\theta + \pi)$. The data on the Fourier images $\hat{\mathcal{K}}(\varkappa)$ at straight lines $\chi \boldsymbol{n}$ with different θ obtained by applying (5.44) to the projections of correlation functions make it possible to reconstruct a (finite) truncated series of (5.46). Having obtained a set of the values of $\hat{\mathcal{K}}_i(\chi) \equiv \hat{\mathcal{K}}(\chi, \theta_i)$ at fixed χ, one arrives at a linear system for determination of $a_n(\chi)$:

$$\sum_n a_n(\chi)\, P_{ni} = \hat{\mathcal{K}}_i(\chi)\ , \tag{5.47}$$

where $P_{ni} \equiv P_n(\theta_i)$. Solving the linear system of not large dimension for $a_n(\chi)$ with different χ yields a reconstruction of the finite-dimensional approximation of spectrum $\hat{\mathcal{K}}(\varkappa)$. Here, we omit the topics of regularization of the solution and solving the system with inaccurate data. Algorithms for solving systems like (5.47) with inaccurately defined (noised) right-hand parts and summation of the Fourier series with inaccurately defined coefficients are worked out quite well [Tikhonov and Arsenin, 1977; Tikhonov et al., 1995]. When solving problems of that kind, owing to the inherently limited number of terms in the series, it is important to choose a suitable polynomial system that gives a good approximation of the desired function by a few polynomials. The choice of polynomial system depends on available a priori information about the presumed spectral structure. In particular, it is often supposed that the ionosphere fluctuations' correlation function is symmetrical about the direction of the local geomagnetic field. In this case, after processing the experimental data, it is convenient to pick out this axis of symmetry and to measure angles θ from it. Then, besides satisfying the central symmetry condition, the polynomials will also meet the condition $P_n(\theta) = P_n(-\theta)$. In the capacity of such polynomials, for example, Legendre polynomials $P_n(\cos\theta)$ can be used, with odd polynomials only remaining in expansion (5.46). A proper choice of polynomial system will yield a system (5.47) of small dimension. There are no limitations on the shape of the function $a_n(\chi)$. Note that in statistical inverse problems, correlation functions and spectra are estimated from finite length data. Therefore, spectral analysis should be used here, including multidimensional analysis [Jenkins and Watts, 1969; Marple Jr., 1987] in processing the data and spectral estimation.

5.5 Tomography of Randomly Inhomogeneous Anisotropic Media

In the previous section, it was assumed that the medium contains isotropic irregularities. However, it is well known that small-scale irregularities in the real ionosphere, particularly in the high latitudes, are markedly anisotropic. Therefore let us analyze the relationship between the parameters of the scattered field and the parameters of three-dimensional anisotropic scattering irregularities and consider the applicability of the tomographic approach to the investigation of such irregularities.

First, seek an average field in the medium containing random anisotropic irregularities. Equation (5.30) in this case can be written as follows:

$$\langle E(\boldsymbol{r})\rangle = E_0(\boldsymbol{r})\,\exp\left(-\frac{1}{8k^2}\right)\int_{z_u}^{z}\sigma^2(z')\,K(0,\boldsymbol{z}')\,\mathrm{d}z'\,. \tag{5.48}$$

As K depends on both the absolute value of z and the orientation of the $0z$ axis in an anisotropic medium, one should use here a vector argument $\boldsymbol{z}'$ of K. Expressing K in terms of the spatial spectral density of electron density fluctuations $\Phi_{\mathrm{N}}(\boldsymbol{k},\boldsymbol{z}')$, we arrive at the following expression for the logarithmic relative field:

$$\ln\frac{\langle E(\boldsymbol{r})\rangle}{E_0(\boldsymbol{r})} = -\frac{\lambda^2 r_{\mathrm{e}}^2}{8\pi^2}\int \Phi_{\mathrm{N}}(\boldsymbol{k},\boldsymbol{z}')\Big|_{k_z=0}\,\mathrm{d}k_x\mathrm{d}k_y\mathrm{d}z'\,.$$

Assuming that the generating mechanism of the irregularities remains the same within the whole ionospheric volume of interest, the spectral density can be written in the form $\Phi_{\mathrm{N}}(\boldsymbol{k},\boldsymbol{z}') = \sigma^2(z')\,F(\boldsymbol{k},\boldsymbol{e}_{z'})$ where $\sigma^2(z')$ is, as previously, the variance of electron density fluctuations depending only on $\boldsymbol{z}'$ absolute values, $\boldsymbol{e}_{z'}$ is the unit vector in the direction of the $0z'$ axis, and $F(\boldsymbol{k},\boldsymbol{e}_{z'})$ is the normalized spectral density depending on the orientation of $\boldsymbol{z}'$. Then,

$$\ln\frac{\langle E(\boldsymbol{r})\rangle}{E_0(\boldsymbol{r})} = -\frac{\lambda^2 r_{\mathrm{e}}^2}{8\pi^2}\int I(z')\,\sigma^2(z')\,\mathrm{d}z'\,, \tag{5.49}$$

where weighting factor $I(z')$ is

$$I(z) = \int_{-\infty}^{\infty}\int_{-\infty}^{\infty} F(\boldsymbol{k},\boldsymbol{e}_z)\Big|_{k_z=0}\,\mathrm{d}k_x\mathrm{d}k_y\,. \tag{5.50}$$

Equation (5.49) means that an average field measured along z yields information about the integral variance of electron density fluctuations along the ray from the satellite to the receiver. This situation is similar to the usual tomographic problem complicated by the weighting factor $I(z)$ that depends, in particular, on the angle between the ray and the local geomagnetic field. However, as we will see below, this is not a critical impediment for the reconstruction of the variance of electron density fluctuations.

It is believed that ionospheric irregularities are strongly elongated in the direction of the geomagnetic field and perhaps in some field-perpendicular direction. In this case, it is convenient to use the spectrum as a function of the quadric surface of the wave vector [Secan et al., 1997; Fremouw and Secan, 1984]:

$$F(\boldsymbol{k}, \boldsymbol{e}_z) = \Phi\left[\left(k_{y\perp}^2 + \beta^2 k_{x\perp}^2 + \alpha^2 k_{\parallel}^2\right) \frac{L_0^2}{(2\pi)^2}\right] , \tag{5.51}$$

where $x_\perp$ and $y_\perp$ are components in a plane perpendicular to the local geomagnetic field, the sign $\parallel$ denotes a field-aligned component, and α and β are relative along-field and cross-field elongations of the irregularities, L_0 is an outer scale of irregularities, and $\boldsymbol{k}$ is a wave vector. Irregularities symmetrical in a plane perpendicular to the geomagnetic field are described by

$$F(\boldsymbol{k}, \boldsymbol{e}_z) = \Phi\left[\left(k_{x\perp}^2 + \alpha^2 k_{\parallel}^2\right) \frac{L_0^2}{(2\pi)^2}\right] , \tag{5.52}$$

where $k_\perp^2 = k_x^2 + k_y^2$.

For calculation of (5.50) with spectral density in the form (5.51), let us specify the coordinate system. Suppose that the z-axis is directed from the receiver to the satellite and the vector of geomagnetic field $\boldsymbol{B}$ contained in the plane $(y, 0, z)$ makes an angle $\theta(z)$ with the $0z$ direction of the $(y, 0, z)$ plane. Assume that in the plane perpendicular to the geomagnetic field, the irregularities are elongated in the direction $\boldsymbol{e}_\beta$ defined by angles θ_β and φ_β in spherical coordinates. Then, the condition of $\boldsymbol{e}_\beta$ orthogonality to the geomagnetic field brings

$$\sin\theta(z) \sin\theta_\beta \sin\varphi_\beta + \cos\theta(z) \cos\theta_\beta = 0 . \tag{5.53}$$

Specifying φ_β and calculating $\theta(z)$, one can find θ_β from (5.53) and thus define the orientation of the cross-field anisotropy of irregularities $\boldsymbol{e}_\beta$.

With the spectrum F in the form (5.51), one can calculate $\ln \frac{\langle E(\boldsymbol{r})\rangle}{E_0(\boldsymbol{r})}$ using (5.49) and (5.50):

$$\begin{aligned}\ln \frac{\langle E(\boldsymbol{r})\rangle}{E_0(\boldsymbol{r})} = &-\frac{\pi r_e^2}{2}\left(\frac{\lambda}{L_0}\right)^2 \int_0^\infty \mathrm{d}y\, \Phi(y) \\ &\times \int_{z_d}^{z_u} \mathrm{d}z\, \sigma_N^2(z) \left\{\left[1 + (\beta^2 - 1)\cos^2\psi\right]\right. \\ &\times \left.\left[1 + (\alpha^2 - 1)\sin^2\theta(z) + \frac{(\beta^2 - 1)\sin^2\psi \cos^2\theta(z)}{1 + (\beta^2 + 1)\cos^2\psi}\right]\right\}^{-1/2} ,\end{aligned} \tag{5.54}$$

where $\sin\psi = \sin\theta_\beta \sin\varphi_\beta \cos\theta(z) - \cos\theta_\beta \sin\theta(z)$, $\cos\psi = \sin\theta_\beta \cos\varphi_\beta$.

Thus, relative measurements of $\ln \frac{\langle E(\boldsymbol{r})\rangle}{E_0(\boldsymbol{r})}$ do not depend on the explicit form of the spectrum. Specific spectral shape $\Phi(y)$ defines only the scaling factor that remains constant and does not depend on the direction of the ray from the satellite to the receiver. In particular, for the power-law spectral shape [Secan et al., 1997]

$$\Phi(\boldsymbol{k}) = \frac{\alpha\,\beta\,L_0^3\,\Gamma(p/2)}{2\pi\Gamma(3/2)\,\Gamma((p-3)/2)} \times \left[1 + \left(\frac{L_0}{2\pi}\right)^2 \left(k_{y\perp}^2 + \beta^2 k_{x\perp}^2 + \alpha^2 k_\parallel^2\right)\right]^{-p/2}, \tag{5.55}$$

we obtain

$$\int_0^\infty \mathrm{d}y\,\Phi(y) = \frac{\alpha\,\beta\,L_0^3\,\Gamma(1/2)\,\Gamma[(p-2)/2]}{\pi^{3/2}\Gamma[(p-3)/2]},$$

where Γ is the gamma function; p is the power index; and $k_\parallel$, $k_{x\perp}$, $k_{y\perp}$ are the components of vector $\boldsymbol{k}$ in the directions parallel and perpendicular to the geomagnetic field. The axes x and y are chosen in such a way that the cross-field anisotropy of irregularities is oriented in the x direction.

Consider the case when the irregularity is elongated in a direction perpendicular to the plane $(y, 0, z)$, that is, $\boldsymbol{e}_\beta = \boldsymbol{e}_x$. Then $\varphi_\beta = 0$, $\theta_\beta = 90°$, $\sin\psi = 0$, $\cos\psi = 1$, and

$$\ln\frac{\langle E(\boldsymbol{r})\rangle}{E_0(\boldsymbol{r})} = -\frac{r_\mathrm{e}^2}{2}\left(\frac{\lambda}{L_0}\right)^2 \frac{\alpha\,L_0^3\,\Gamma[(p-2)/2]}{\pi^{1/2}\,\Gamma[(p-3)/2]} \times \int_{z_\mathrm{d}}^{z_\mathrm{u}} \frac{\sigma^2(z')}{\left[1+(\alpha^2-1)\sin^2\theta(z')\right]^{1/2}}\mathrm{d}z'. \tag{5.56}$$

Just the same result would be obtained if the spectral function (5.52) were used that describes irregularities symmetrical about the geomagnetic field. It conforms to the geometry of the experiment when signals from polar-orbiting satellites are measured at a meridional receiving chain and the elongation of irregularities in the west–east direction corresponds to that in the x direction. Therefore, the simpler equation (5.56) allowing for a field-aligned only asymmetry of irregularities in certain cases may serve as a good approximation for experiments with navigational satellites. The above results were obtained for a spectral shape obeying the power law. Similar conclusions follow from considerations based on the Gaussian spectral shape, too; the only difference is the scaling factor $\int \mathrm{d}y\,\Phi(y)$ contained in (5.54) [Tereshchenko et al., 1998].

Consider the specificity of the use of the average field in satellite ionospheric tomography. The above computations prove the applicability of the measurements of an average field to tomographic reconstruction of the variance of electron density fluctuations in the ionosphere. Basic equations were obtained within the scope of the Markov approximation. The latter does not imply small fluctuations of amplitude and allows theoretical tomographic analysis of the data, even for strong scintillations. However, in a real experiment, it is impossible to measure the field in the absence of irregularities. Instead of this idealized quantity, in practice one can use the measured field averaged over some interval. Let us analyze the changes in (5.56) caused by substitution, instead of $E_0(\boldsymbol{r})$, of $E_{0,\mathrm{e}}(\boldsymbol{r})$ obtained by the approximation of the experimental amplitude. Since phase fluctuations of the scattered field are zero-mean Gaussian (normally distributed) variables, one comes to

$$E_{0,\mathrm{e}}(\boldsymbol{r}) = E_0(\boldsymbol{r})\,\frac{\langle A\rangle}{A_0}\,. \tag{5.57}$$

Mean amplitude $\langle A\rangle$ is linked with the amplitude of undisturbed field (in the absence of irregularities) A_0 by the equation [Rytov et al., 1989]

$$\langle A\rangle = A_0\,\exp\left(-\frac{1}{2}\,\sigma_\chi^2\right)\,, \tag{5.58}$$

where σ_χ^2 is the variance of the logarithmic relative amplitude of the scattered field

$$\sigma_\chi^2 = \left\langle \left(\ln\frac{A}{A_0} - \langle \ln\frac{A}{A_0}\rangle\right)^2 \right\rangle\,. \tag{5.59}$$

Therefore, the following relationship takes place

$$\ln\frac{\langle E(\boldsymbol{r})\rangle}{E_{0,\mathrm{e}}(\boldsymbol{r})} = \ln\frac{\langle E(\boldsymbol{r})\rangle}{E_0(\boldsymbol{r})} + \frac{1}{2}\,\sigma_\chi^2\,. \tag{5.60}$$

Thus, owing to the substitution of $E_{0,\mathrm{e}}$ for E_0, the experimental estimate of the average field contains the function of the variance of the logarithmic relative amplitude in its explicit form. This means that, for analyzing experimental data on an average field, one needs to obtain analytical expressions not only for the logarithmic relative field $\ln\frac{\langle E(\boldsymbol{r})\rangle}{E_0(\boldsymbol{r})}$ but also for the variance of logarithmic relative amplitude σ_χ^2. In the literature on scattering of radio waves in randomly inhomogeneous media, no analytical expressions for the variance of the logarithmic relative amplitude obtained in terms of the Markov approximation are referred to. The available formulas suitable for amplitude tomography based on the variance of amplitude have been derived by the method of smooth perturbations within the scope of Rytov's approximation. This approach is applicable to weak scattering beneath the limiting value of the variance $\sigma_\chi^2 = 0.3$. Experimental data measured in subauroral and auroral regions show that the observed amplitude scintillations very seldom exceed this level. Therefore, the Rytov approach can be used in the analysis of scattering by small-scale irregularities in the high-latitude ionosphere

Thus, we cannot take the advantage of the Markov approximation in analyzing experimental data on an average field, and so we have to manage within the scope of the Rytov approach. Hence, instead of using the average field function, the tomographic reconstruction of small-scale irregularities can be based on the variance of the logarithmic relative amplitude that has an experimental estimate equal to its theoretical value.

Consider the possibility of tomographic reconstruction based on amplitude data only. Let us show that the variance of the logarithmic relative amplitude contains the same information on the properties of scattering irregularities as the function of the average field.

For this purpose, let us seek the relationship between the amplitude of the wave scattered by small-scale ionospheric irregularities and the parameters of the irregularities. In terms of the Rytov approach, the following equation can be obtained for the variance of the logarithmic relative amplitude [Rytov et al., 1989]:

$$\sigma_\chi^2 = \frac{\lambda^2 r_e^2}{4\pi^2} \int_{z_d}^{z_u} \iint\limits_{-\infty}^{\infty} \Phi_N(\boldsymbol{k}, \boldsymbol{z})\,|_{k_z=0} \sin^2 \frac{R_F^2\left(k_x^2 + k_y^2\right)}{4\pi}\, \mathrm{d}k_x \mathrm{d}k_y \mathrm{d}z\,, \quad (5.61)$$

where $R_F = [\lambda\, z(z_0 - z)/z_0]^{1/2}$ is the Fresnel radius. With the spectrum in the form $\Phi_N(\boldsymbol{k}, \boldsymbol{z}) = \sigma^2(z)\, F(\boldsymbol{k}, \boldsymbol{e}_x)$, as used previously, (5.61) will be written as follows:

$$\sigma_\chi^2 = \frac{\lambda^2 r_e^2}{4\pi^2} \int_{z_d}^{z_u} \sigma^2(z) \iint\limits_{-\infty}^{\infty} F(\boldsymbol{k}, \boldsymbol{e}_z)\,|_{k_z=0} \sin^2 \frac{R_F^2\left(k_x^2 + k_y^2\right)}{4\pi}\, \mathrm{d}k_x \mathrm{d}k_y \mathrm{d}z\,. \quad (5.62)$$

In contrast to the analysis of an average field where it was enough to define the spectrum as a generalized quadric form, in the case in question, the spectral shape should be specified for further calculations. In most investigations of the spectra of ionospheric irregularities, the power-law spectrum is believed the most suitable approximation for describing the irregularities. Therefore, one can use the spectral shape given by (5.55). Note also that in the experiments using VHF satellite signals, the Fresnel radius is less than a kilometer, whereas the reported outer scale L_0 of the irregularities is of the order of tens of kilometers [Aarons, 1982]. Due to the fact that the ratio of the Fresnel radius to the outer scale is small, in the calculation of (5.62), the original spectrum can be replaced by

$$\begin{aligned} F_0(\boldsymbol{k}, \boldsymbol{e}_z) = {} & \frac{\alpha\,\beta\,\Gamma(p/2) L_0^3}{2\pi\,\Gamma(3/2)\Gamma\left[(p-3)/2\right]} \\ & \times \left[\left(\frac{L_0}{2\pi}\right)^2 \times \left(\alpha^2 k_\parallel^2 + \beta^2 k_{x\perp} + k_{y\perp}\right)\right]^{-p/2}. \end{aligned} \quad (5.63)$$

Integration of (5.62) with respect to k_x and k_y yields the equation for the variance of the logarithmic relative amplitude of a wave scattered by anisotropic irregularities, valid for $0 < p < 4$:

$$\sigma_\chi^2 = \frac{\lambda^2 r_e^2 \, \alpha \, \beta \, L_0^{3-p} \, \pi^{(p-1)/2}}{2^{-p/2+3} \sin\left[\frac{\pi}{4}(p-2)\right] \Gamma[(p-3)/2]} \int_{z_d}^{z_u} \frac{\sigma^2(z)}{\sqrt{1+\gamma(z)}} R_F^{p-2}(z)$$
$$\times F\left[1-\frac{p}{2}, \frac{1}{2}, 1, \frac{\gamma(z)}{1+\gamma(z)}\right] \left[C(z) - \sqrt{A^2(z)\, B^2(z)}\right]^{-p/2} \mathrm{d}z \,, \tag{5.64}$$

$$\gamma(z) = \frac{2\sqrt{A^2(z)+B^2(z)}}{C(z) - \sqrt{A^2(z)+B^2(z)}} \,,$$

$$A = \frac{1}{2}\left[(\alpha^2-1)\sin^2\theta(z) + (\beta^2-1)(\sin^2\psi\cos^2\theta(z) - \cos^2\psi)\right] \,,$$

$$B = (\beta^2-1)\sin\psi\cos\psi\cos\theta(z) \,,$$

$$C = 1 + \frac{1}{2}\left[(\alpha^2-1)\sin^2\theta(z) + (\beta^2-1)(\sin^2\psi\cos^2\theta(z) + \cos^2\psi)\right] \,.$$

Here, ψ is the orientational angle of the anisotropy of irregularities in a plane perpendicular to the local geomagnetic field, $\theta(z)$ is the angle between the ray (from the satellite to the receiver) and the vector of the local geomagnetic field, and $F\left(1-\frac{p}{2}, \frac{1}{2}, 1, \frac{\gamma(z)}{1+\gamma(z)}\right)$ is a hypergeometric function.

For cross-field isotropic irregularities ($\beta = 1$), (5.64) reduces to

$$\sigma_\chi^2 = \frac{\lambda^2 r_e^2 \, \alpha \, L_0^{3-p} \, \pi^{(p-1)/2}}{2^{-p/2+3} \sin\left[\frac{\pi}{4}(p-2)\right] \Gamma[(p-3)/2]} \int_{z_d}^{z_u} \frac{\sigma^2(z)}{\sqrt{1+(\alpha^2-1)\sin^2\theta(z)}}$$
$$\times R_F^{p-2}(z)\, F\left[1-\frac{p}{2}, \frac{1}{2}, 1, \frac{(\alpha^2-1)\sin^2\theta(z)}{1+(\alpha^2-1)\sin^2\theta(z)}\right] \mathrm{d}z \,. \tag{5.65}$$

Equation (5.65) is similar to (5.56), except for the additional exponent of the Fresnel radius and the hypergeometric function contained in the weighting factor. The hypergeometric function reaches its maximum at $\sin\theta(z) = 0$. Moving off the geomagnetic zenith, the function decreases sharply (at $\alpha \gg 1$) quickly arriving at its asymptotic value so that its influence is significant only within a narrow cone around the zenith ray. In the rest of the region, the hypergeometric function can be replaced by its asymptotic value

$$\Gamma\left(1-\frac{p}{2}, \frac{1}{2}, 1, 1\right) = \frac{\Gamma((p-1)/2)}{\Gamma(p/2)\, \Gamma(1/2)} \,,$$

and then (5.65) for σ_χ^2 will differ from (5.56) only by the slowly varying function $R_F^{p-2}(z)$ contained in the integrand. This function may be easily calculated from the experimental data and included in the program for reconstructing the variance of electron density fluctuations in the ionosphere.

The above reasoning and the conclusions are valid also for irregularities anisotropic in a plane perpendicular to the geomagnetic field.

Figure 5.9 portrays the latitudinal dependencies of normalized variance $\sigma_\chi^2/(\sigma_\chi^2)_{\max}$ of the logarithmic relative amplitude calculated for cross-field isotropic and cross-field anisotropic irregularities when $\theta_{\min} = 0.2°$ (nearly zenith satellite pass, upper panel), $\theta_{\min} = 4°$ (middle panel), and $\theta_{\min} = 10°$. The actual orbital parameters of navigational satellites with an orbital inclination of 83° and flight altitude of about 1000 km are used in the calculations. To manifest the effect of the aspect-due enhancement of scintillations, the ionospheric irregularities are assumed to make a statistically homogeneous layer with a constant variance of electron density fluctuations. The lower boundary of the layer is $z_{\mathrm{d}} = 250$ km, and its upper boundary $z_{\mathrm{u}} = 350$ km. The power index is $p = 3.5$. Calculations were made for the receiving point located at 69°01′ N, 35°42′ E, which is one of the regular receiving sites where satellite radio signals were measured. Results for $\alpha = 30$ and $\beta = 1$ are shown by crosses, for $\alpha = 30$ and $\beta = 4$ by dots, and for $\alpha = 30$ and $\beta = 10$ by solid lines.

The orientation of the cross-field anisotropy of the irregularities in all three cases was 103° (counted clockwise from the north). As seen from the figure, everywhere outside a comparatively narrow sector about the geomagnetic zenith, the weighting coefficient is a smooth and slowly varying function. For $\theta_{\min}$ less than 2°, the wings of the curve are practically

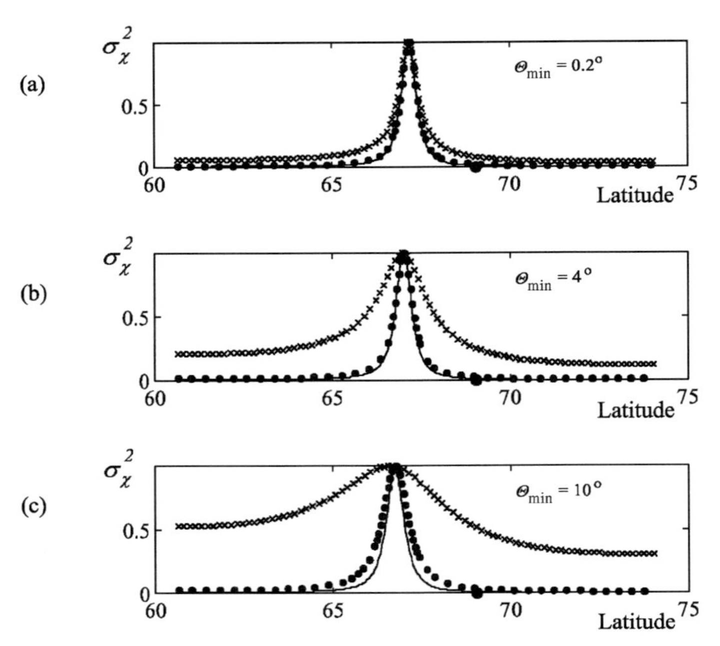

Fig. 5.9. Behavior of the normalized variance of the logarithmic relative amplitude portraying the effect of the weighting function

independent of β within the range of $\beta = 2 - 10$. For $\theta_{\min}$ higher than $2°$, the off-zenith side parts of the curve only faintly depend on β, too. Thus, when making a tomographic reconstruction, the weighting coefficients caused by the anisotropy of irregularities should be taken into account only if probing rays intersect the ionospheric volume of interest close to geomagnetic zeniths of the receiving sites.

To verify the applicability of the chosen spectral model to the analysis of experimental data on amplitude scintillation of satellite radio signals, a comparison has been made of the theoretical variance of the logarithmic relative amplitude and that calculated from measured data. The values of the logarithmic relative amplitude were calculated as a function of the location of the satellite. Experimental data were measured in a series of experiments carried out from 1997–1998 in the Murmansk region (northwest Russia) and in 1995 and 1997 in north Scandinavia. Receiving sites were located in a latitudinal sector from 68 to 70° N and covered different longitudinal regions of the auroral zone.

In the experiments, radio signals from Russian navigational satellites were measured by three or four receivers arranged in a chain oriented close to the ground projection of either northward or southward satellite flight. The length of registration was about 12 minutes that corresponds to the time interval when the satellite was in the radio visibility range. The data were sampled every at 0.02 s. For each satellite pass, the first amplitude was calculated from measured data and then from the variance of the logarithmic relative amplitude. In calculating the amplitude variance, the whole body of amplitudes measured during a given satellite pass was divided into a set of consequent intervals 10 s long shifted by 1 s. Then for each of the 10-s intervals, the variance of the log relative amplitude was calculated. Thus, processed experimental data were plotted versus satellite position. Obtained experimental dependencies were then compared with the theoretical curves calculated from (5.64) with different values of anisotropy parameters (relative elongation of the irregularities along and perpendicular to the geomagnetic field and the orientation of their cross-field anisotropy). The power index p contained in (5.64) was obtained from the high-frequency approximation of the amplitude spectrum of the experimental data. For calculating theoretical curves, it is necessary to define the model of the spatial distribution of electron density fluctuations in the ionosphere. The simplest model of the distribution of electron density fluctuations in a faintly disturbed ionosphere is a statistically homogeneous layer spreading horizontally within a wide enough region. The layer was assumed 100 km thick; the height of its center in different models varied from 250 to 350 km.

Analysis of a large data set showed that in weak ionospheric disturbances, the experimental variance of the logarithmic relative amplitude can be properly reproduced by the model of power-law spectrum small-scale scattering irregularities. Some results of the comparison are shown below. Figure 5.10a portrays experimental (solid line) and theoretical (crosses) curves of the logarithmic relative amplitude measured in April 1997 in the Verkhnetulomsky (68°35′ N, 31°45′ E), Murmansk, region. The x-axis is the geographical latitude of the satellite. As seen from the figure, the experimental curve has a single maximum caused by an enhancement of scintillation that takes place near the geomagnetic zenith of the receiving site. The shape and the location of the maximum depend on the anisotropic parameters of the irregularities and also on the geometry of the experiment (the position of the satellite with respect to the vector of the geomagnetic field). In the case shown in the figure, the satellite trajectory lay very close to the geomagnetic zenith (the minimum angle between the line of sight and the direction of the local geomagnetic field was 0.25°. With this geometry, the orientation of the cross-field anisotropy of the irregularities has almost no effect on the shape and the position of the maximum, so that experimental curves may be simulated by a model of irregularities symmetrical about the geomagnetic field. The best fit of the theoretical curve to the experimental one in the present case is observed at $\alpha = 45$.

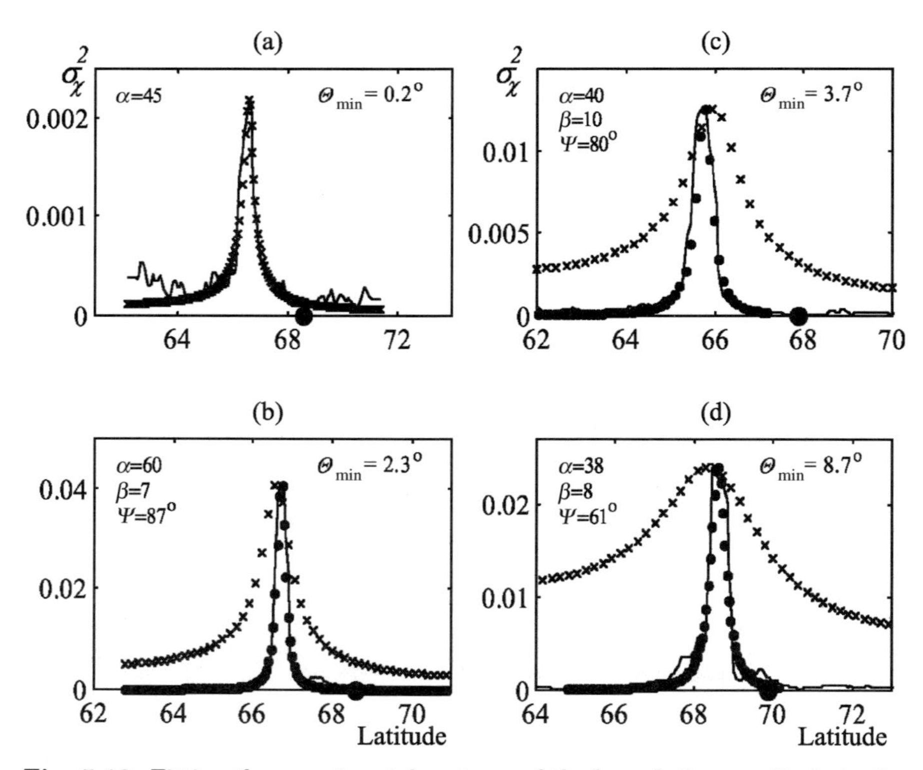

Fig. 5.10. Fitting the experimental variance of the log relative amplitude to the model of three-dimensional anisotropic irregularities with the power-law spectrum: (**a**) zenith satellite pass, (**b**, **c**, **d**) off-zenith satellite passes

Figure 5.10b,c,d displays experimental and fitted theoretical curves calculated from the data measured in April 1997 in Verknetulomsky (b) and in November 1995 in Esrange, Sweden (67.89° N, 21.12° E, c) and Tromsø, Norway (69.59° N, 19.22° E, d). Theoretical curves calculated for the model of cross-field isotropic irregularities are shown by crosses, and those calculated for cross-field anisotropic irregularities are shown by dots. Satellite trajectories in the cases shown did not cross the regions of the geomagnetic zeniths of the receiving sites: the minimum angle between the line of sight and the geomagnetic field was $\theta_{\min} = 2.3°$ in (b), $\theta_{\min} = 3.7°$ in (c), and $\theta_{\min} = 8.7°$ in (d). Here the experimental variance of the log relative amplitude can no longer be reproduced by the model of irregularities symmetrical in a plane perpendicular to the geomagnetic field since the cross-field anisotropy of irregularities has an increasing effect on the amplitude curves with growing $\theta_{\min}$. The effect becomes noticeable starting from $\theta_{\min} \sim 1$–$1.5°$. The width of the maximum depends mostly on the values of field-aligned and cross-field elongation of irregularities, and the location of the maximum is mainly determined by the orientation of the cross-field anisotropy of irregularities. In (b), the best fit of the model to the experimental curve gives $\alpha = 60$, $\beta = 7$, $\psi = 87°$, where ψ is the angle between the ground projection of the cross-field anisotropy and geographical north, measured eastward. In (c) the best fit is obtained at $\alpha = 40$, $\beta = 10$, $\psi = 80°$; in (d), the best approximation of experimental curves is reached with $\alpha = 38$, $\beta = 8$, $\psi = 61°$. Note that the position of the maximum depends strongly on the orientation of the cross-field anisotropy of irregularities. Based on this effect, a correlation was established between the orientation of the cross-field anisotropy and the direction of ionospheric convection [Tereshchenko et al., 2000a].

Thus, good agreement between the experimental and fitted data proves the applicability of the chosen spectral model to the analysis of measured amplitude scintillations of satellite radio signals. The power-law spectrum can be taken as a basic model in developing the tomographic approach to the reconstruction of the variance of electron density fluctuations in the ionosphere.

5.6 Reconstruction of the Variance of Electron Density Fluctuations from Amplitude Data

The above analysis of the experimental variance of the logarithmic relative amplitude was carried out under an assumption made on the constant variance of density fluctuations within some ionospheric region. In this case, the dependence of the logarithmic amplitude upon satellite position has a single maximum caused by enhancement of scintillations at some specific position of the ray with respect to the geomagnetic field vector and the orientation of irregularities. Thus, if an experimental curve has a single maximum that can be reproduced by fitting the proposed model, the assumption of constant variance of density fluctuations seems quite acceptable. In simultaneous measurements of satellite radio signals at several receiving sites, the assumption of constant variance of density fluctuations seems plausible if each of the experimental amplitude variance curves has a single aspect-due maximum and all experimental curves can be reproduced by one and the same model.

However, the spatial uniformity of the variance of electron density fluctuations over tens to hundreds of kilometers is not typical of the high-latitude ionosphere. Due to specific physical processes taking place in the near-polar regions, the structure of the high-latitude ionosphere may undergo significant spatial and temporal changes both in the spectrum of irregularities and in the variance of density fluctuations.

The changes in ionospheric parameters taking place during registration of a satellite radio signal can produce multiple extrema of various shapes and magnitudes in the curve of the variance of the logarithmic relative amplitude. Correct interpretation of such data requires determination of physical reasons causing changes in the experimental variance of the log amplitude of the radio wave scattered by the ionospheric layer. As seen from the theory, the spectral parameters and the variance of the logarithmic relative amplitude appear in the expression for the amplitude variance not separately but as a product, which makes it impossible to distinguish between the effects of spectral changes and variations in density fluctuations. However, one could complement the experiment by correlation measurements at one of the receiving sites, which helps to reveal the cause of changes in amplitude variance.

Within the region of a constant spectrum of irregularities, two factors may cause variations in the variance of the logarithmic relative amplitude. First, extrema in the amplitude variance curves can be produced by changes in the variance of electron density fluctuations reflecting a nonuniform structure within the ionospheric volume. It is just spatial distribution of the variance of electron density fluctuations which is the subject of amplitude tomography. Another factor contributing to the variance of the logarithmic relative amplitude, as shown above, is aspect enhancement of scintillations caused by scattering from small-scale irregularities evenly distributed in the ionosphere.

In tomographic experiments, satellite radio signals are measured by several receivers arranged in a receiving chain that is usually oriented along the ground projection of the northward or southward satellite path. With this geometry, each ionospheric volume in a vertical plane containing the satellite path and the chain is consecutively illuminated by the rays from the moving satellite to the receivers. If some ionospheric volume illuminated by the rays contains an irregularity of electron density, then the variance of electron density fluctuations within this volume can be tomographically reconstructed from the set of amplitude data measured along the probing rays intersecting the volume. Note that the set of multireceiver data each containing nearly sole aspect-due maximum is not well suited for tomographic reconstruction (Fig. 5.11). This problem is equivalent to the reconstruction of a spherically symmetrical ionosphere (in ray tomography) that is difficult in practical implementation. Besides this, the regions with maximum signal-to-noise ratios corresponding to different receiving sites are spaced out somewhat apart; hence, they do not form a system of intersecting rays (lower panel in Fig. 5.11a). In the right side of Fig. 5.11b, a situation is sketched that fits well for tomographic reconstruction.

Thus, the data suitable for tomographic reconstruction of statistical structure of the ionosphere must meet the following two requirements:

(1) Extrema of amplitude variance measured at different receiving sites should correspond to the same ionospheric volume containing irregularities of electron density fluctuations, that is, in other words, the probing rays should intersect within the volume with irregularities.

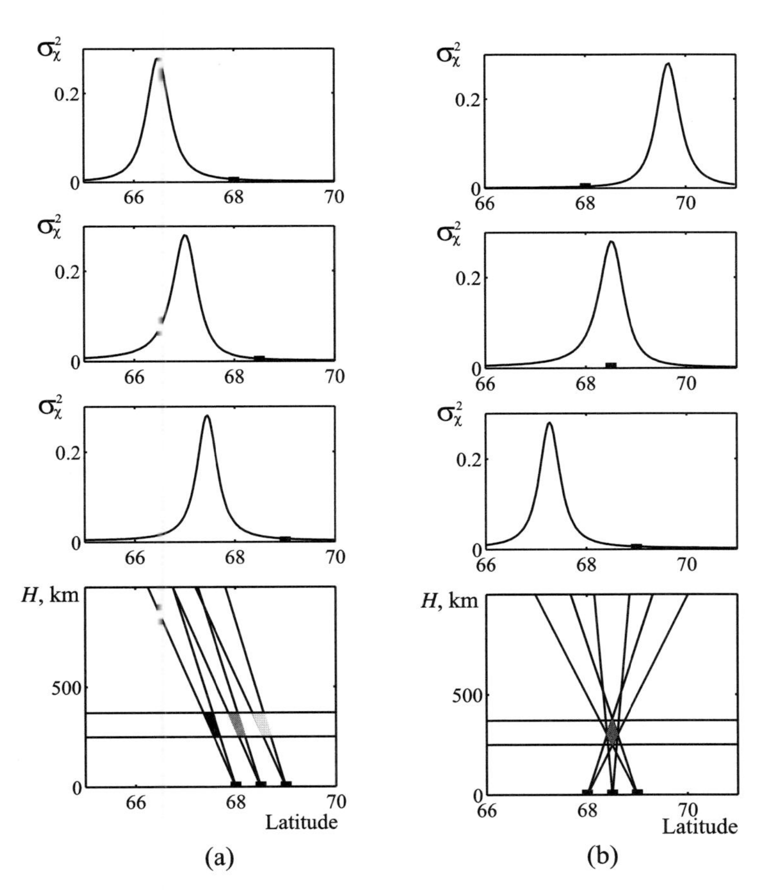

Fig. 5.11. Set of aspect-due maxima badly suited for tomographic reconstruction (**a**) and those produced by scattering from the same ionospheric volume (**b**) as an example favorable for tomographic reconstruction

(2) The extrema should come from the regions with a high enough signal-to-noise ratio.

These conditions are not too strict and, as the analysis of experimental data shows, they are met in practice.

Figure 5.12 shows experimental results obtained at 18:20–18:30 UT on February 2, 1996 at Kola peninsula. In this experiment, amplitudes and phases of satellite radio signals were measured at three receiving sites – Umba, Lovozero, and Tumanny. The tomographic receiving chain was oriented along the ground projection of northward circuits of orbits of the Russian navigational satellites.

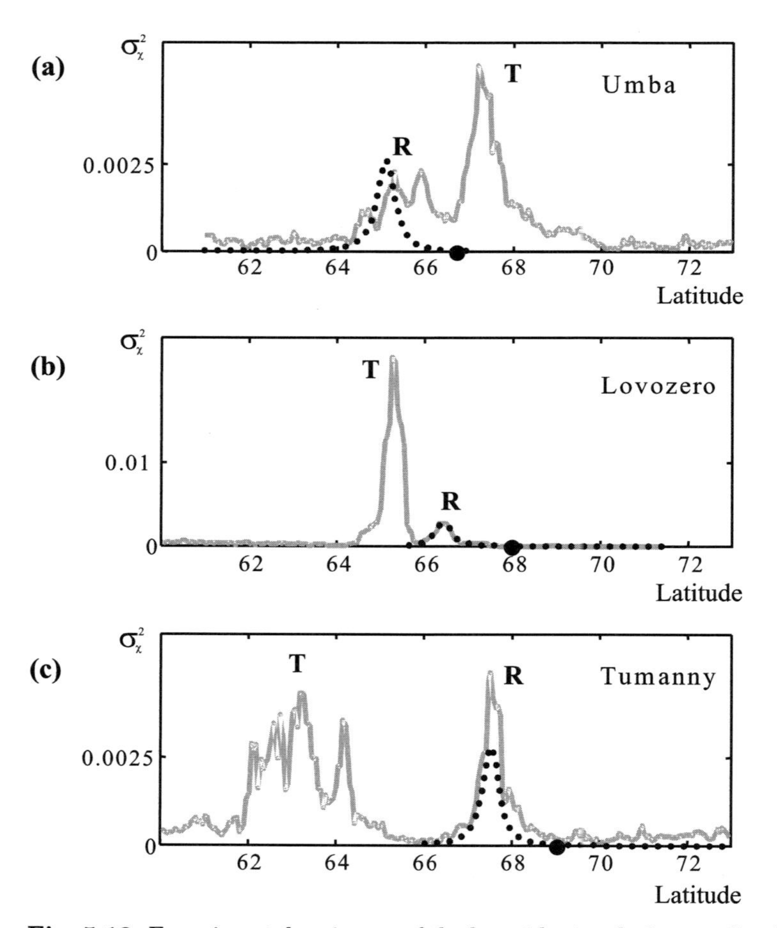

Fig. 5.12. Experimental variances of the logarithmic relative amplitude and model aspect-due maxima in Umba, Lovozero, and Tumanny (February 2, 1996, 18:20–18:30 UT)

Gray lines in the figure show experimental curves of the variance of the logarithmic relative amplitude of satellite radio signals measured in Umba (upper panel), Lovozero (middle panel), and Tumanny (lower panel).

Each experimental curve contains several extrema. The most distinct peak in Umba is located at 67.3° latitude somewhat north of the receiving site. A few other maxima of lower magnitude are observed south of the main maximum within an interval from 65° to 66.5°. In Lovozero, there are two separate maxima in the amplitude variance curve seen south of the receiving site. A larger maximum is observed at about 65.4°; a smaller one is at approximately 66.5°. Both extrema are smoother than those in Umba and Tumanny. The curve in Tumanny contains an isolated maximum at 67.5° south of the receiving site and also a rather wide zone of increased amplitude variance within the latitudinal range from 62° to 64°. The northernmost separate extremum is comparatively smooth but the southern area of increased amplitude variance is sharply indented.The magnitude of amplitude variance is nearly the same in both regions of maxima.

Each curve contains a maximum somewhat south of the receiving site that is likely to be caused by geometrical enhancement of scintillations close to the local geomagnetic zenith. These peaks are marked by "R" in the figure. The "aspect-due" maxima in all curves are reproduced, with some accuracy, by the same model of irregularities evenly distributed within a spherical layer 200 km thick centered at 280 km in height. The fitted anisotropic parameters are $\alpha = 40$, $\beta = 9$, and $\psi = 100°$.

As seen from the figure, the aspect-produced experimental maximum in Lovozero is properly reproduced by the theoretical curve. Similarly shaped experimental and theoretical aspect maxima in Tumanny are nearly collocated, however, their magnitudes do not match quite well. Less agreemeint is observed in Umba where the theoretical maximum does not match exactly any of experimental peaks, but it falls well into an interval of increased amplitude variance. All of these discrepancies do not speak for constant variance of electron density fluctuations in the ionosphere.

There is another set of regions with increased amplitude variance (marked by "T") observed in all three experimental curves that correspond to scattering from the same ionospheric volume. Once the spectrum of irregularities is determined from the aspect enhancement of scintillations, the spatial distribution of small-scale irregularities within this volume can be reconstructed from the amplitude variance data by the same procedure as in ray tomography, with σ^2 used instead of N.

Plate 15a maps the results of reconstruction of σ^2 relative changes in the ionosphere calculated from experimental data shown in Fig. 5.12 [Tereshchenko et al., 2002]. Note that the reconstructed results are insensitive to the changes in spectral shape outside the region of aspect enhancement of amplitude scintillations.

6. Further Development of Tomographic Methods

6.1 Radio Tomography Based on Quasi-Tangential Sounding

In the previous chapters, various modifications of radio tomography with ground-based receiving systems have been discussed. The use of satellite-borne receiving systems makes it possible to sound the ionosphere in the satellite-to-satellite direction and allows obtaining information about the ionosphere over a set of quasi-tangential rays. In particular, nowadays, most promising for ionospheric and atmospheric investigations are the systems mounted in low- and middle-orbiting satellites that receive radio signals from navigational systems such as GPS/GLONASS.

For a long time, the radio occultation method has been known as a source of information about the height profile of the refractive index. In [Tatarskii, 1968; Phinney and Anderson, 1968; Fjeldbo and Eshleman, 1965], forward and inverse problems of radio sounding Earth's atmosphere are considered. In the simplest scheme of the occultation technique for investigating the ionosphere and atmosphere of the planet, one satellite is observing the radio rise (or radio set) of another one carrying a transmitter of electromagnetic waves. Thus, the movement of satellites provides a section of the media being sounded, and the ionosphere and atmosphere can be investigated separately. In the experiment, an amplitude is measured as well as the Doppler refractive frequency shift of the received signal that is linked with the refractive angle of the probing wave in the medium. For spherically stratified media under probing, a vertical profile of the refractive angle $\epsilon(p)$ is related to the refractive index profile $n_i(x)$ at probing frequency f_i by the Abel transform (being a special case of the Radon transform here [Natterer, 1986]) by the following expression:

$$n_i(x) = \exp\left[\frac{1}{\pi}\int_x^H \frac{\epsilon(p)\ dp}{\sqrt{p^2 - x^2}}\right] , \qquad x = rn(r) . \tag{6.1}$$

From (6.1), using formula (6.2) for the refractive index in the propagation medium (atmosphere and ionosphere) and gas state equations, using a barometric formula, and making some additional assumptions, one can obtain height profiles of density, temperature T [K], pressure p [mbar], and the partial pressure of steam p_w [mbar] in the atmosphere and electron density N_e [electrons/m^3] in the ionosphere:

$$n = 1 + 77.6\frac{p}{T} + 0.373\frac{p_w}{T^2} - 40.3\frac{N_e}{f_i^2}, \tag{6.2}$$

where f_i [Hz] is the probing frequency in hertz.

In [Phinney and Anderson, 1968] for the first time, various aspects of solving the inverse problem of radio sounding have been considered, and in [Tatarskii, 1968], possibilities for application of these techniques to studies of Earth's atmosphere have been first analyzed. The analysis showed that, for technical reasons at that time (1965–1968), the method was quite rough and little suitable for Earth investigation or for meteorological applications. Here, we should stress the value of occultation techniques in obtaining primary information about planetary atmospheres and ionospheres. Basic data on atmospheres and ionospheres of solar system planets (Mars [Kliore et al., 1969; Stewart and Hogan, 1973], Venus [Fjeldbo et al., 1975; Kolosov et al., 1979], Jupiter [Kliore et al., 1976; Lindal et al., 1981], the Saturn and Titan system [Tyler et al., 1981; Hudson and Tyler, 1983], Uranus [Lindal et al., 1987], and Neptune [Lindal, 1992]) have been derived just by this method.

Considered apart should be radio probing investigations of Earth. In contrast to exploration of unstudied planets where prompt acquisition of any reliable information about the atmosphere is helpful, in Earth studies, the occultation techniques should be verified by comparison of the data obtained by other methods (e.g., in meteorological or ionospheric networks) and should have advantages over the latter. Of primary interest in studying Earth's atmosphere are variations of the parameters of its atmosphere and ionosphere, which result in the necessity for high accuracy determination of the parameters themselves [Yakovlev et al., 1995; Kursinski et al., 1996; Ware et al., 1996].

The development of GPS/GLONASS satellite navigational systems with quantum stabilities of radiated carrier frequencies together with the use of new techniques for determining the kinematic parameters of both satellites made it possible to come over the technical difficulties inherent in earlier investigations. Successful experiments at the "MIR" station [Yakovlev et al., 1995] and MICROLAB satellite (GPS/MET experiment) [Kursinski et al., 1996; Ware et al., 1996; Hoeg et al., 1995] demonstrated the principal possibility for applying the approaches initially developed for exploring solar system planets to investigating Earth's atmosphere. Note that during the recent period, obtaining information by the occultation method turned from single successful experiments into standard techniques, although it had its own difficulties in interpreting the data obtained [Kursinski et al., 1995, 1997; Rocken et al., 1997]. The radio eclipse system of monitoring the atmosphere and ionosphere can be made more efficient if one uses a specialized system with several satellites performing radio probing of the upper atmosphere above different Earth regions. European countries have already taken this approach: two specialized satellites – CHAMP and ODIN – have been already launched for the purpose of investigating the Arctic polar atmosphere that is climate-producing in this region.

From the statement of the problem, it is quite clear that the source of difficulties is the influence of irregular structures in the media of signal propagation (protonosphere, ionosphere, and atmosphere), which violate the condition of quasisphericity, on the accuracy of the reconstruction of regular parameters. The dynamics of perturbations in these media is superimposed on atmospheric variations, that is, in "unfavorable" conditions, "shading" of atmospheric structures by ionospheric ones is possible as well as the appearance of artifacts in atmospheric profiles [Zakharov and Kunitsyn, 1998]. As shown a series of cases, multipath propagation of the sounding radiation may arise [Zakharov and Kunitsyn, 1999].

Let us briefly expound the results of the earlier analysis of simulation of the influence of irregular structures in various geophysical conditions and multipath propagation in different situations on the accuracy of reconstructing the parameters of propagation media in occultation experiments. The simulation technique is discussed in detail in [Zakharov and Kunitsyn, 1998, 1999; Andreeva et al., 2000a].

In numerical "end-to-end" experiments in atmosphere sounding, cases of horizontally inhomogeneous atmosphere were considered with several hundred kilometer to thousand kilometer scale sizes of irregularities which is almost always typical of the atmosphere. It was found that reconstructed profiles in many cases had no specific features that had been set in the atmosphere model. The results of reconstruction show that the recovered profiles are spatially "averaged" with some certain weight. The path length of the sounding ray between radii R and $R+h$ is, with $h \ll R$, $\sqrt{2Rh}$. Hence, the typical path length of a ray in the atmosphere ($R \sim 6400\,\text{km}$, $h \sim 5$–$10\,\text{km}$) is $L_\text{a} = 400$–$600\,\text{km}$, and structures with sizes less than 300 km are noticeably distorted – they get "smeared" along the ray length in the atmosphere. Images of irregularities with scale sizes larger than 300 km bear more resemblance to the original, but also in this case, locally the profiles can be significantly distorted, and errors may exceed 10–15%.

Similar analysis shows that as well the atmospheric front turns out smeared horizontally. Smooth fronts – with the front surface inclined by less than 0.02 rad – are reconstructed with much resemblance to the original. The limiting value of the front tilt is an angle of the order of the regular refraction in the atmosphere (for Earth $\sim 20\,\text{mrad}$), which has a clear geometric interpretation. Steeper and localized meteofronts (with the front thickness $L_\text{f} = 50$–$100\,\text{km}$) are recovered with significant distortion [Zakharov and Kunitsyn, 1998, 1999; Andreeva et al., 2000a; Berbeneva et al., 2001]. For instance, the slope angle of the frontal surface of the reconstructed image decreases to a ratio of the order of L_a/L_f times, and reconstruction errors depend on the amplitude of the disturbance and its dimensions. For a set of typical atmospheric values, the errors may reach 1–10% for various characteristic front parameters.

The positions of reconstructed equipotential surfaces within irregularities and the atmospheric front at ground level are somewhat shifted with respect to those given in the model. The shift value, as the analysis shows, is faintly dependent on the disturbance amplitude within a range of typical atmospheric values and is defined by the displacement of the ground projection of the ray perigee point to which the profile is reduced. For meteorological heights, this regular shift is up to 50 km.

Standing apart are errors caused by the influence of dispersive media of signal propagation on the profile being reconstructed. Within the frequency range of GPS signal transmissions, such media are the protonosphere and ionosphere. The known two-frequency method for correction of the ionosphere influence [Lindal et al., 1981; Tyler et al., 1981] works well for large and smooth perturbations that can be considered spherically symmetrical within the reconstruction domain. Irregular structures like a thin ionospheric sporadic layer give rise to additional nonsystematic errors since the most important factor in their emergence is various "hits" and passing by of rays through one or several irregularities in cases of their different orientation, which is probably typical in the practice of radio sounding [Zakharov and Kunitsyn, 1998, 1999].

In the regular ionosphere, the value of the radial shift of a ray at frequency f_1 with respect to the ray f_2 may be, according to estimates, up to about 1.5 km in typical ionospheric situations. Practically, situations are possible when a ray at one frequency hits a thin irregular structure while another one passes it by or goes "on the edge." Such an influence of irregular ionospheric structures cannot be eliminated by the two-frequency correction based on compensation for the regular, quite smooth influence of the ionosphere on the atmospheric refractive angle.

These considerations are illustrated in Fig. 6.1 [Berbeneva et al., 2001] portraying the errors of reconstructing height profiles $n(h)$ in the regular atmosphere in the presence of ionospheric perturbation with the scale length along the sounding ray equal to 100 km, thickness 10 km, located at 100 km in altitude. The amplitude of perturbation is 0.5 times the maximum of the F2 layer, and the angular position changes in such a way that in the experimental geometry the probing rays pass by the irregularity (curve 1), go near the edge (curve 2), and one of rays comes along the irregularity through its center (curve 3). The data obtained show that the errors of reconstructing the height profiles of meteoparameters within the altitude interval 5–35 km due to the presence of ionospheric structures like thin sporadic layers may reach 0.5–1.5% decreasing to 0.1% in regular cases. The increase in these errors leads to an error in reconstruction, e.g., of the temperature profile of not less than 1 K at atmospheric hieghts. Therefore, ionospheric irregularities

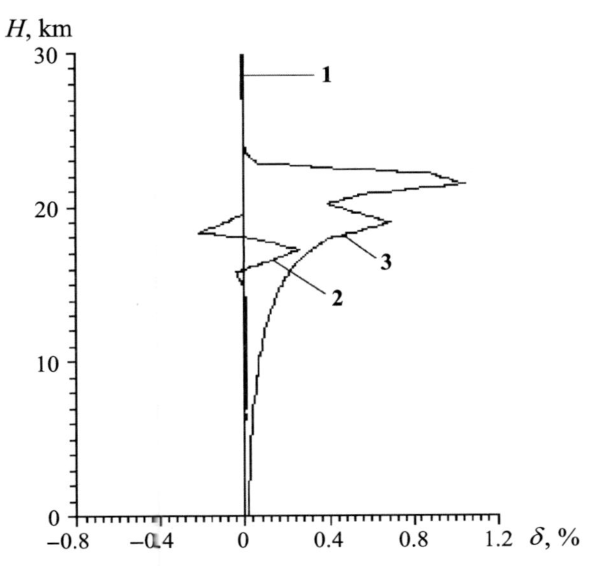

Fig. 6.1. Reconstruction errors (in %) of height profiles $n(h)$ for the regular atmosphere in the presence of a strong irregularity in the E-layer ionosphere, with different positions of the irregularity: the sounding rays pass by the irregularity (*curve 1*), lie close to the edge (*curve 2*), and one of the rays passes along the irregularity through its center (*curve 3*)

may "shade" atmospheric variations of the refractive index caused by minor (by a few percent) changes in meteoparameters or lead to artifacts emerging in the profiles. These cases are subjected to specific processing because for the sliding incidence of a ray, the appearance of interference and diffraction effects is possible [Zakharov and Kunitsyn, 1998, 1999; Andreeva et al., 2000a; Berbeneva et al., 2001].

Perhaps a good way out of this situation could be the use of excessive data for the purpose of investigating a single region or applying tomographic principles together with occultation techniques for monitoring the ionospheric disturbances discussed below.

The protonosphere is not spherically symmetrical and has a complex envelope structure. In general, the protonospheric contribution to the integral electron concentration (along the ray of propagation) is in many cases comparable with that of the ionosphere. According to estimates, the geometric divergence of rays at different frequencies in the protonosphere can achieve 10% of the divergence in the ionosphere, and the refractive angle in the protonosphere can be 1–10% of the ionospheric angle. Therefore, the above considered peculiarities are also valid if the protonosphere is taken into account, the mentioned estimate being allowed for.

Correction for the ionospheric influence, in accordance with [Vorob'ev and Krasilnikova, 1993], in the case of ionospheric irregularities may produce errors even in reconstructing regular profiles not less than 0.5% which is not less than 1 K in the temperature profile within the height interval 10–35 km. Low horizontal resolution is a source of additional systematic error. With increasing height, reconstruction errors increase due to the relative augmentation of the ionospheric influence and the processes of stratification in the atmosphere; within the altitude interval up to 5–8 km, the humid component is important.

In the numerical "end-to-end" experiment on monitoring the ionosphere by radio occultation techniques, cases were considered of a horizontally inhomogeneous ionosphere with scale sizes of irregularities of several hundred kilometers to a few thousand kilometers typical of the ionosphere. It was found that the reconstructed profiles in some cases reveal no characteristic features set in the model simulation of specific situations in the ionosphere. It is evident that reconstructed profiles are spatially "averaged" with some weight. The typical path length of a ray in the ionosphere is 3000 km so that structures with scale sizes less than 1000 km are practically unresolved. Reconstructed irregularities with extent larger than 3000 km look like the original but also in this case locally the profiles are significantly disturbed.

The analysis of the data obtained shows that the presence of a local irregularity always encumbers interpreting the specified profile by the occultation method. As a low-orbiting satellite moves, the irregularity (assumed stable during the time interval of observation that is about 8 minutes) exhibits itself in the probing data that correspond to different heights, which depend on the position of the irregularity with respect to the perigee of the sounding rays.

The analysis of errors in reconstructing ionospheric electron density profiles from radio occultation data using the GPS navigation system showed that in the presence of irregular structures (such as sporadic layers, troughs, an equatorial anomaly, and so on), local errors may increase significantly, in some particular cases reaching 100%.

As seen from the previous chapters, by RT based on low-orbiting satellites with a network of ground-based receivers, quite high-quality tomographic images of the ionosphere can be obtained. Note that the comparison of RT results with incoherent scatter data [Foster et al., 1994] showed that RT can provide better horizontal resolution. However, in ordinary RT experiments, the family of rays joining the transmitter on board the low-orbiting satellite with the ground-based receiver, which intersect the ionosphere, are limited in elevation. Thus, the rays that propagate along ionospheric layers are missing. Therefore, along with high horizontal resolution, RT has worse vertical resolution, that is, "thin" in altitude and rather extended parallel to Earth surface structures that are poorly reconstructed. In particular, ill reconstructed are quasi-regular stratified structures like the E-layer and the extended sporadic E. In its turn, since in the radio occultation approach a set of quasi-tangential rays is used, this method has low horizontal resolution.

It works well only when reconstructing ionospheric structures larger than several thousand km.

It is evidently expedient to use RT data with ground reception of signals together with data of quasi-tangential sounding from satellite to satellite, that is, to combine RT and radio occultation techniques. It should be mentioned that in phase-difference RT, data on phase differences (Doppler phase shift) are employed; therefore, radio occultation Doppler data are of the same nature as RT data along the rays from the satellite to the ground-based receivers. Hence, radio occultation data over a set of quasi-tangential rays are a projection additional to RT data with ground reception of signals, and the radio occultation method is a particular case of the tomographic phase-difference approach.

The simulation [Andreeva et al., 2000a; Berbeneva et al., 2001] showed that the combination of RT and radio occultation techniques makes it possible to improve noticeably the quality of reconstructions; in particular, vertical resolution becomes considerably higher. Figure 6.2 shows an example of a RT

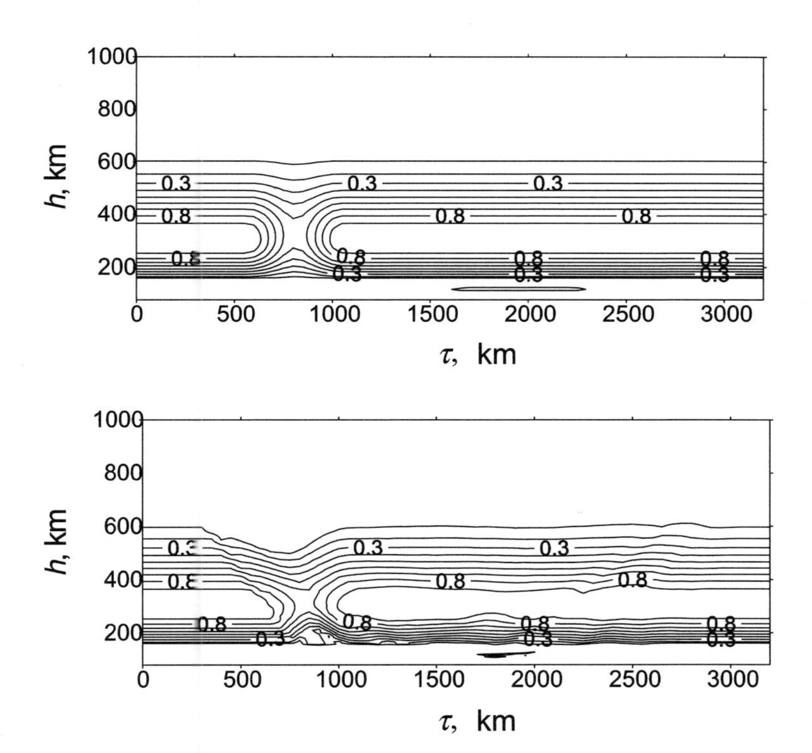

Fig. 6.2. An example of RT reconstruction (*bottom*) using radio occulatation data. The model (*top*) in an ionosphere containing a trough and an E layer; a thin sporadic E_s is centered at 100 km. The minimum concentration in the E_s layer is 25% of the peak F-layer concentration

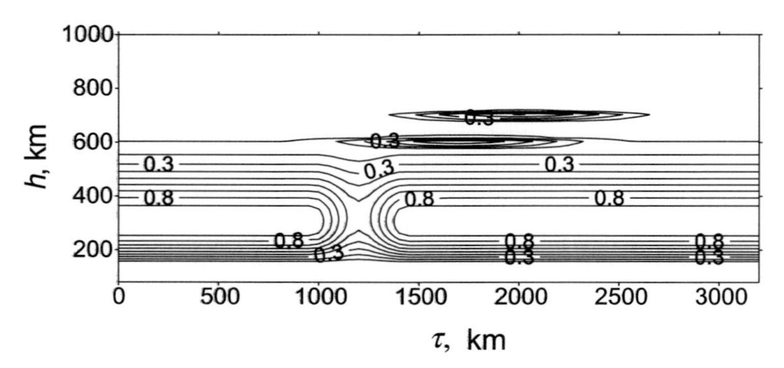

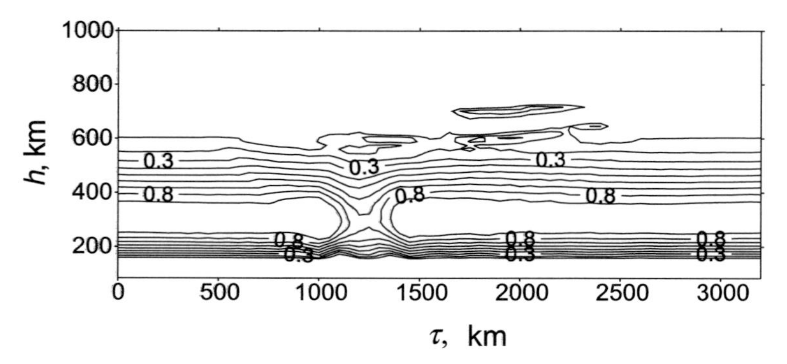

Fig. 6.3. An example of RT reconstruction (*bottom*) using radio occultation data. The model (*top*) is an ionosphere containing a pair of comparatively thin extended layers 50 km thick and 1000 km long located at 600 km and 700 km in altitude

image obtained with the radio occultation data involved. The model used is the ionosphere containing a trough and the E-layer; a thin sporadic E_s 20 km thick and 500 km long is located at 100 km in altitude. The peak electron density in the E_s is 25% of the maximum of the F-layer concentration. Ordinary RT with ground reception recovers well all of the structures except for the thin and extended E_s layer. The use of radio occultation data in addition to RT allows also the E_s layer to be reconstructed. Another example of reconstruction of a pair of comparatively thin and extended layers, each 50 km thick and 1000 km long, located at 600 and 700 km in altitude, is portrayed in Fig. 6.3. RT reconstruction without radio occultation data reveals only some thickening of the main F layer; the high-altitude layers are not resolved. Note that such layers were observed by incoherent scatter radar sounding of the ionosphere [Foster et al., 1994]. On the other hand, the use of the radio occultation method alone would make it possible to distinguish only the greatly extended structures, but the profiles of strong irregularities other than extended layers could not be reconstructed. In particular, a trough-like structure would not be recovered.

6.2 Tomography Using High-Orbiting Satellites. Space-Time Radio Tomography

As seen from the preceding chapters, methods for satellite RT of the ionosphere have been actively developed during recent years. A lot of new experimental RT images of the ionosphere are obtained containing new geophysical information. All of the RT experiments described above were realized with the use of low-orbiting (~1000 km) navigational satellites "Cikada" (Russia) and "Transit" (USA). The progress in low-orbital satellite RT of the ionosphere excites growing interest in applying similar tomographic approach to the investigation of near-earth space (the Earth protonosphere–plasmasphere–magnetosphere) using probing signals from high-orbiting GPS and GLONASS satellite. The satellites fly at about 20,000 km in altitude, and their operating frequencies lie within the range 1.2–1.7 GHz. Of interest is also sounding the ionosphere by the signals from high-orbiting navigational satellites. In this section, we will consider in brief the possibilities for satellite RT of near-Earth space. Also, various geometry schemes of several possible tomography experiments will be analyzed, and examples will be given of reconstructing simple models of plasma distribution in near-Earth space. Estimates of the possibilities of ionosphere diagnostics based on GPS/GLONASS systems will be discussed.

Let us recall the basic relations connecting experimentally measured physical quantities with the parameters of the propagation medium. It is known [Wells et al., 1986; Dixon, 1991] that the differential phase between two L-band carriers at $f_1 = 1.575\,\text{GHz}$ and $f_2 = 1.227\,\text{GHz}$ can be used to make precise relative total electron content (TEC) measurements (in the GPS system and similar in the GLONASS system). The differential group delay between the 10.23 MHz modulation on the two L-band carriers is used to fit these precise relative TEC measurements to an absolute scale.

The difference in simultaneous dual-frequency GPS code measurements P_1, P_2 (in meters) equals the TEC along the transmitter–receiver ray [Wells et al., 1986; Dixon, 1991]:

$$\text{TEC} \equiv \int N\,\mathrm{d}\sigma = S(P_2 - P_1) + e_\text{p}\,, \tag{6.3}$$

where $\int N\,\mathrm{d}\sigma$ denotes integration of the electron density N along the signal path. Merging all bias terms (differential instrumental group delays in the receiver and in the satellite and multipath and observation noise) yields the error e_p (m^{-2}). The scaling factor S converts units of distance (m) to units of TEC (electrons/m^2) and is derived from the GPS frequencies with

$$S = 0.0225\frac{f_1^2 \cdot f_2^2}{f_1^2 - f_2^2} = 9.52 \cdot 10^{16}\ \text{m}^{-3}\,.$$

The observation equation for the TEC from phase measurements Φ_1, Φ_2 (m) looks similar. The error e_Φ (m^{-2}) include the differential instrumental biases of the phases and an additional bias term due to the carrier phase ambiguities M_1, M_2,

$$\mathrm{TEC} \equiv \int N \, \mathrm{d}\sigma = S(\Phi_2 - \Phi_1) + S(M_1\lambda_1 - M_2\lambda_2) \,, \qquad (6.4)$$

where λ_1, λ_2 denote the wavelengths of the sounding waves.

Thus, the phase and group delay measurements made with the use of modern navigational GPS/GLONASS systems allow determination of the linear integrals of the electron concentration. Further consideration of RT problems by means of GPS/GLONASS systems requires the following fact to be taken into account. TEC can be measured with a relative accuracy of approximately $3 \cdot 10^{14}$ electrons/m^2, and absolute values can be determined to approximately (1–2)$\cdot 10^{16}$ electrons/m^2, plus any unknown GPS satellite differential code offsets [Klobuchar et al., 1994; Snow and Romanowski, 1994; Wanniger et al., 1994].

From this, it follows that the methods are capable of high accuracy determination of the relative value of the TEC, but they lead to rather big errors in obtaining the absolute value of the TEC. Therefore, it is expedient not to use the absolute values of the TEC themselves but their differences as input data for RT reconstruction. For this purpose, the most suitable will be the method of phase-difference tomography described above (Sect. 2.1) which involves information only about linear integral differences but not about their absolute values.

Let us consider some possible schemes for the geometry of tomography experiments with low- and high-orbiting satellites. Numerical simulation revealed the considerable importance of experimental geometry: the relative position of transmitters and receivers, the mode in which the rays intersect the reconstructed region, and the range of angles between the rays. Various combinations of ground-based receivers and measuring systems on board the satellites (within VHF/ULF frequency range) were analyzed, including those with elliptical and circular orbits such as "Cikada," "Transit," GPS, GLONASS, Microlab, etc. The use of high-orbiting navigational systems such as GPS/GLONASS with only ground-based receivers within the frame of the traditional RT approach (no temporal changes taken into account) for the purposes of RT of the ionosphere and protonosphere seems hardly promising. First, it is impossible to separate the contribution of the ionosphere to signal variations from those of the protonosphere–plasmasphere. Second, the low angular velocity of high-orbiting satellites noticeably worsens the geometry of the experiment and the RT possibilities. The angular velocity of GPS satellites is 0.0001456 rad/s that is one-seventh the angular velocity of satellites like "Transit" (0.0009966 rad/s). Such satellites pass through only small angles during dozens of minutes. For example, the altitude of the GPS satellite orbit is 20,200 km, and within 30 minutes the satellite position along the orbit will change by only 15°. In conventional RT, it is impossible to use the data from a longer series of measurements (of the order of a few hours) because as a rule the electron density in the region under study changes considerably during such a time interval. Let us illustrate this by an example of the geometry of an RT experiment with five ground-based receivers spaced 300 km apart (Fig. 6.4). The receivers are assumed to be located in the plane containing the path of high-orbiting satellites. Recall that GPS satellites are arranged in fours in six planes and the GLONASS satellites in eights in three planes; therefore, one ground receiver can "see" only two satellites from each GPS plane and three satellites from each GLONASS plane. Figure 6.4a displays the geometry of an experiment where five ground-based receivers are measuring signals from two GPS satellites during an hour-long period. Even after an hour of continuous measurements, considerable "holes" still remain in the reconstruction plane where the rays do not reach. The situation with GLONASS satellites is a little better (Fig. 6.4b); however, only after an hour and longer measurements.

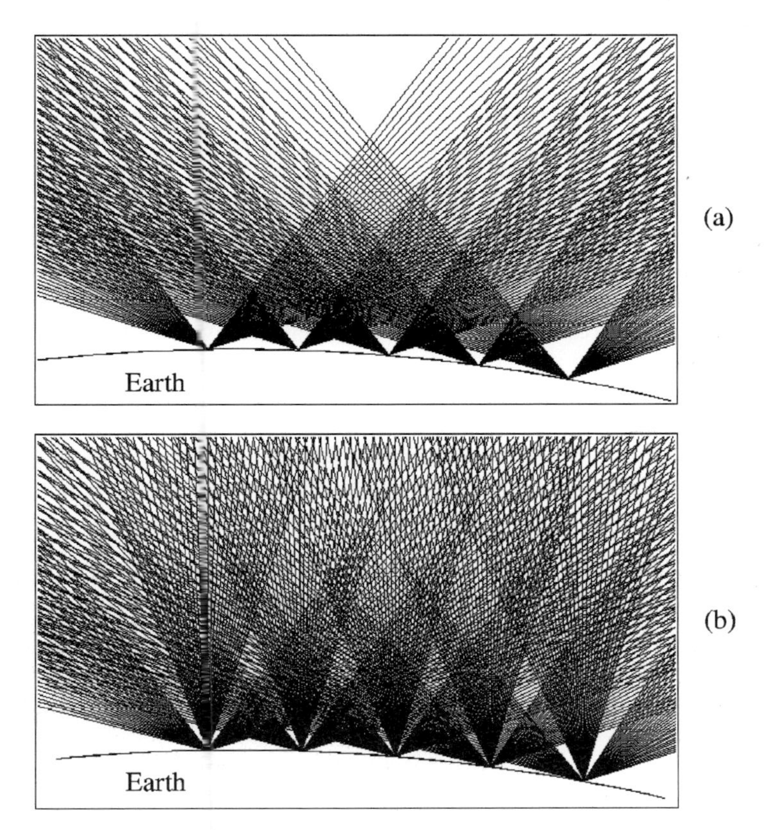

Fig. 6.4. RT experimental geometries ($T_{\rm m}$ = 3600 s): (**a**) five ground-based receivers and two GPS satellites; (**b**) five ground-based receivers and three GLONASS satellites

Nevertheless, the use of a rather dense (with about 100 km spacing of the receivers) network of ground-based GPS/GLONASS receivers will make it possible to implement RT of the ionosphere (up to 1000 km altitudes) as both the reconstruction of a set of 2-D projections and restoration of the 3-D ionospheric structure above the region of observations. However, realization of this scheme of ionospheric RT (with recording time intervals of 30–60 minutes and longer) involves development of new methods allowing for temporal variations of the recovered structures.

The space-time RT approach was proposed in [Kunitsyn et al., 1997]. The process of constructing the forward problem approximate operator (projection operator) that relates the electron concentration values and its derivatives (with high order approximations) in the nodes of a certain grid with the measured data has been described above (Sect. 2.2). Doppler or phase-difference measurements require higher order interpolation than piecewise-constant representation of the recorded function. Therefore, for these measurements, one should construct an approximation of the projection operator (a matrix) that would provide linear integral continuity with respect to the satellite angle. If, in addition, one tries to take into account also the time dependency of the desired function, the appearance of one more variable – time – increases the dimensionality of the tomography problem. In other words, the plane problem of ray tomography becomes three-dimensional. The time dependency of the sought function in between two instants can be taken into account in a similar way, that is, can be expressed in terms of an appropriate polynomial. Then, the rays intersecting this structure will cross it each in its own time. Figure 6.5 portrays such a transfer from 2-D plane tomography to 3-D tomography, different time of data (ray) recording and possible temporal dependency of the sought function $F(h, \tau, t)$ being taken into account. As a rule, 40–60 minute time intervals are sufficient for the purposes of ionospheric RT; it is usually quite enough to use linear, square, or cubic temporal dependency of the desired function. Avoiding bulky formulas, we will give here the relations for the simplest linear temporal interpolation. In this particular case, the construction procedures and the formulas for spatial coefficients are similar to those in ordinary "stationary" RT; they only "split," together with the variables, since not a single value of the function but two values are to be found within each spatial interval. For example, with linear temporal dependency of the values of the "beginning" F^{b} and the "end" F^{e} of the sought function, the temporal evolution $F \rightarrow (F^{\mathrm{e}} - F^{\mathrm{b}})T + F^{\mathrm{b}}$. Here $T = t/T_{\mathrm{m}}$ is normalized (to the total time of measurements T_{m}) time. Evidently, the matrix elements will also bifurcate, and the procedure for constructing each element will remain similar to (2.21) but with corresponding recalculation in the temporal dependency of the ray passage T: $a_{mn} \rightarrow (a^{\mathrm{e}}_{mn} - a^{\mathrm{b}}_{mn})T + a^{\mathrm{b}}_{mn}$. Therefore, when forming the matrix elements, in integration along each ray intersecting the medium at time T,

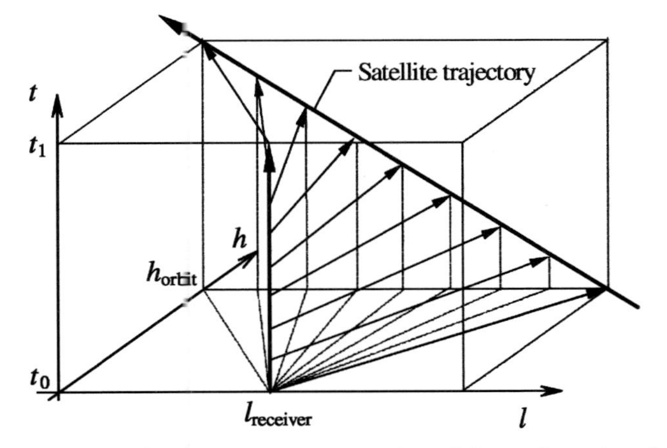

Fig. 6.5. A schematic representation of the trajectories of a satellite and a receiver in three-dimensional space-time tomography and the rays connecting positions of the satellite and the receiver at different moments

the value $a_{mn}(1-T)$ refers to the "beginning" element a^{b}_{mn}, and the value $a_{mn}T$ refers to the "end" element a^{e}_{mn}.

The results of numerical simulation of space-time RT proved the possibility of realizing such schemes, with recording time intervals of the order of 40-60 minutes. RT reconstruction of the temporal evolution of the ionosphere according to the scheme shown in Fig. 6.4b yielded the following results. As a model, an ionosphere with 15% variation in total electron concentration during an hour and change in the trough structure was used. The five stages of temporal evolution in the model ionosphere are shown in Fig. 6.6, left. In the interim interval, the ionosphere was changing linearly, but the rate of temporal evolution at each spatial point was different. The right column of plots in Fig. 6.6 illustrates quite an acceptable result of the reconstruction of the given temporal evolution from data of hour-long measurements. Note that phase RT is hardly applicable in such cases. The relative contribution of the protonosphere to the whole ionosphere–protonosphere TEC varies within wide limits and depends on many factors: the geographical position of the receiver, the season, time, and solar activity [Davies, 1980; Breed et al., 1997; Ciraolo and Spalla, 1997; Lunt et al., 1999a,b]. The relative contribution of the protonosphere often reaches 10% in the daytime and 50% in the nighttime.

Thus the contribution of the protonosphere electron content to the TEC cannot be neglected. It follows that it is impossible to separate the contributions of the ionosphere and the protonosphere using the data on TEC (changes in phases and group delays). The situation is different in measurements of phase differences (either in time or in space). When a ray intersects an irregularity, characteristic phase differences would be proportional to variations of the electron content in the irregularity and independent of the irreg-

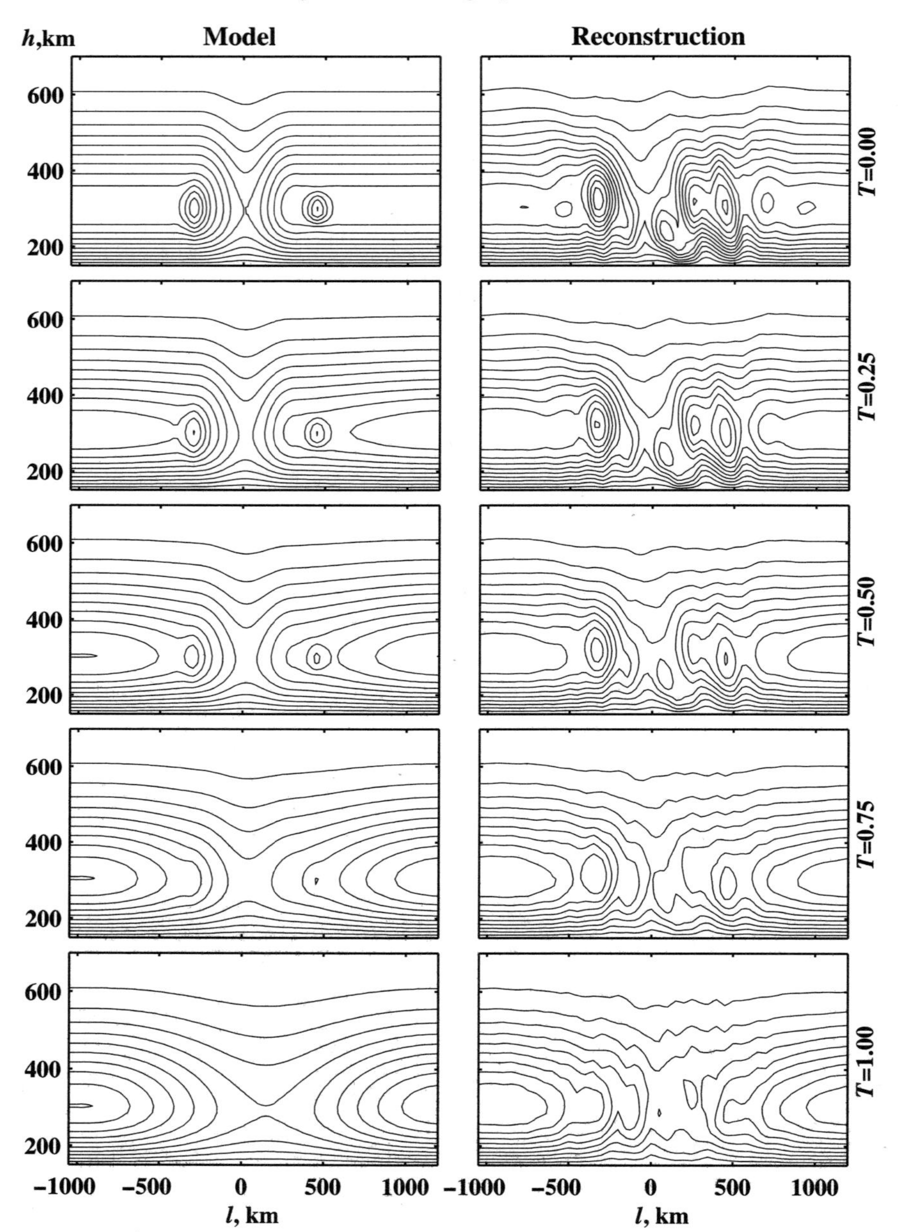

Fig. 6.6. Temporal evolution of the ionosphere model (*left side*) and computer simulation results (*right side*); T is the normalized time

ularity dimensions. Thereby, characteristic variations in phase difference are proportional to $\Delta N \sim 10^9\,\mathrm{m}^{-3}$ in the protonosphere and to $\Delta N \sim (10^{11}\text{–}10^{12})\,\mathrm{m}^{-3}$ in the ionosphere. Therefore, as a rule, the contribution of the protonosphere in phase-difference measurements is much less than 1% of the ionospheric contribution. Consequently, it is possible to determine the ionosphere structure in the first step by using data of phase-difference measurements, and then in the second step, by separating the ionospheric contribution to determine the protonosphere structure by phase measurements. Thus, phase (group delays) measurements and phase-difference measurements make it possible to separate the data and to reconstruct the structure of the ionosphere and the protonosphere separately.

For the purposes of satellite RT of near-Earth space, it is necessary to use combinations of various constellations, in particular, the scheme when the receivers of the signals from high-orbiting satellites are installed on board middle- and low-orbiting satellites. However, the use of only one low orbiter (with an orbit altitude of 500–2000 km) does not provide good geometry for the experiment. Figure 6.7a displays a sketch of the experiment (A) with one low-orbiting satellite combined with a few high-orbiting satellites – one Transit and three GLONASS (angular positions at the initial instant were 15°, 60°, 105°). From the figure, one can see how small is the domain of ray intersections. With this scheme, it is possible to study only that region of the protonosphere where rays are intersecting if it contains strong enough local irregularities and the contribution from other protonospheric regions is negligible.

A little better is the geometry of measurements by means of several middle-orbiting satellites (flying in one orbit) combined with a high orbiter. Figure 6.7b shows the scheme of the experiment (B) with three satellites like "Cikada" (angular positions at the initial time instant 0°, 30°, 60°) and one GPS satellite (15). It is seen that half-an-hour-long measurements make it possible to recover the structure of the protonosphere within a wide enough strip. Although this geometry is worse than the typical geometry of experiments with low-orbiting satellites (as the range of rays intersection angles is much narrower); however, as will be shown below, it still allows determination of RT images of the protonosphere.

Good geometry for an RT experiment with fan beams of intersecting rays is provided by a few middle-orbiting satellites revolving in the direction opposite to that of a high-orbiting satellite. Figure 6.7c shows a scheme of such an experiment (C) with one GPS and three Microlab-like satellites (with the altitude of about 600 km and initial angular positions 90°, 120°, and 150°) revolving opposite to the GPS revolution direction. After recording the signals for 20 minutes, the ray structure consisting of three beams is satisfactory for RT reconstruction. Experiments with filling the studied region by quasi-tangential rays (the radio occultation technique) considered previously are illustrated by the sketch in Fig. 6.7d.

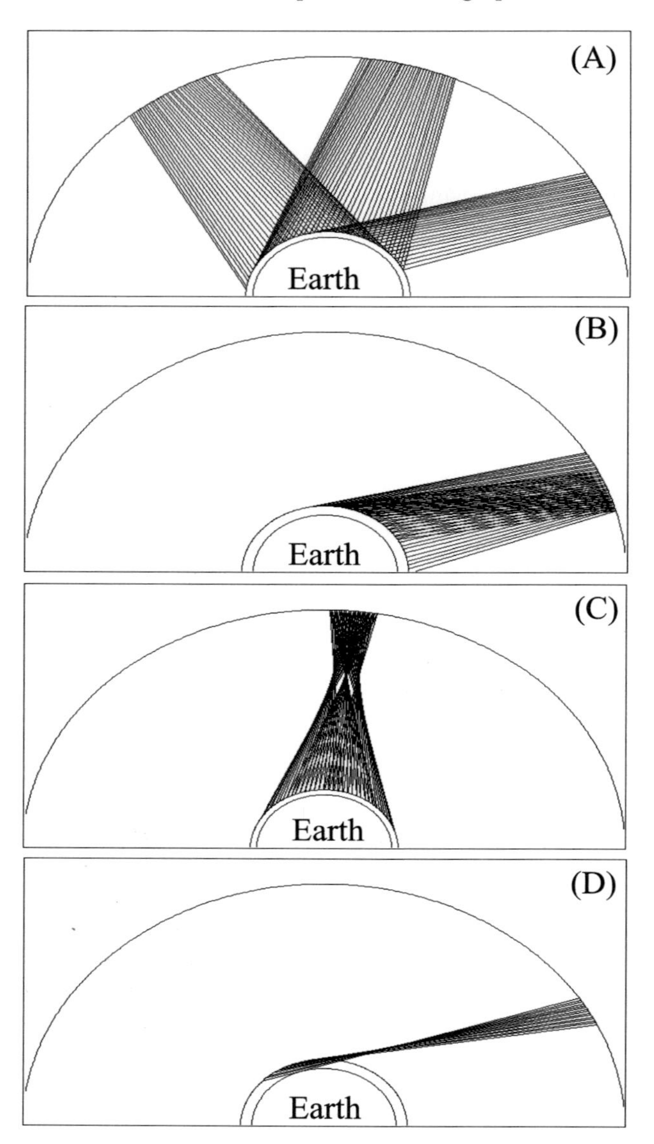

Fig. 6.7. RT experimental geometries. (A) One middle orbital satellite in a 1000 km orbit and three GLONASS satellites; $T_{\mathrm{m}} = 2400\,\mathrm{s}$; (B) three Transit satellites in 1000 km orbit and one GPS satellite: $T_{\mathrm{m}} = 1800\,\mathrm{s}$; (C) one GPS satellite and three satellites in 1000 km orbit rotating in the direction opposite to the GPS; $T_{\mathrm{m}} = 1200\,\mathrm{s}$; (D) one satellite in 1000-km orbit and one GPS satellite using the data from quasi-tangential rays; $T_{\mathrm{m}} = 900\,\mathrm{s}$

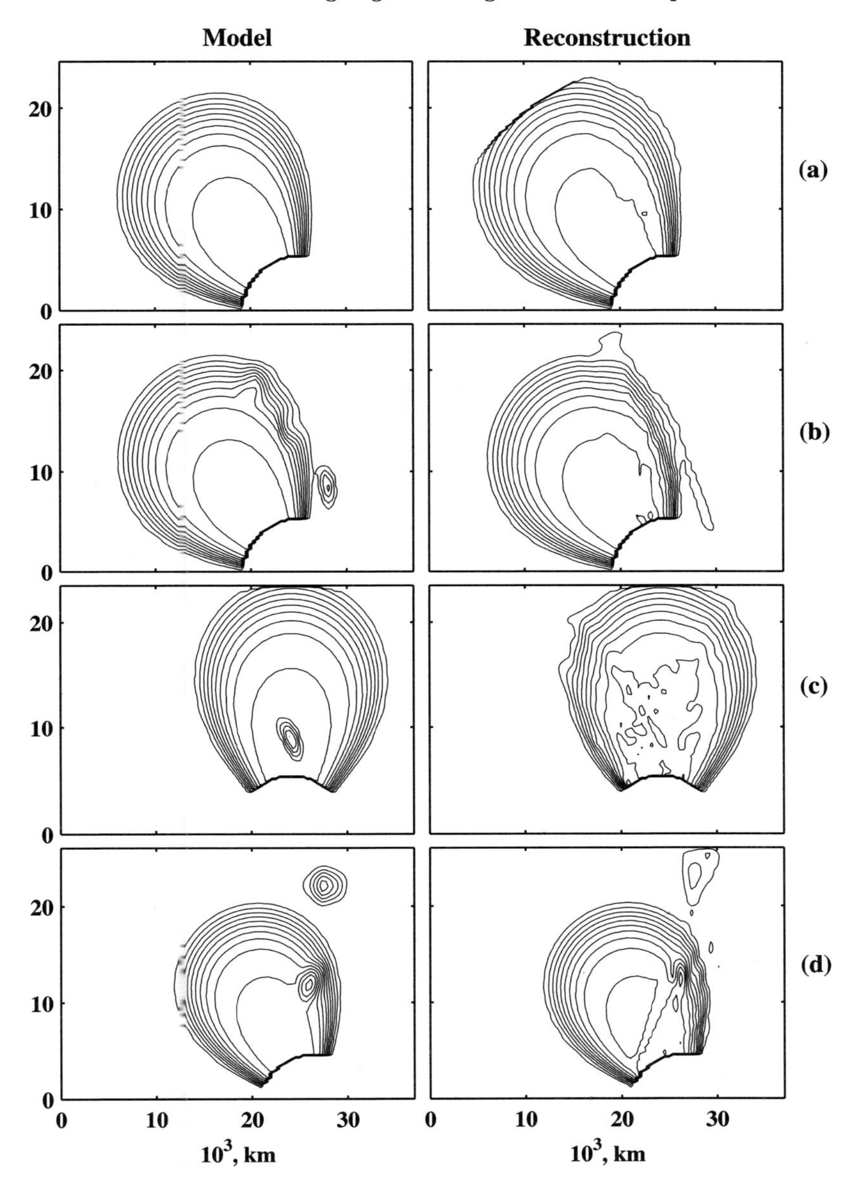

Fig. 6.8. Results of computer simulation of near-space RT reconstruction in units 10^{12} el m^{-3}. (**a**) The model (*left side*) and the reconstruction (*right side*) by scheme B with initial guess error (+10% in evaluation of the L-shell dimension); (**b**) the model and the reconstruction by scheme B; (**c**) the model and the reconstruction by scheme A; and (**d**) the model and the reconstruction by scheme C

Let us present some results of numerical simulations [Kunitsyn et al., 1997]. RT reconstructions of the plasmaspheric structure or of the L-shell position give good enough results for all three schemes, provided that we have a priori information about the electron density value, the location, and the size of the structures under study. However, even low accuracy of the required a priori information (of the order of 30–50%) is quite enough here. By varying different initial guesses and comparing the reconstructed TEC (restored from the reconstruction) with that really measured, one can choose the best initial guess and obtain the final result with acceptable reconstruction quality. An example is shown in Fig. 6.8a of reconstructing the protonosphere model according to scheme (B) with 10% error in the initial guess in estimating the L-shell dimensions. The quality of reconstruction of the structures containing localized irregularities is much poorer. The results of reconstructing the plasmaspheric model [in accordance with the scheme (B)] containing an additional localized inhomogeneity are shown in Fig. 6.8b. Localized irregularities are distorted and stretch when reconstructed. Such poor quality of reconstruction is caused by geometry unsuitable for the experiment. The angles of intersection of the sounding rays in this scheme are small, which leads to "stretching" and distortion of the localized object in the reconstruction. Scheme (A) also strongly distorts the localized irregularities (Fig. 6.8c). The results of RT reconstruction with scheme (C) are somewhat better. Figure 6.8d shows the results of RT reconstruction of the protonospheric model with localized irregularities. In this case, as usual, the localized inhomogeneity is revealed quite well in the reconstruction, although distortions and artifacts are still visible. The higher quality of reconstruction using scheme (C) is due to better experimental geometry (Fig. 6.7) where large angles of the probing ray intersections are seen and only a small region of nonintersecting rays is observed at the edge of the region of reconstruction. Of course, the existence of upper and lower ray beams causes corresponding distortions and artifacts in RT reconstructions. However, by developing optimum methods and algorithms for RT reconstruction as applied to such particular schemes, one can significantly improve the results.

At present, methods for ionospheric RT with high-orbiting satellites are being developed; therefore, estimation of the proposed methods and results obtained seems premature. Here, a series of works should be mentioned developing this new field of investigation [Hajj et al., 1994; Kuklinski, 1997; Rius et al., 1997; Hajj and Romans, 1998; Ruffini et al., 1998; Hernández-Pajares et al., 2000; Howe et al., 1998], however, good practical results still need time.

Thus, based on the analysis carried out, from the numerical simulation and the results presented, one can infer the following:

- The use of high-orbiting navigational systems GPS/GLONASS with only ground-based receivers for the purposes of near-space tomography (protonosphere – plasmasphere – magnetosphere of Earth) does not seem promising;

- The use of combinations of various constellations, in particular, those with receivers of signals from high-orbiting satellites like GPS/GLONASS installed on board low-orbiting satellites, will make it possible to realize RT of near space, although the results depend strongly on the experimental geometry;
- Ionospheric RT using high-orbiting systems is possible on the basis of the phase-difference (to separate the contributions of the ionosphere and protonosphere) and space-time approaches (for determination of temporal evolution).

6.3 Frequency-Space Radio Tomography

In many cases, the object under study in tomography is considered a definite function of the coordinates, and the tomographic inversion consists of the determination of this definite function from a set of integral projections or "strips" through the medium. In this section, we consider the medium as a statistical ensemble where only certain moments of the function of interest are well defined. It is therefore not meaningful to invert separate realizations of the medium, only some of the statistical properties of the medium. In random radio waves from a source, it is only the mean power as a function of direction (and frequency) which is meaningful. In radar astronomy of a planet, it is only power as a function of frequency and time delay which is interesting; for an early example, see Hagfors et al. [Hagfors et al., 1968].

The problem of a medium that varies both spatially and temporally is more often encountered. An example of tomography applied in such a case has been considered by Hagfors and Tereshchenko [Tereshchenko and Hagfors, 1997]. Rather than discussing such problems with time or frequency as one of the coordinates, it is possible to approach the problem in a general way, where the time/frequency coordinate is just one of the generalized coordinates of n dimensional space ($x \in R^n$) and where the unknown function is to be derived from k dimensional aggregates in R^k, where $k < n$. An early approach along these lines was discussed by Radon [Radon, 1917]. Following this procedure, we can formulate the two-dimensional problem ($n = 2, k = l$) of frequency-space tomographic reconstruction.

Let $f(u, v)$ be a function of the two variables u, v that are dimensionless coordinates in time and space, respectively. We shall consider a statistically stationary process and introduce the following quantities related to the temporal correlation function f:

$$R_f(u_2 - u_1, v) = \langle f(u_2, v) f^*(u_1, v) \rangle , \qquad (6.5)$$

where the angle brackets denote statistical averaging.

The Fourier time spectrum is related to the correlation function through the Wiener–Khinchine theorem,

$$S_f(\Omega, v) = \int R_f(u, v) \exp(2\mathrm{i}\pi\Omega u)\, \mathrm{d}\, u\,. \tag{6.6}$$

$S_f(\Omega, v)$ is related to the power spectrum through

$$\begin{aligned} S_f(\Omega, v) &= \lim_{T\to\infty} \frac{1}{T} \langle |s_{f\mathrm{T}}(\Omega, v)|^2 \rangle \\ &= \lim_{T\to\infty} \frac{1}{T} \langle P_{f\mathrm{T}}(\Omega, v) \rangle \end{aligned} \tag{6.7}$$

and where

$$s_{f\mathrm{T}}(\Omega, v) = \int_{-T/2}^{+T/2} f(u, v) \exp(2\mathrm{i}\pi\Omega u)\, \mathrm{d}\, u$$

is the Fourier transform of the signal itself.

In practice, however, the measurements are made over a finite time interval or over a finite region of space. When the time interval is limited to $-u_0/2 \le u \le u_0/2$, we can define a new time function $F(u, v)$ by

$$\begin{aligned} F(u, v) &= f(u, v) \prod\left(\frac{2u}{u_0}\right), \\ \prod(x) &= \begin{cases} 1\,, |x| \le 0.5 \\ 0\,, |x| > 0.5\,. \end{cases} \end{aligned} \tag{6.8}$$

Let us determine a function $W(\Omega_1, v_1)$ as a symbolic integral as follows:

$$W(\Omega_1, v_1) = \int P_F(\Omega, v)\, \mathrm{d}L\,, \tag{6.9}$$

where L is the family of aggregates and $\mathrm{d}L$ is an element of L. The problem of reconstructive tomography then amounts to the determination of the local characteristics of the object characterized, either in terms of the spectral function S_f or the correlation function R_f. To transform (6.9), we shall apply Fourier synthesis as first discussed by Bracewell [Bracewell, 1956]. The Fourier transform of the strip integrals is performed along central sections, i.e., along lines all of which intersect the coordinate origin.

Let us discuss the particular case where the family of linear aggregates consists of straight lines in the (Ω, v) space, as shown in Fig. 6.9.

Next, we introduce the new coordinate system (Ω', v') related to the integration line L. Denote the angle between the Ω axis and the Ω' axis as ψ, and let r be the distance between the origin and the line L. Following the Fourier synthesis method in (6.9), we change from the variables Ω, v to the variables r, ψ, which for any point on the line are related as follows:

$$\begin{aligned} \Omega &= r\cos\psi - v'\sin\psi\,, \\ v &= r\sin\psi + v'\cos\psi\,. \end{aligned} \tag{6.10}$$

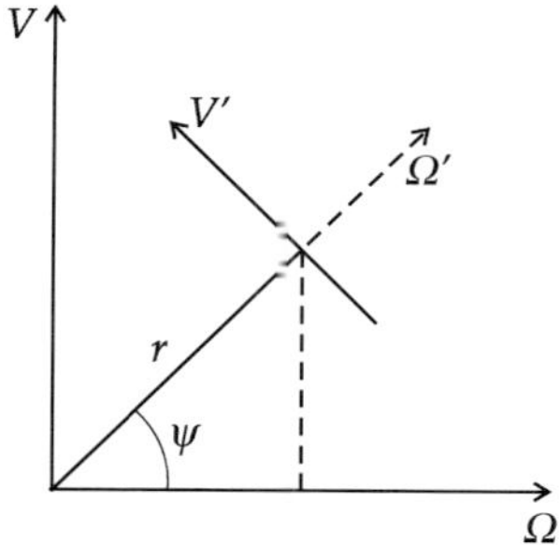

Fig. 6.9. Coordinate system

Using the variables r and v', (6.9) can be rewritten as follows:

$$w(r,\psi) = \int P_F(r\cos\psi - v'\sin\psi, r\sin\psi + v'\cos\psi)\,\mathrm{d}v'\,. \tag{6.11}$$

The spectrum of $F(u,v)$ can be related to the spectrum of $f(u,v)$ by the convolution theorem:

$$s_F(\Omega, v) = \int \frac{\sin\pi u_0(\Omega - \Omega')}{\pi(\Omega - \Omega')}\, s_f(\Omega', v)\,\mathrm{d}\Omega'\,. \tag{6.12}$$

Squaring this, taking the ensemble average, and making use of the following relation for a statistically homogeneous medium:

$$\langle s_f(\Omega', v)\, s_f^*(\Omega'', v)\rangle = \delta(\Omega' - \Omega'')\, S_f(\Omega', v)\,, \tag{6.13}$$

we arrive at the relationship

$$P_F(\Omega, v) = \int \frac{\sin^2\pi u_0(\Omega - \Omega')}{[\pi(\Omega - \Omega')]^2}\, S_f(\Omega', v)\,\mathrm{d}\Omega'\,. \tag{6.14}$$

We now substitute (6.14) in (6.11) and find for the line integrals

$$W(r,\psi) = \int\!\!\int S_f(\Omega', r\sin\psi + v'\cos\psi) \times \frac{\sin^2\pi u_0(\Omega' - r\cos\psi + v'\sin\psi)}{[\pi(\Omega' - r\cos\psi + v'\sin\psi)]^2}\,\mathrm{d}v'\,\mathrm{d}\Omega'\,. \tag{6.15}$$

Changing the variables of integration, $\Omega' = \Omega$ and $v' = -r\tan\psi + +v/\cos\psi$, we obtain this in the slightly simpler form,

$$W(r,\psi) = \int\!\!\int S_f(\Omega)\, \frac{\sin^2\pi u_0\left(\dfrac{r}{\cos\psi} - \Omega - v\tan\psi\right)}{|\cos\psi|\left[\left(\dfrac{r}{\cos\psi} - \Omega - v\tan\psi\right)\right]^2}\,\mathrm{d}v\,\mathrm{d}\Omega\,. \tag{6.16}$$

This can also be rewritten in terms of the correlation function R_f as

$$W(r,\psi) = u_0/|\cos\psi| \int\int \Lambda\left(\frac{u}{2u_0}\right) R_f(u,v) \times \exp\left[\mathrm{i}2\pi u\left(\frac{r}{\cos\psi} - v\tan\psi\right)\right] \mathrm{d}u\,\mathrm{d}v\,. \quad (6.17)$$

The Λ function, which is the convolution of our Π function with itself, is defined by

$$\Lambda(x) = \begin{cases} 1-2|x|\,, & |x| \le 0.5 \\ 0\,, & |x| > 0.5\,. \end{cases}$$

Taking the Fourier transform with respect to r on both sides of (6.17), integrating over u, bringing the factor containing ρ to the other side of the equation, introducing $\tilde{S}_f(\Omega, v)$, and Fourier transforming, the result finally leads to

$$\tilde{S}_f(\Omega,v) = \int_0^{2\pi} \frac{\Delta r}{\cos\psi} \times \int_0^{1/\Delta r} \frac{\rho}{1-\rho\Delta r} \exp\left[\mathrm{i}2\pi\rho\left(\Omega\cos\psi + v\sin\psi\right)\right] \times \int W(r,\psi)\exp(-\mathrm{i}2\pi r\rho)\,\mathrm{d}r\,\mathrm{d}\rho\,\mathrm{d}\psi\,, \quad (6.18)$$

where $\Delta r = |\cos\psi|/u_0$. Here, we must observe that $\tilde{S}_f(\Omega,v) = S_f(\Omega,v)$, only provided the autocorrelation function $R_f(u,v) = 0$, whenever $|u| \le 0$. When this is not the case, $\tilde{S}_f(\Omega,v)$ is related to $S_f(\Omega,v)$ by the relation

$$\tilde{S}_f(\Omega,v) = \int S_f(\Omega',v)\frac{\sin 2\pi u_0\,(\Omega-\Omega')}{\pi\,(\Omega-\Omega')}\,\mathrm{d}\Omega'\,. \quad (6.19)$$

There is a slight difference between (6.18) and the result of [Hagfors and Tereshchenko, 1991], equation (8). The reason for the discrepancy stems from the fact that the authors compare the chirp reconstruction result (6) of their paper with the reconstruction of an image from data obtained with constant strip width. Actually, under the finite time interval assumption, one should perform the comparison by tomography with strip width varying with angle as the cosine of the inclination angle y, as seen from (6.18).

We can also see this by introducing new integration variables x, y in (6.16) using the relationships

$$\Omega = x\cos\psi - y\sin\psi\,,$$
$$v = x\sin\psi + y\cos\psi\,. \quad (6.20)$$

Taking into account that the Jacobian is equal to 1, we obtain

$$W(r,\psi) = \iint \frac{u_0^2}{|\cos\psi|} S_f(x\cos\psi - y\sin\psi, x\sin\psi + y\cos\psi) \times \frac{\sin^2 \dfrac{\pi(r-x)}{\Delta r}}{\left[\dfrac{\pi(r-x)}{\Delta r}\right]^2} \, dx \, dy \tag{6.21}$$

with the same definition of the strip width as in (6.18). The inversion problem stated in the beginning is therefore solved by (6.18).

Application of this approach to ionospheric research is discussed in [Tønsager and La Hoz, 1999].

6.4 Optical Tomography of the Ionosphere

In contradistinction to the main purpose of radio tomography to reconstruct the distribution of electron density in the ionosphere, the aim of optical tomography is to recover the distribution of the volume emission rate within some ionospheric volume. Several reasons may cause atmospheric luminescence, either artificial (due to anthropogenic effects as, for example, rocket launching) or natural (for instance, those producing polar auroras).

The first attempts to apply the tomographic approach to studying auroras (auroral tomography) were made more than 10 years ago. These investigations were based on experimental data obtained by satellite-borne scanning photometers [Solomon et al., 1985], rocket photometers [McDade and Llewellyn, 1991; McDade et al., 1991], and ground-based photometers [Vallance et al., 1991; Aso et al., 1990]. Although such investigations are still quite active nowadays (e.g., ARIES, ALIS, COTIF projects), only a few tomographic results have been obtained up to now. This is mainly caused by measuring equipment not ideal; another reason is the complexity of data processing. However, auroral tomography now draws growing interest extending toward not only polar region investigations but also midlatitude studies. At present, advanced methods for processing traditional experimental data are developing [Nygrén et al., 1997]. As well, new large-scale experiments are being carried out, such as, e.g. those using satellite-borne scanning spectrographs together with simultaneous satellite transmission of radio signals [Kamalabadi et al., 1999].

One of the purposes of auroral tomography is reconstruction of the two-dimensional distribution of volume emission rates in specific auroral forms such as arcs, auroral belts, or their combinations. This problem can be solved by performing measurements at a chain of ground-based scanning photometers arranged in a line, in particular, along a geomagnetic meridian. Then, from ground-based photometer measurements of auroral brightness in a series of emissions at various photometer elevations, optical tomograms for each emission can be recovered from the data obtained at the receiving chain. The most suitable for this purpose are atomic oxygen emissions at 557.7 and 630.0 nm and molecular nitrogen emission at 427.8 nm.

Reconstruction of the data in auroral tomography is similar to that in ray radio tomography previously discussed in Chap. 2. Although the subject in auroral tomography – brightness – is different from that of radio tomography, the reconstruction algorithms of ray radio tomography may be adjusted to the inversion of optical data. Naturally, due to the specificity of the physical subjects investigated, the reconstruction procedure will be slightly different. In particular, to provide appropriate spatial resolution in the tomographic method, different spatial grid steps should be used in reconstructing radio and optical data when dividing the atmospheric region of interest into geometric elements (pixels). Specific regularization should be used in optical tomography to control both vertical and horizontal variations of auroral brightness, and some other changes should be made in the inversion software as well. Reducing the dimension of the grid cell opens an opportunity to investigate fine auroral structures, but, on the other hand, this results in a great increase in time required for processing the data. Therefore, in practice, a compromise should be found between these two factors.

Let us start demonstrating the application of the tomographic approach to the reconstruction of optical data with the cases when a stable and bright enough diffuse arc is observed from all receiving sites at high enough elevations. Such a situation is portrayed in Fig. 6.10. In this figure, the results are shown of optical measurements made at Kola peninsula by three spaced scanning photometers installed in Apatity (67°34′ N, 33°24′ E), Verkhnetulomsky (68°35′ N, 31°45′ E), and Korzunovo (69°24′ N, 30°59′ E). The total length of the receiving chain was 226 km. Photometers measured the auroral brightnesses in three emissions – 557.7, 630.0, and 427.8 nm. The photometer spin rate was four complete rotations per minute and the elevation angle sampling interval was 0.5°. Figure 6.10 displays the results of measuring the auroral brightness in Rayleigh (R) in emission of 557.7 nm as a function of the direction of the reception of the emission so that 90° corresponds to the zenith, 180° to the northward direction, and 0° to the southward direction.

The reconstruction of the optical volume emission rate recovered from auroral brightness data is shown in Plate 16. The pixel size of 2.5 km × 2.5 km corresponds to the spatial resolution of the measuring instruments used. The specific feature of the recovered two-dimensional distributions is a distinct tendency toward decreasing height of both the bottom and the peak heights within the auroral form from south to north. This speaks for northward increasing the hardness of precipitating electrons. Another peculiarity seen in the reconstruction is the presence of two patches with contrary changes of

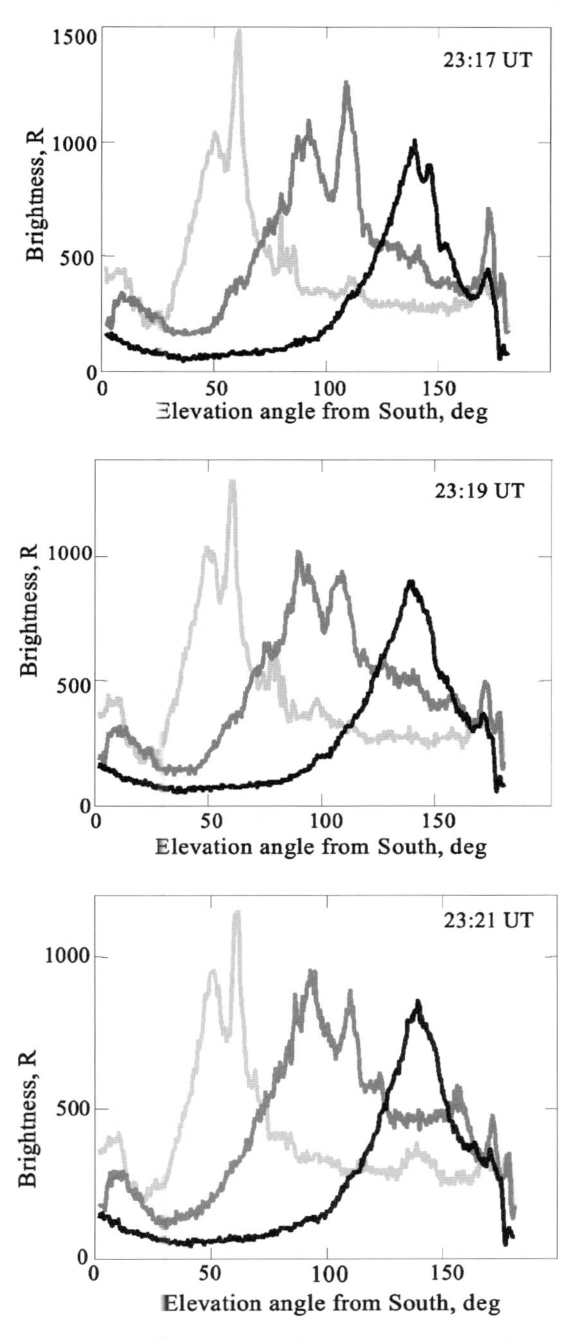

Fig. 6.10. The 557.7-nm brightness versus elevation angle measured at three receiving sites at 23:17 (upper panel), 23:19 (middle panel), and 23:21 UT (bottom panel) on February 10, 1999. The brightness measured in Apatity is shown by a black line, in Verkhnetulomsky by a gray line, and in Nikel by a light-gray line

brightness in them: during the time interval from 23:17 to 23:21, the brightness in the southern patch increases while it decreases in the northern one.

The existence of spatially and temporally stable (during a few minutes) auroral forms makes it reasonable to carry out simultaneous experiments in optical and radio tomography. Therefore, simultaneously with optical measurements made at the receiving sites described above, in 1999 also satellite radio signals were measured at a chain of radio tomographic receivers located in Kem (64°57′ N, 34°36′ E), Polyarnye Zori (67°37′ N, 32°45′ E), Verkhnetulomsky, and Korzunovo. The distance between Apatity and Polyarnye Zori is small so that, as a first-order approximation, both types of measurements can be considered as providing a two-dimensional distribution of the recovered parameters in the same vertical plane. Figure 6.11 portrays the collocation of the receiving sites mentioned above. The solid line shows the ground projection of the satellite path as it moves from north to south; the projection is close to the geomagnetic meridian $\Lambda = 125°$. Dense siting of the satellite radio receivers at Kola Peninsula provides a system of rays intersecting not only in the F-layer ionosphere but also at lower altitudes of the E layer which allows recovering the distribution of electron density both in the F- and E-layer ionosphere. Therefore, it becomes possible to compare the structure of observed auroras with the distribution of electron density. Note that reconstructions of electron density and auroral volume emission rates for the same time period may be used not only for qualitative analysis but also for

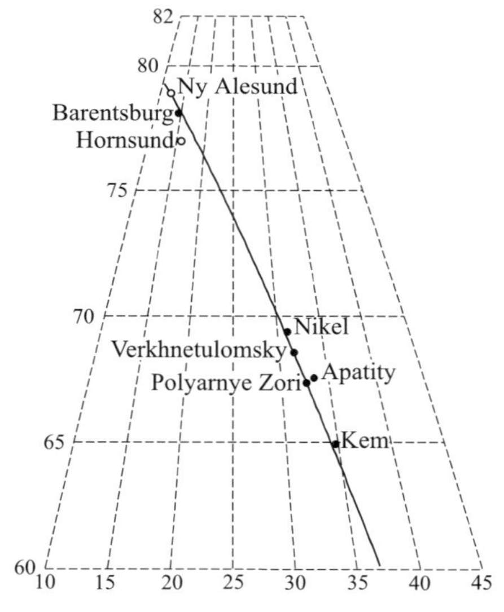

Fig. 6.11. A tomographic chain of PGI experiments. Presently operational RT receiving sites at Kem, Polyarnye Zori, Verkhnetulomsky, Nikel, Barentsburg. Hornsund, and Ny Alesund are planned to start operation in 2002

quantitative comparison of the data recovered by satellite (ray) and auroral tomography.

Plate 15b (top panel) portrays the two-dimensional distribution of electron density obtained from a radio tomogram synthesized for the period from 00:52 to 00:55 UT. A distinct enhancement of electron density south of Verkhnetulomsky is clearly seen in the reconstruction, whereas no increase in electron concentration is observed north of Verkhnetulomsky. TV images obtained by an all-sky camera in the Loparskaya observatory as well as scanning photometer data showed practically no aurora in the region north of Verkhnetulomsky till 00:56 UT, i.e., during the interval of the synthesized radio tomogram. Therefore, one can infer that a correlation exists between the observed precipitation and the appearance of enhanced electron density in the E-layer ionosphere. Plate 15b (middle panel) and (bottom panel) portray the reconstruction of the volume emission rate observed at 00:54 UT in the E-layer ionospheric region of enhanced electron density seen in the radio tomographic reconstruction.

Summary

The theoretical fundamentals of ionospheric radio tomography (RT), including diffraction, ray, and statistical RT, are expounded in the present monograph; experimental results are shown to illustrate the possibilities of tomographic approaches. The theoretical methods developed and the software and special equipment designed for RT investigations made an essential base for practical realization of satellite RT of the ionosphere.

The first tomographic images of the ionosphere were obtained in 1990; further scientific investigations by the ray RT method yielded a lot of new geophysical information about the structure of the ionosphere. During the last 10 years, ray RT studies of the ionosphere drew great interest in the geophysical community; more than 10 scientific groups and laboratories are working in this field. In the monograph, examples are given of investigations by ray RT carried out during recent years in Europe, America, and Asia in cooperation with scientists from several universities and research centers. The existing low-orbiting navigational systems were used as sounding sources. Examples are shown of RT sections of the most interesting ionospheric structures (troughs, the equatorial anomaly, traveling ionospheric disturbances, quasi-wave structures, ionosphere bends, fingerlike forms, etc.) Also given are examples of RT images of artificial ionospheric disturbances caused by anthropogenic factors, in particular, by rocket launching, industrial explosions, and powerful HF radiation.

Investigations of the irregular structure of the ionosphere by diffraction and statistical RT are presented, which have no analogues till now. Other methods of ionospheric studies are not capable of obtaining such information about irregularities (the structure of localized inhomogeneities, the section of the spectrum of electron density fluctuations, the distribution of the ionospheric plasma fluctuations intensity). Optical tomography of the ionosphere is described.

The experiments carried out demonstrated the fine possibilities of the RT techniques; nevertheless, that is evidently only the very beginning, and further advances in satellite RT will open wider prospects for its employment in monitoring the ionosphere and near-Earth space. In the monograph, the possibilities and outlook for the combined application of RT and occultation techniques are considered; also prospects are discussed of using combined systems comprising various constellations and ground reception of signals for the purposes of RT of the ionosphere and near space. Progress in the widespread adoption of radio tomographic sounding methods will involve a lot of both theoretical and technical problems to be solved. However, the development of different satellite RT systems will undoubtedly be of great benefit for practice and will yield new basic knowledge about the ionosphere and near-Earth space.

Color Plates

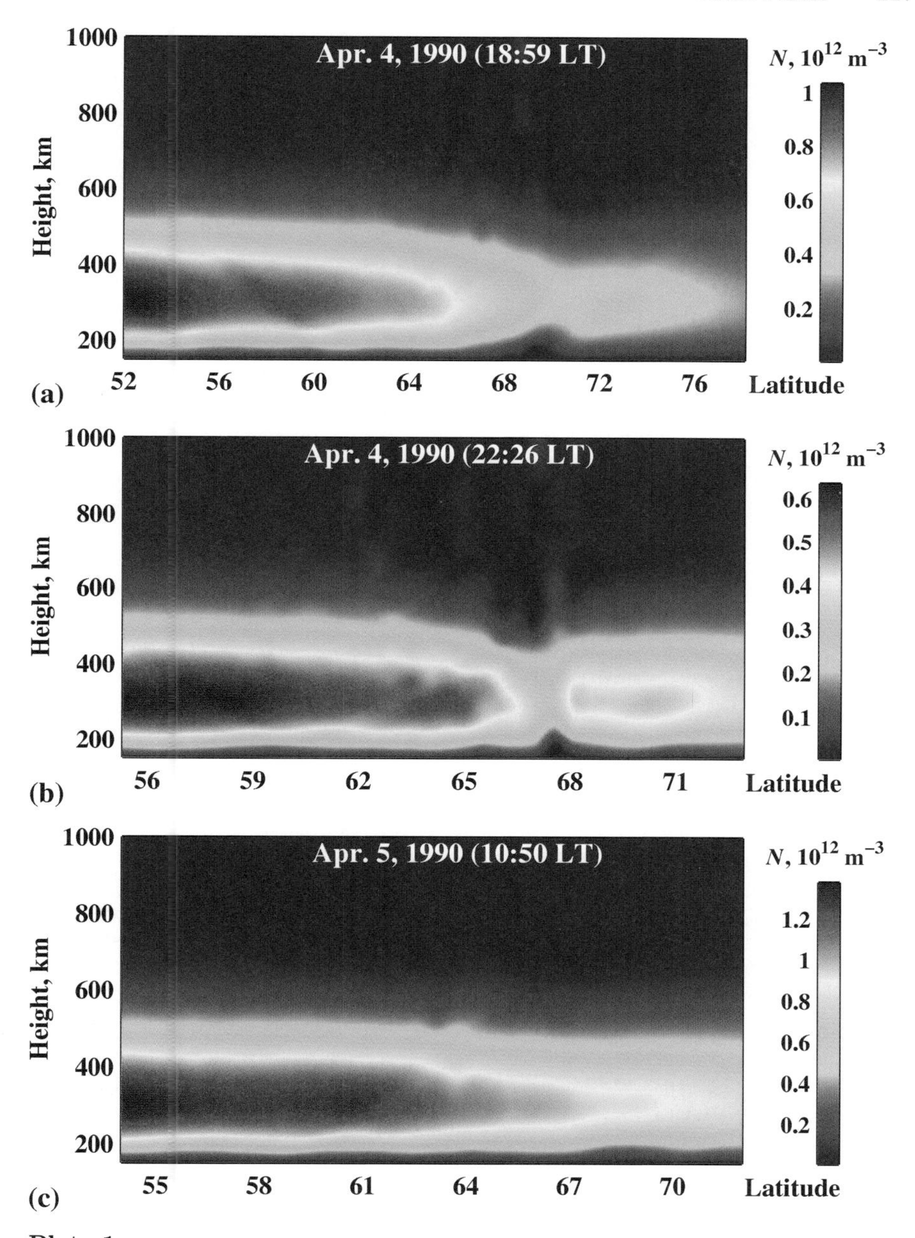

Apr. 4, 1990 (18:59 LT)
Height, km
1000
800
600
400
200
52
56
60
64
68
72
76
Latitude
N, 10^{12} m^{-3}
1
0.8
0.6
0.4
0.2
(a)
Apr. 4, 1990 (22:26 LT)
Height, km
1000
800
600
400
200
56
59
62
65
68
71
Latitude
N, 10^{12} m^{-3}
0.6
0.5
0.4
0.3
0.2
0.1
(b)
Apr. 5, 1990 (10:50 LT)
Height, km
1000
800
600
400
200
55
58
61
64
67
70
Latitude
N, 10^{12} m^{-3}
1.2
1
0.8
0.6
0.4
0.2
(c)

Plate 1.

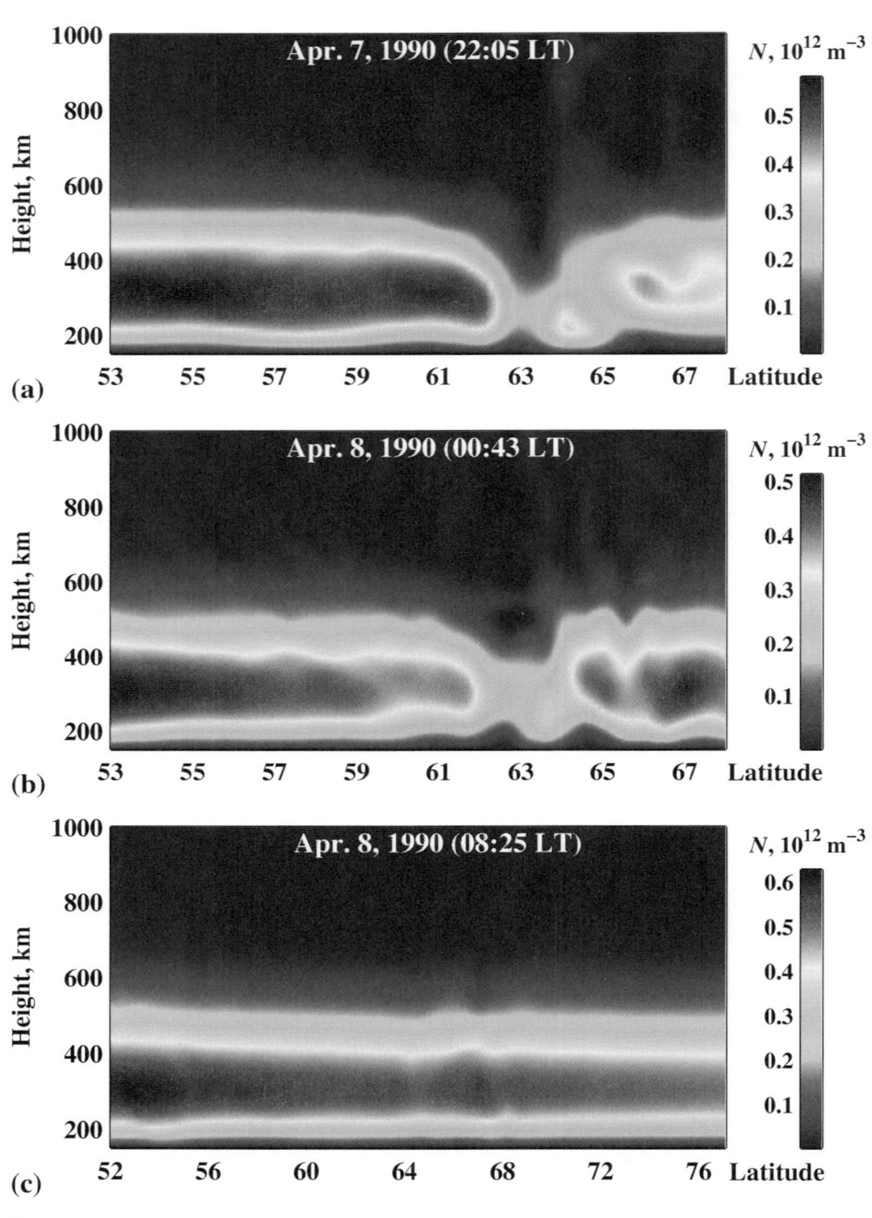
Apr. 7, 1990 (22:05 LT)
Height, km
1000
800
600
400
200
53
55
57
59
61
63
65
67
Latitude
N, 10^{12} m^{-3}
0.5
0.4
0.3
0.2
0.1
(a)
Apr. 8, 1990 (00:43 LT)
Height, km
1000
800
600
400
200
53
55
57
59
61
63
65
67
Latitude
N, 10^{12} m^{-3}
0.5
0.4
0.3
0.2
0.1
(b)
Apr. 8, 1990 (08:25 LT)
Height, km
1000
800
600
400
200
52
56
60
64
68
72
76
Latitude
N, 10^{12} m^{-3}
0.6
0.5
0.4
0.3
0.2
0.1
(c)

Plate 2.

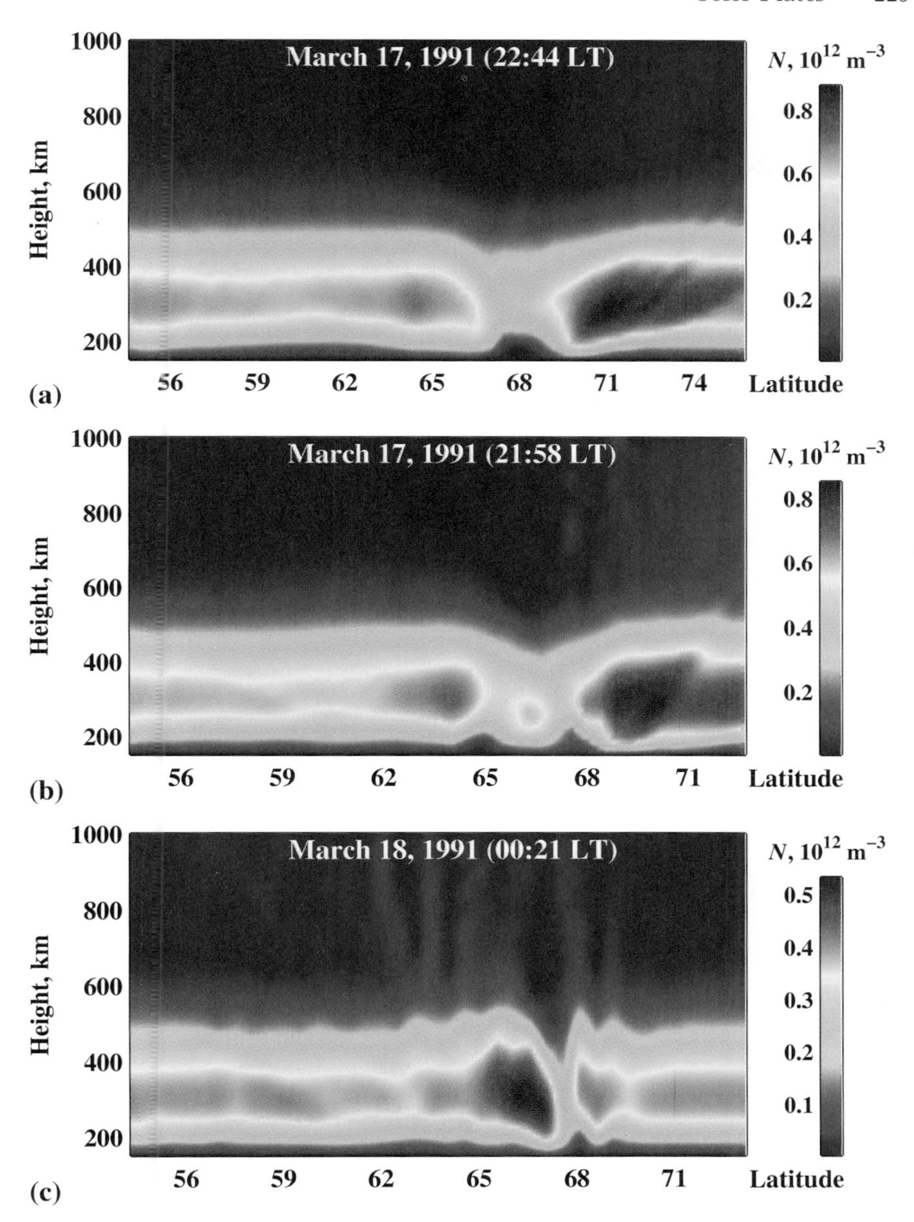
March 17, 1991 (22:44 LT)
Height, km
1000
800
600
400
200
56
59
62
65
68
71
74
Latitude
N, 10^{12} m^{-3}
0.8
0.6
0.4
0.2
(a)
March 17, 1991 (21:58 LT)
Height, km
1000
800
600
400
200
56
59
62
65
68
71
Latitude
N, 10^{12} m^{-3}
0.8
0.6
0.4
0.2
(b)
March 18, 1991 (00:21 LT)
Height, km
1000
800
600
400
200
56
59
62
65
68
71
Latitude
N, 10^{12} m^{-3}
0.5
0.4
0.3
0.2
0.1
(c)

Plate 3.

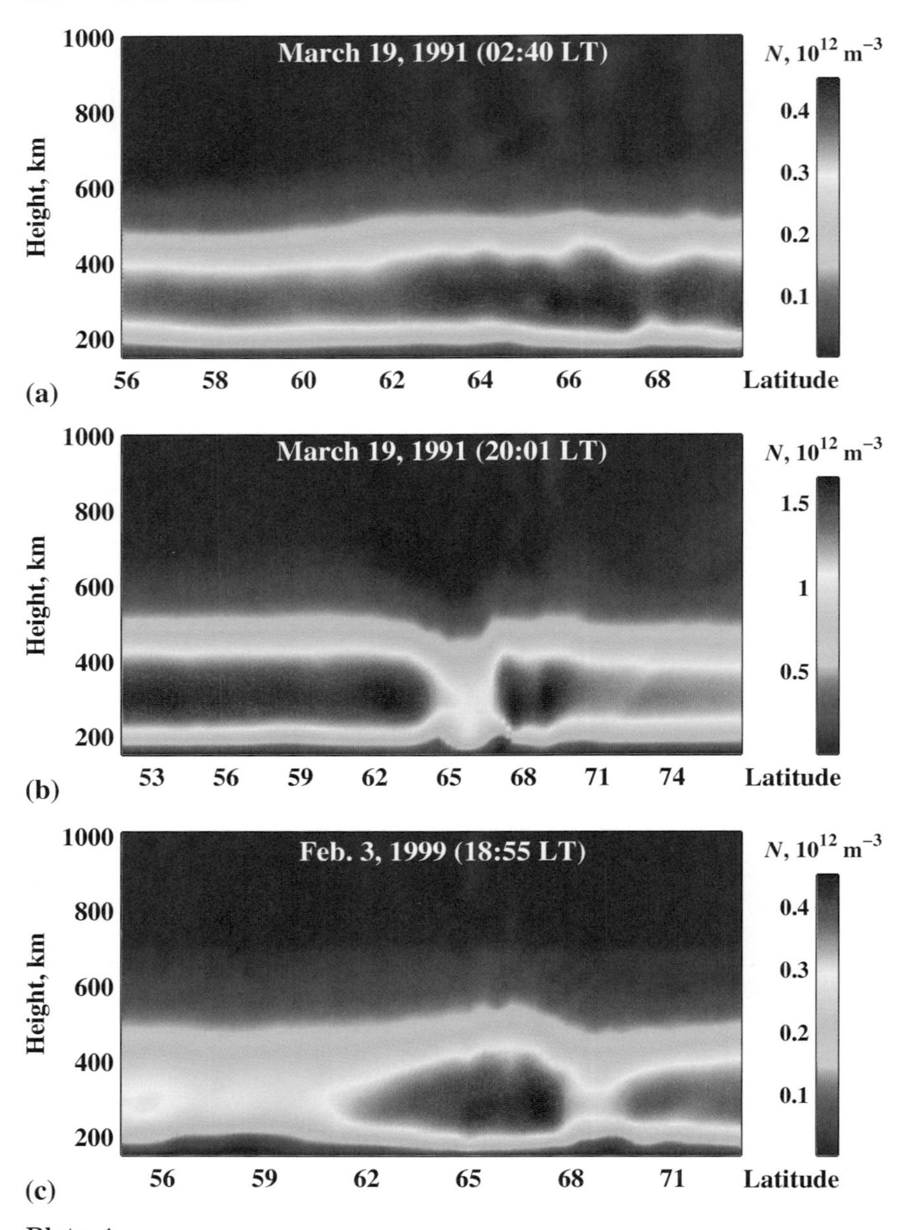

March 19, 1991 (02:40 LT)
Height, km
1000
800
600
400
200
56
58
60
62
64
66
68
Latitude
N, 10^{12} m^{-3}
0.4
0.3
0.2
0.1
(a)
March 19, 1991 (20:01 LT)
Height, km
1000
800
600
400
200
53
56
59
62
65
68
71
74
Latitude
N, 10^{12} m^{-3}
1.5
1
0.5
(b)
Feb. 3, 1999 (18:55 LT)
Height, km
1000
800
600
400
200
56
59
62
65
68
71
Latitude
N, 10^{12} m^{-3}
0.4
0.3
0.2
0.1
(c)

Plate 4.

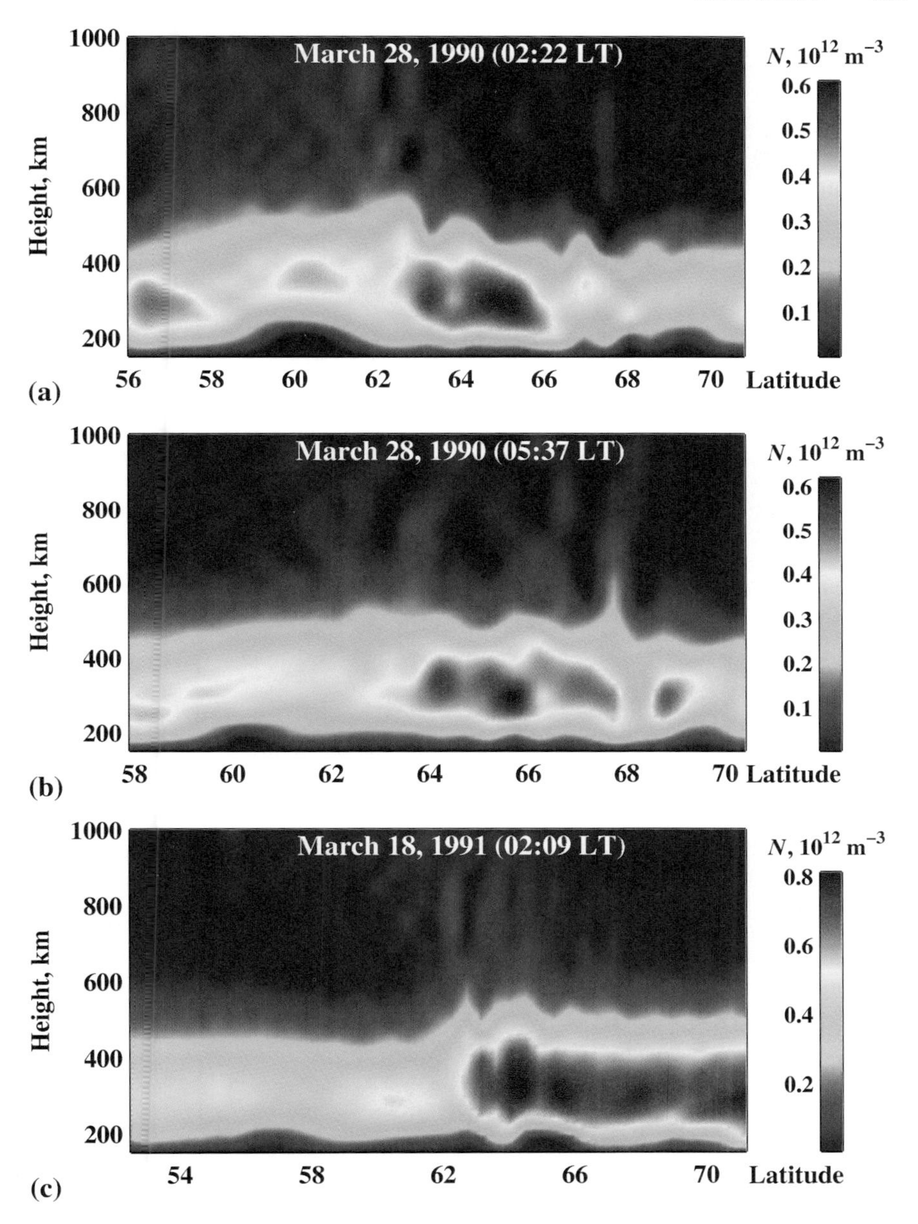

March 28, 1990 (02:22 LT)
Height, km
1000
800
600
400
200
56
58
60
62
64
66
68
70
Latitude
N, 10^12 m^-3
0.6
0.5
0.4
0.3
0.2
0.1
(a)
March 28, 1990 (05:37 LT)
Height, km
1000
800
600
400
200
58
60
62
64
66
68
70
Latitude
N, 10^12 m^-3
0.6
0.5
0.4
0.3
0.2
0.1
(b)
March 18, 1991 (02:09 LT)
Height, km
1000
800
600
400
200
54
58
62
66
70
Latitude
N, 10^12 m^-3
0.8
0.6
0.4
0.2
(c)

Plate 5.

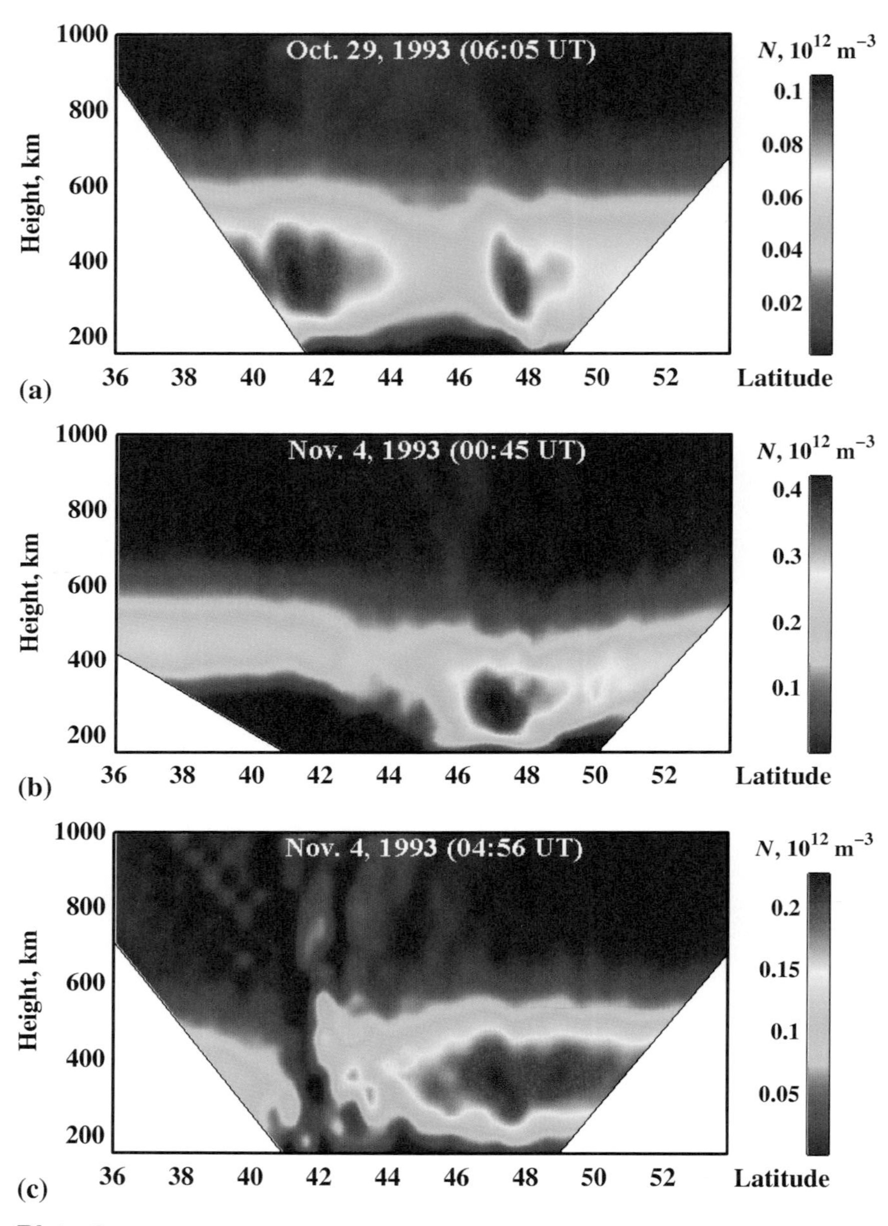
Oct. 29, 1993 (06:05 UT)
Height, km
1000
800
600
400
200
36 38 40 42 44 46 48 50 52 Latitude
N, 10^12 m^-3
0.1
0.08
0.06
0.04
0.02
(a)
Nov. 4, 1993 (00:45 UT)
Height, km
1000
800
600
400
200
36 38 40 42 44 46 48 50 52 Latitude
N, 10^12 m^-3
0.4
0.3
0.2
0.1
(b)
Nov. 4, 1993 (04:56 UT)
Height, km
1000
800
600
400
200
36 38 40 42 44 46 48 50 52 Latitude
N, 10^12 m^-3
0.2
0.15
0.1
0.05
(c)

Plate 6.

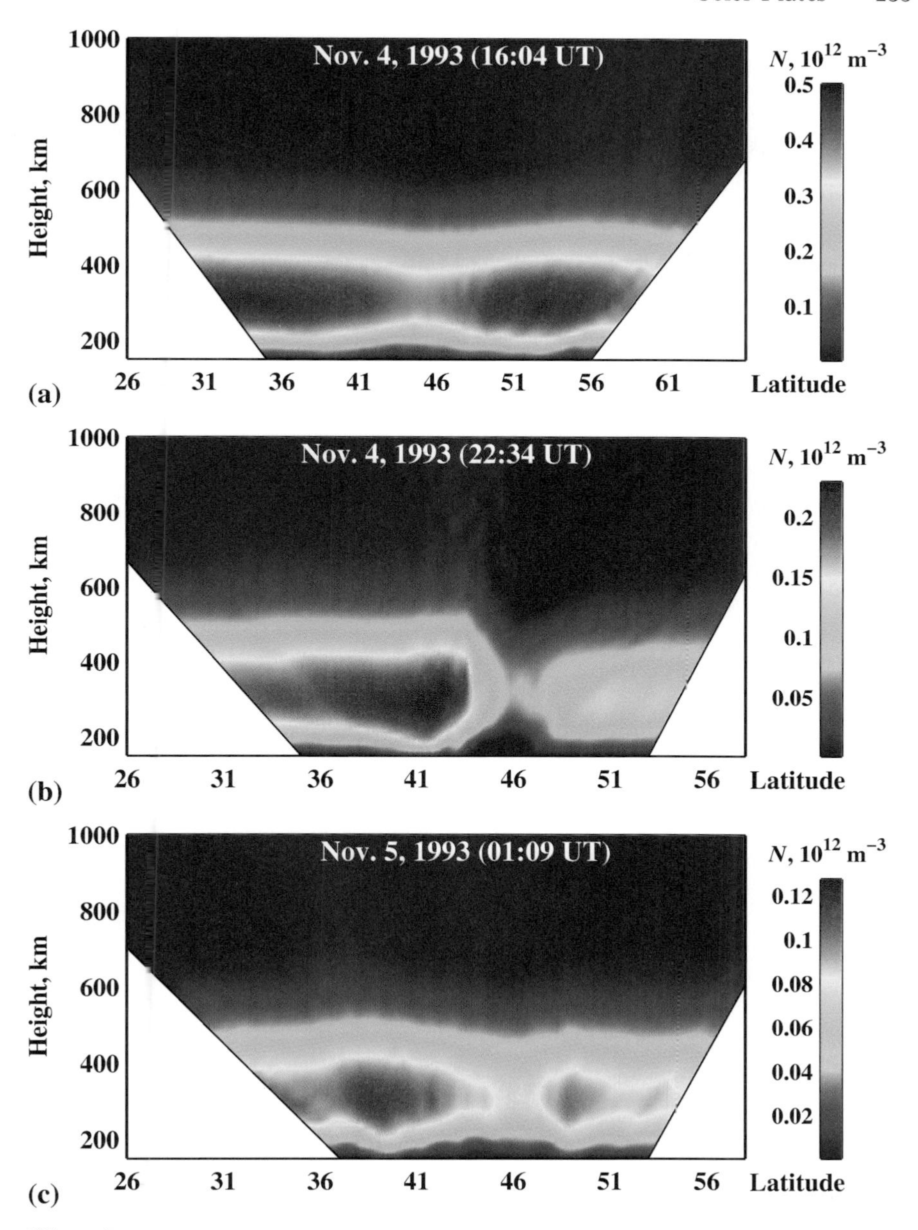
Nov. 4, 1993 (16:04 UT)
Height, km
1000
800
600
400
200
26 31 36 41 46 51 56 61 Latitude
N, 10^12 m^-3
0.5
0.4
0.3
0.2
0.1
(a)
Nov. 4, 1993 (22:34 UT)
Height, km
1000
800
600
400
200
26 31 36 41 46 51 56 Latitude
N, 10^12 m^-3
0.2
0.15
0.1
0.05
(b)
Nov. 5, 1993 (01:09 UT)
Height, km
1000
800
600
400
200
26 31 36 41 46 51 56 Latitude
N, 10^12 m^-3
0.12
0.1
0.08
0.06
0.04
0.02
(c)

Plate 7.

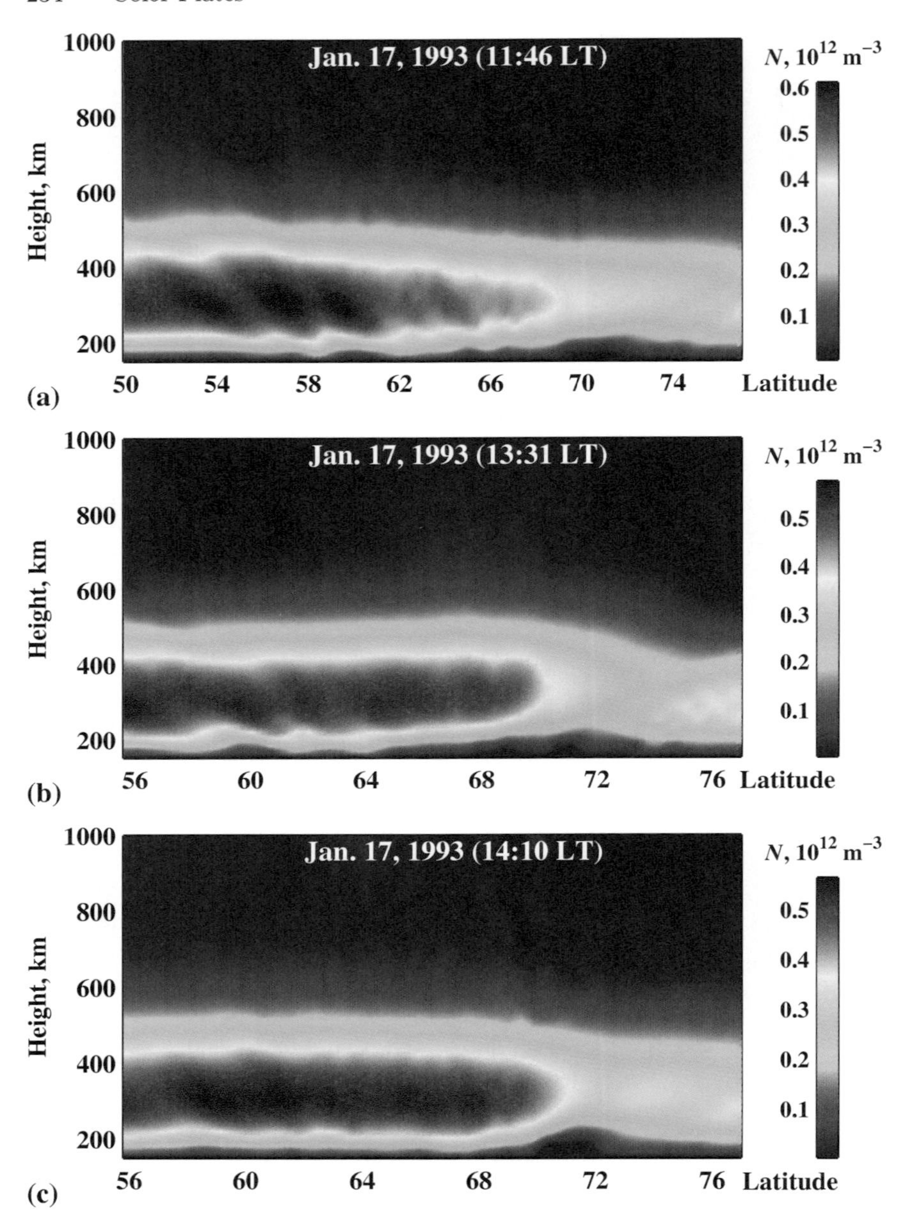
Jan. 17, 1993 (11:46 LT)
Height, km
1000
800
600
400
200
50 54 58 62 66 70 74 Latitude
N, 10^12 m^-3
0.6
0.5
0.4
0.3
0.2
0.1
(a)
Jan. 17, 1993 (13:31 LT)
Height, km
1000
800
600
400
200
56 60 64 68 72 76 Latitude
N, 10^12 m^-3
0.5
0.4
0.3
0.2
0.1
(b)
Jan. 17, 1993 (14:10 LT)
Height, km
1000
800
600
400
200
56 60 64 68 72 76 Latitude
N, 10^12 m^-3
0.5
0.4
0.3
0.2
0.1
(c)

Plate 8.

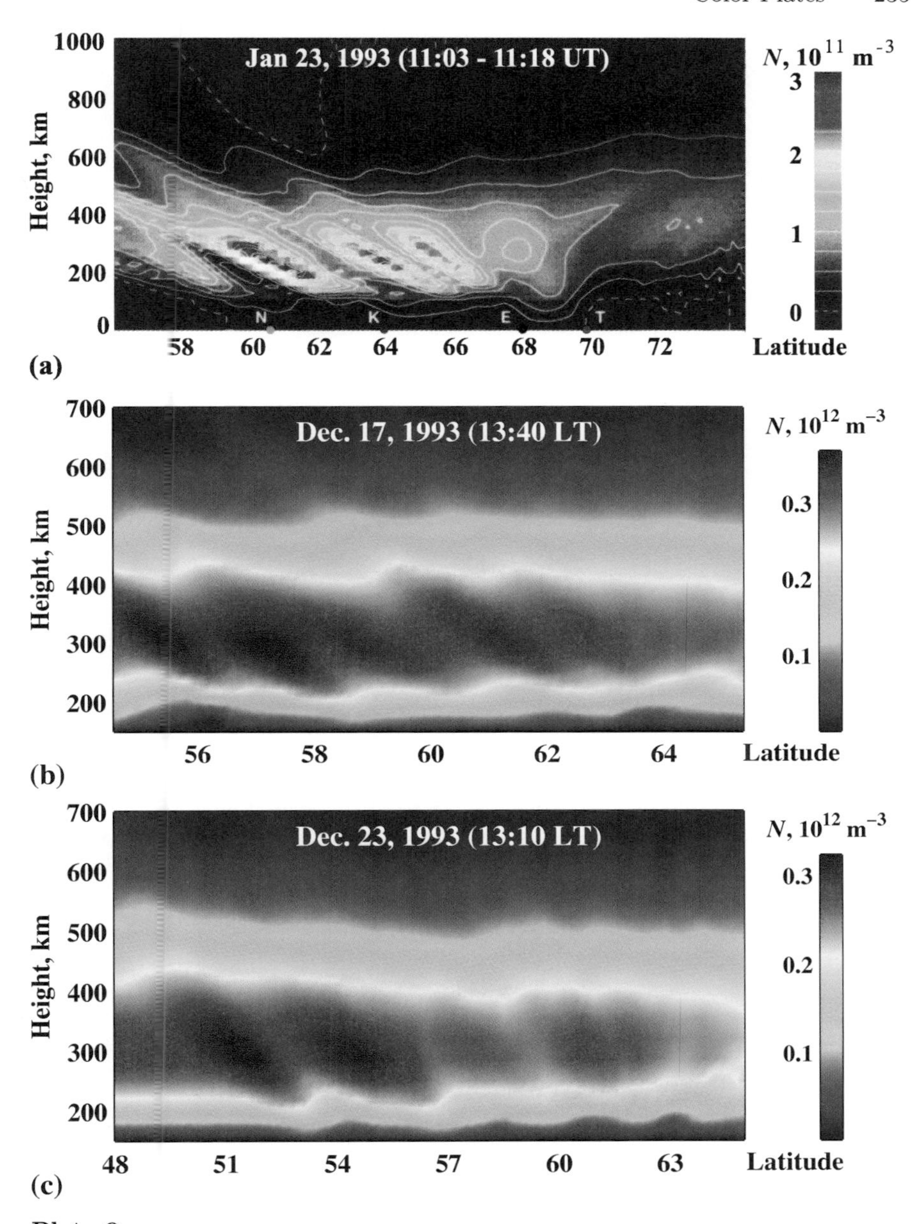
Jan 23, 1993 (11:03 - 11:18 UT)
Height, km
1000
800
600
400
200
0
N
K
E
T
58
60
62
64
66
68
70
72
Latitude
N, 10^11 m^-3
3
2
1
0
(a)
Dec. 17, 1993 (13:40 LT)
Height, km
700
600
500
400
300
200
56
58
60
62
64
Latitude
N, 10^12 m^-3
0.3
0.2
0.1
(b)
Dec. 23, 1993 (13:10 LT)
Height, km
700
600
500
400
300
200
48
51
54
57
60
63
Latitude
N, 10^12 m^-3
0.3
0.2
0.1
(c)

Plate 9.

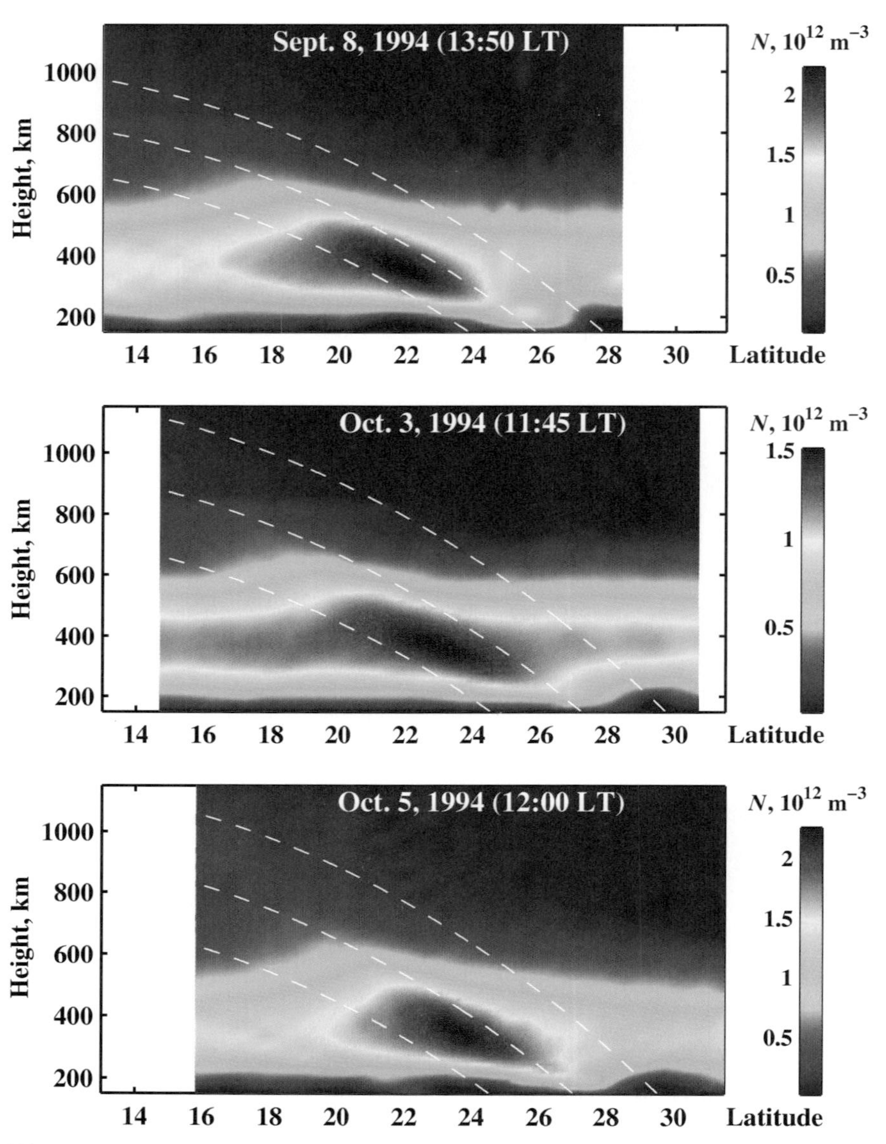
Sept. 8, 1994 (13:50 LT)
Height, km
1000
800
600
400
200
14 16 18 20 22 24 26 28 30 Latitude
N, 10^12 m^-3
2
1.5
1
0.5
Oct. 3, 1994 (11:45 LT)
Height, km
1000
800
600
400
200
14 16 18 20 22 24 26 28 30 Latitude
N, 10^12 m^-3
1.5
1
0.5
Oct. 5, 1994 (12:00 LT)
Height, km
1000
800
600
400
200
14 16 18 20 22 24 26 28 30 Latitude
N, 10^12 m^-3
2
1.5
1
0.5

Plate 10.

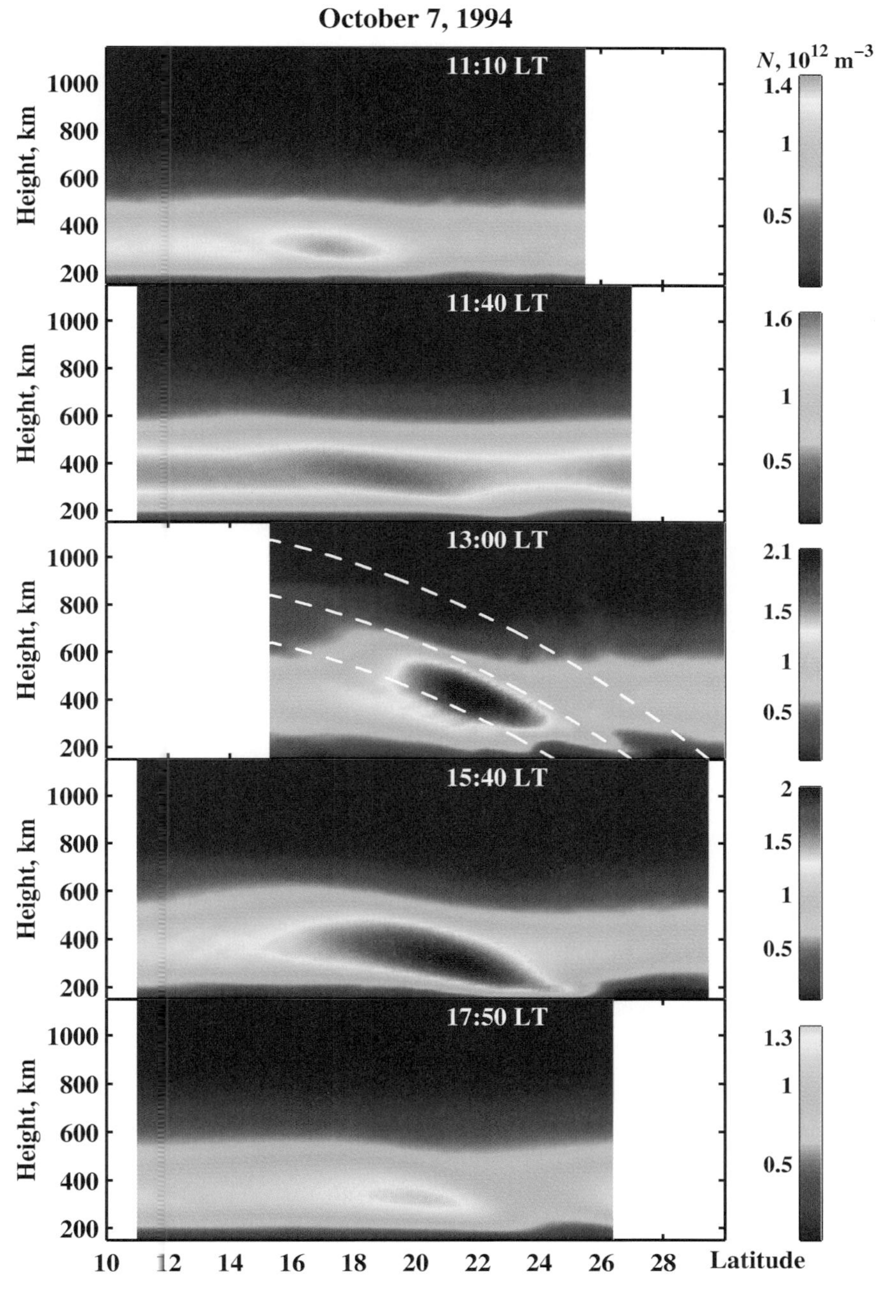
October 7, 1994
11:10 LT
11:40 LT
13:00 LT
15:40 LT
17:50 LT
Height, km
1000
800
600
400
200
N, 10^{12} m^{-3}
1.4
1
0.5
1.6
2.1
1.5
2
1.3
10
12
14
16
18
20
22
24
26
28
Latitude

Plate 11.

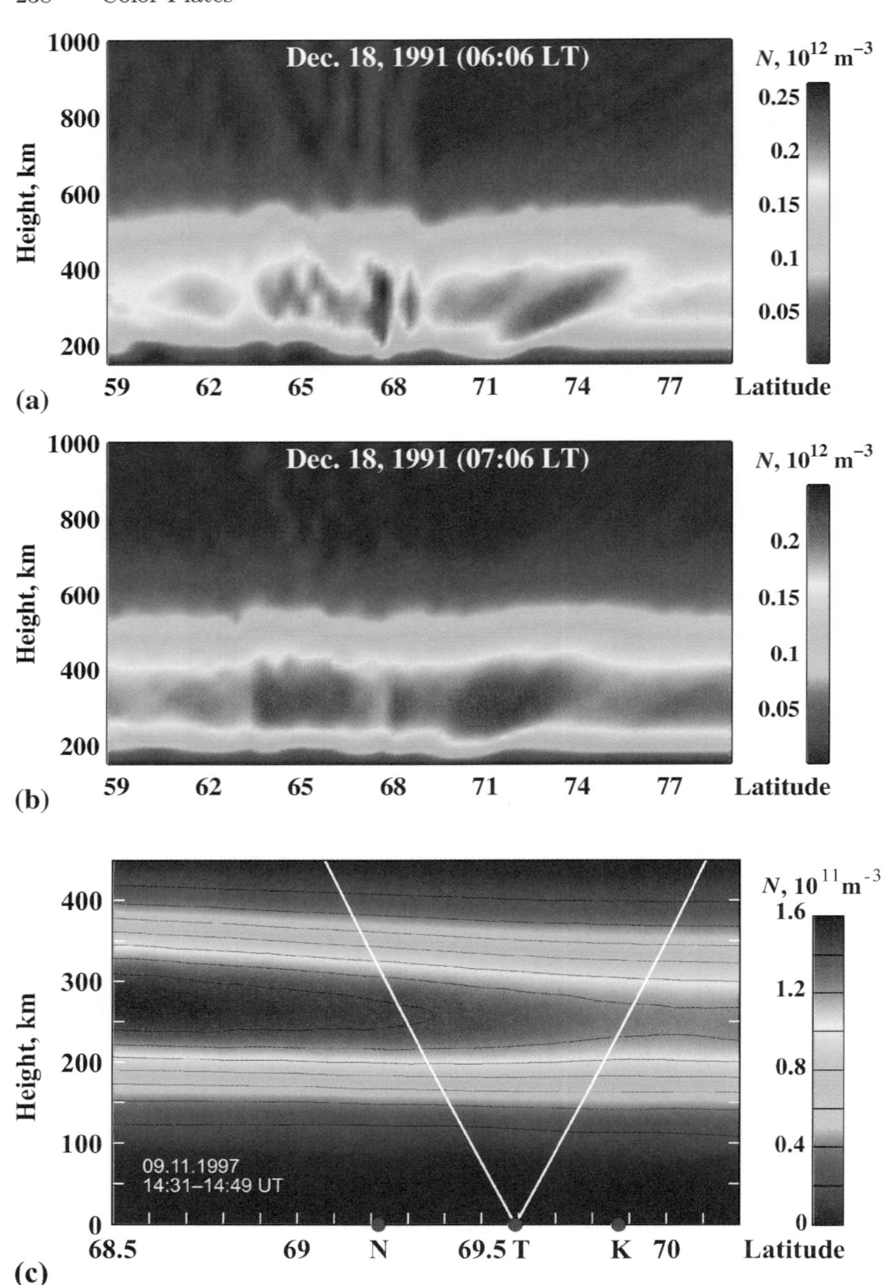
Dec. 18, 1991 (06:06 LT)
Height, km
1000
800
600
400
200
59 62 65 68 71 74 77 Latitude
N, 10^12 m^-3
0.25
0.2
0.15
0.1
0.05
(a)
Dec. 18, 1991 (07:06 LT)
Height, km
1000
800
600
400
200
59 62 65 68 71 74 77 Latitude
N, 10^12 m^-3
0.2
0.15
0.1
0.05
(b)
Height, km
400
300
200
100
0
09.11.1997
14:31–14:49 UT
68.5 69 N 69.5 T K 70 Latitude
N, 10^11 m^-3
1.6
1.2
0.8
0.4
0
(c)

Plate 12.

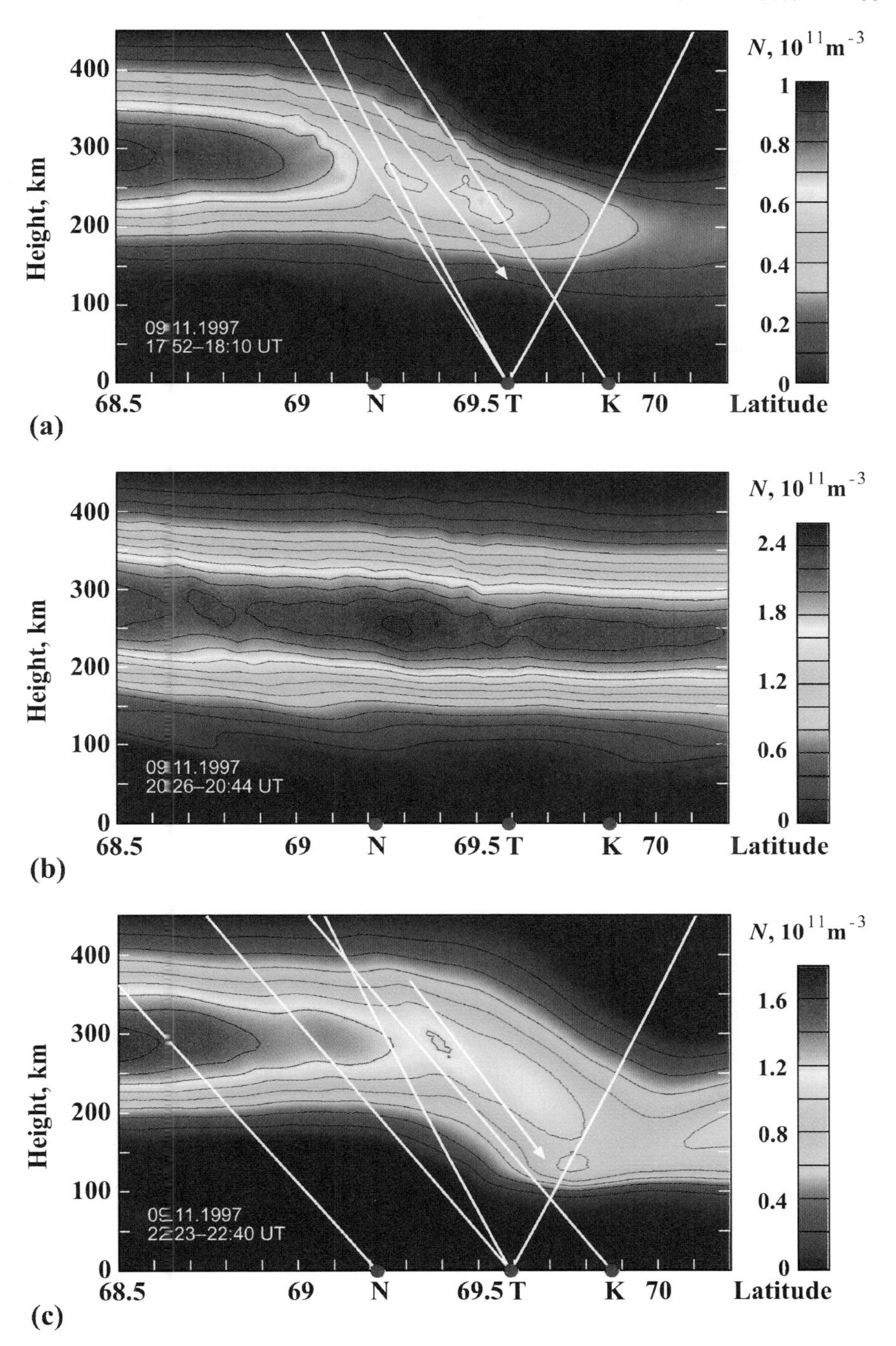

N, 10^{11}m^{-3}
Height, km
09.11.1997
17:52–18:10 UT
20:26–20:44 UT
22:23–22:40 UT
68.5
69
N
69.5 T
K
70
Latitude
(a)
(b)
(c)

Plate 13.

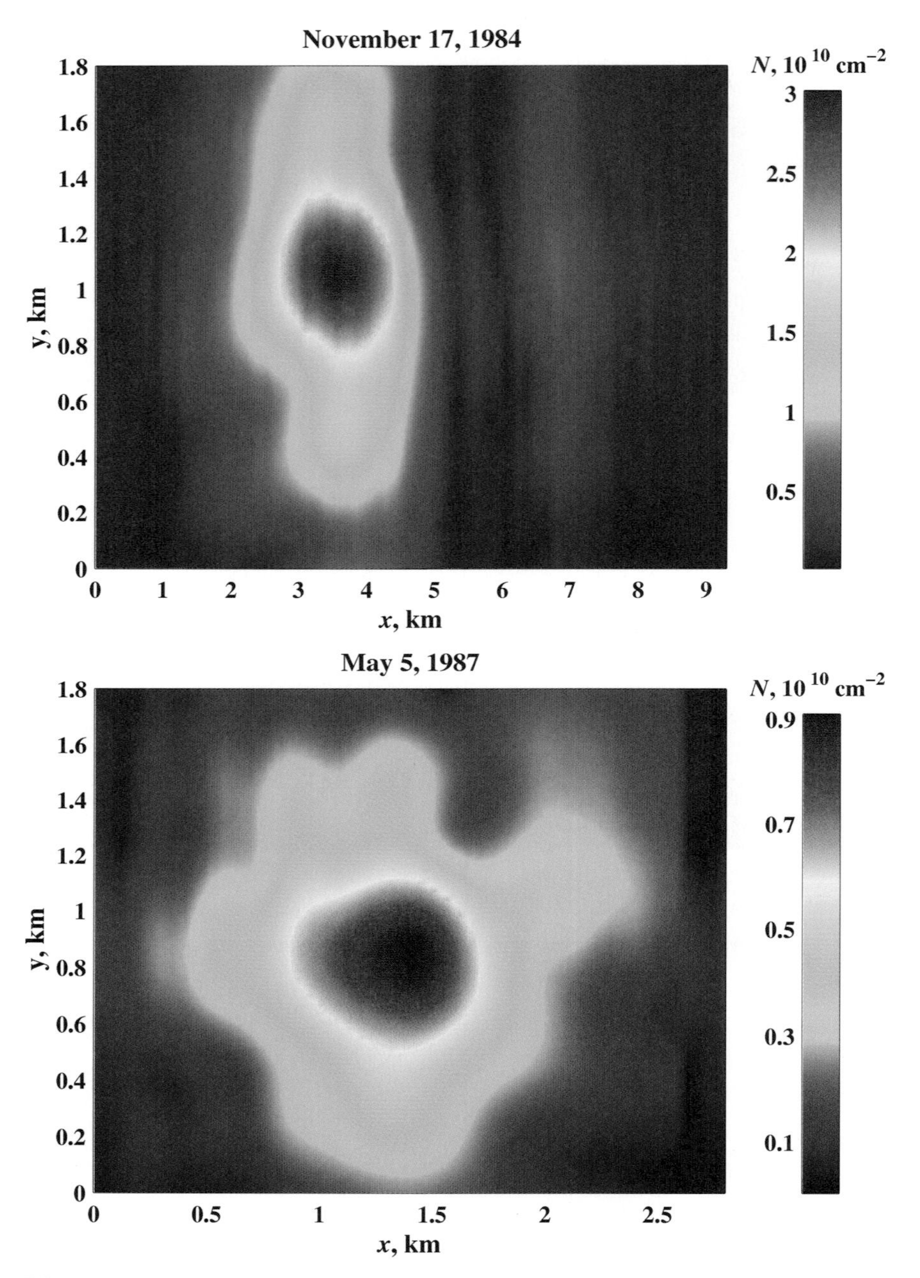
November 17, 1984
1.8
1.6
1.4
1.2
1
0.8
0.6
0.4
0.2
0
y, km
0 1 2 3 4 5 6 7 8 9
x, km
N, 10 10 cm−2
3
2.5
2
1.5
1
0.5
May 5, 1987
1.8
1.6
1.4
1.2
1
0.8
0.6
0.4
0.2
0
y, km
0 0.5 1 1.5 2 2.5
x, km
N, 10 10 cm−2
0.9
0.7
0.5
0.3
0.1

Plate 14.

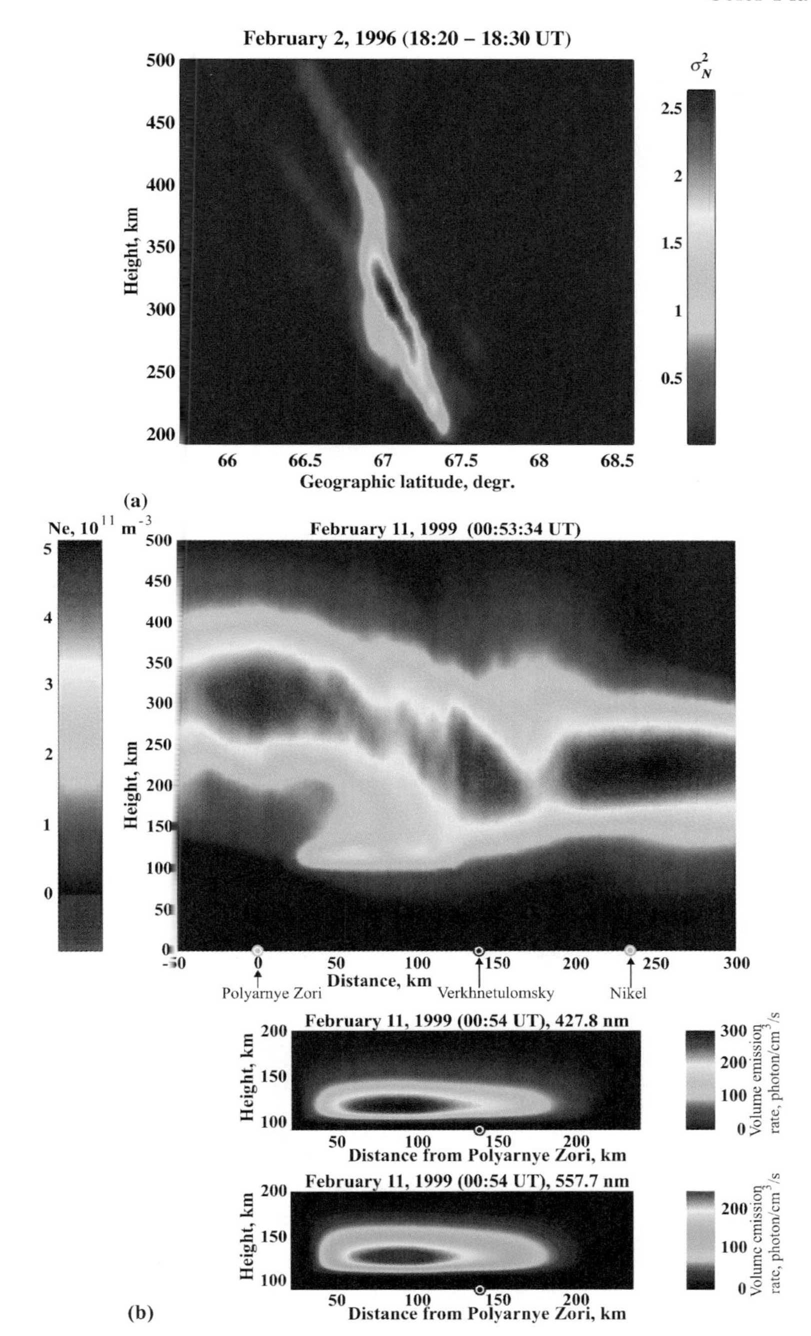

February 2, 1996 (18:20 – 18:30 UT)
Height, km
Geographic latitude, degr.
σ_N^2
(a)
Ne, 10^{11} m^{-3}
February 11, 1999 (00:53:34 UT)
Height, km
Distance, km
Polyarnye Zori
Verkhnetulomsky
Nikel
February 11, 1999 (00:54 UT), 427.8 nm
Height, km
Distance from Polyarnye Zori, km
Volume emission rate, photon/cm^3/s
February 11, 1999 (00:54 UT), 557.7 nm
Height, km
Distance from Polyarnye Zori, km
Volume emission rate, photon/cm^3/s
(b)

Plate 15.

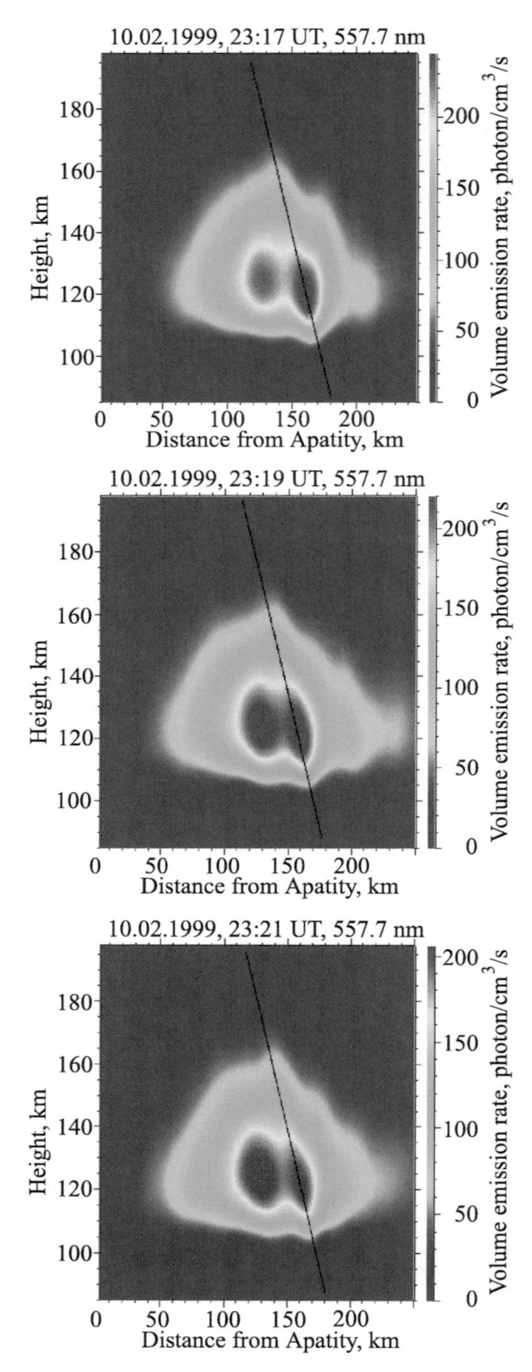
10.02.1999, 23:17 UT, 557.7 nm
Height, km
Distance from Apatity, km
Volume emission rate, photon/cm^3/s
10.02.1999, 23:19 UT, 557.7 nm
Height, km
Distance from Apatity, km
Volume emission rate, photon/cm^3/s
10.02.1999, 23:21 UT, 557.7 nm
Height, km
Distance from Apatity, km
Volume emission rate, photon/cm^3/s

Plate 16.

References

[Aarons, 1982] Aarons, J. (1982). *Proc. IEEE*, 70(4):360–378.

[Aarons et al. 1971] Aarons, J., Whitney, H., and Allen, R. (1971). *Proc. IEEE*, 59(2):54–66.

[Afraimovich et al., 1989] Afraimovich, E., Pirog, O., and Terekhov, A. (1989). *Preprint Sib. IZMIR*, 19-89:1–12 (in Russian).

[Akasofu, 1974] Akasofu, S.-I. (1974). *Space Sci. Rev.*, 16(5/6):617.

[Anastasio and Pan, 2001] Anastasio, M. and Pan, X. (2001). *Appl. Opt.*, 40(20):3334–3345.

[Andreeva et al., 1999] Andreeva, E., Berbeneva, N., Kunitsyn, V., and Popov, A. (1999). *Geomagnetism and Aeronomy*, 39(6):109–114.

[Andreeva et al., 2000a] Andreeva, E., Berbeneva, N., Zakharov, V., and Kunitsyn, V. (2000a). *Telecommun. Radio Eng.*, 1:74–80.

[Andreeva et al., 2000b] Andreeva, E., Franke, S., Yeh, K., and Kunitsyn, V. (2000b). *Geophys. Res. Lett.*, 27(16):2465–2468.

[Andreeva et al., 2001a] Andreeva, E., Franke, S., Yeh, K., Kunitsyn, V., and Nesterov, I. (2001a). *Radio Sci.*, 36(2):299–309.

[Andreeva et al., 1990] Andreeva, E., Galinov, A., Kunitsyn, V., Mel'nitchenko, Y., Tereshchenko, E., Filimonov, M., and Chernyakov, S. (1990). *J. Exp. Theor. Phys. Lett.*, 52:145–148.

[Andreeva et al., 2001b] Andreeva, E., Gokhberg, M., Kunitsyn, V., and other (2001b). *Space Res.*, 39(1):13–17.

[Andreeva and Kunitsyn, 1989] Andreeva, E. and Kunitsyn, V. (1989). *Propagation and Diffraction of Waves in Heterogeneous Media*, pp. 9–14. MPTI, Moscow (in Russian).

[Andreeva et al., 1992] Andreeva, E., Kunitsyn, V., and Tereshchenko, E. (1992). *Ann. Geophys.*, 10:849–855.

[Anger, 1990] Anger, G. (1990). *Inverse Problems in Differential Equations.* Plenum, London.

[Aso et al., 1990] Aso, T., Hashimoto, T., Abe, M., Ono, T., and Ejiri, M. (1990). *J. Geomagn. Geoelectr.*, 42:579.

[Austen et al., 1988] Austen, J., Franke, S., and Liu, C. (1988). *Radio Sci.*, 23:299–307.

[Baker, 1981] Baker, P. (1981). *Radio Sci.*, 17(2):213–217.

[Bakhrakh and Kurochkin, 1979] Bakhrakh, L. and Kurochkin, A. (1979). *Golographiya v mikrovolnovoy tekhnike.* Sovetskoye Radio, Moskva.

[Baron, 1986] Baron, M. (1986). *J. Atmos. Terr. Phys.*, 49(9–10):767–772.

[Berbeneva et al., 2001] Berbeneva, N., Kunitsyn, V., Razinkov, O., and Zakharov, V. (2001). *Phys. Chem. Earth (A)*, 26(3):131–138.

[Berezanskiy, 1955] Berezanskiy, Y. (1955). *Dokl. AN SSSR*, 105:197.

[Bernhardt et al., 1997] Bernhardt, P., Dymond, K., Picone, J., Cotton, D., Chakrabarti, S., Cook, T., and Vickers, J. (1997). *Radio Sci.*, 32(5):1965–1972.

[Bernhardt et al., 1998] Bernhardt, P., McCoy, R., Dymond, K., Picone, J., Meier, R., Kamalabadi, I., Cotton, D., Charkrabarti, S., Cook, T., Vickers, J., Stephan, A., Kersley, L., Pryse, S., Walker, I., Mitchell, C., Straus, P., Na, H., Biswas, C., Bust, G., Kronschnabl, G., and Raymund, T. (1998). *Phys. Plasmas*, 5(5):2010–2021.

[Biswas and Na, 2000] Biswas, C. and Na, H. (2000). *Radio Sci.*, 35(3):905–920.

[Bojarski, 1981a] Bojarski, N. (1981a). *Radio Sci.*, 16(6):1025–1028.

[Bojarski, 1981b] Bojarski, N. (1981b). *J. Math. Phys.*, 22(8):1647–1650.

[Bolomey, 1989] Bolomey, J. (1989). *Trans. Microwave Theory Techn.*, 37(12):2109–2117.

[Booker, 1958] Booker, H. (1958). *Proc. IRE*, 46:610–613.

[Booker et al., 1950] Booker, H., Ratcliffe, G., and Shinn, D. (1950). *Philos. Trans. R. Soc.*, A(242):579–607.

[Born and Wolf, 1970] Born, M. and Wolf, E. (1970). *Principles of Optics*. Pergamon Press, New York, 4th ed.

[Bracewell, 1956] Bracewell, R. (1956). *Aust. J. Phys.*, 9:198–217.

[Breed et al., 1997] Breed, A., Goodwin, G., and Vandenberg, A. (1997). *Radio Sci.*, 32(4):1635–1643.

[Breit and Tuve, 1926] Breit, G. and Tuve, M. (1926). *Phys. Rev.*, 28:554.

[Brekke, 1997] Brekke, A. (1997). *Physics of the Upper Polar Atmosphere*. John Wiley & Sons associated with Praxis, Chichester.

[Briggs et al., 1950] Briggs, B., Philips, G., and Shinn, D. (1950). *Proc. Phys. Soc.*, 63:106–125.

[Budden, 1961] Budden, K. (1961). *Radio Waves in the Ionosphere*. Cambridge University Press, Cambridge.

[Bukhshtaber and Maslov, 1990] Bukhshtaber, V. and Maslov, V. (1990). *Voprosi kibernetiki. Matematicheskiye problemi tomographii*, pp. 7–60. Nauka, Moscow.

[Burov et al., 1986] Burov, V., Goryunov, A., Saskovetz, A., and other (1986). *Akust. zhurnal*, 32(4):439–449.

[Burov et al., 1989] Burov, V., Richagov, M., and Saskovetz, A. (1989). *Vestnik MGU. Ser.3. Phizika, astronomiya*, 30(1):44–48.

[Bust et al., 2001] Bust, G., Coco, D., and Gaussiran, T. (2001). *Radio Sci.*, 36(6):1599–1605.

[Bust et al., 2000] Bust, G., Coco, D., and Makela, J. (2000). *Geophys. Res. Lett.*, 27(8):2849–2852.

[Bust et al., 1994] Bust, G., Cook, J., Vasicek, C., Ward, S., and Kronschnabl, G. (1994). *Int. J. Imaging Syst. Technol.*, 5:160–168.

[Caorsi et al., 1988] Caorsi, S., Cagnani, G., and Pastorino, M. (1988). *Radio Sci.*, 23(6):1094–1106.

[Censor, 1983] Censor, Y. (1983). *Proc. IEEE*, 71:409–419.

[Chadan and Sabatier, 1989] Chadan, K. and Sabatier, P. (1989). *Inverse Problems in Quantum Scattering Theory*. Springer-Verlag, New York.

[Chen et al., 2000] Chen, B., Stamnes, J., and Stamnes, K. (2000). *Appl. Opt.*, 39(17):2904–2911.

[Ciraolo and Spalla, 1997] Ciraolo, L. and Spalla, P. (1997). *Radio Sci.*, 32(3):1071–1080.

[Colton and Kress, 1992] Colton, D. and Kress, R. (1992). *Inverse Acoustic and Electromagnetic Scattering Theory*. Appl. Math. Sc. 93. Springer-Verlag, New York.

[Cook and Close, 1995] Cook, J. and Close, S. (1995). *Ann. Geophys.*, 13:1320–1324.

[Crane, 1977] Crane, R. (1977). *Proc. IEEE*, 65(2):180–199.

[Dadzhion and Mersero, 1988] Dadzhion, D. and Mersero, R. (1988). *Tzifrovaya obrabotka mnogomernykh signalov.* Mir, Moskva.

[Datta and Bandyopadhyay, 1986] Datta, A. and Bandyopadhyay, B. (1986). *Proc. IEEE.*, 74(4):604–606.

[Davies, 1969] Davies, K. (1969). *Ionosphere Radio Waves.* Blaisdel, Watham.

[Davies, 1980] Davies, K. (1980). *Space Sci. Rev.*, 25(4):356–430.

[Davies, 1990] Davies, K. (1990). *Ionospheric Radio.* Peter Peregrinus, London.

[Devaney, 1978] Devaney, A. (1978). *J. Math. Phys.*, 19:1526–1531.

[Devaney, 1982] Devaney, A. (1982). *Proc. SPIE Int. Soc. Opt. Eng.*, 358:10–16.

[Devaney, 1984] Devaney, A. (1984). *IEEE Trans. Geosci. and Remote Sensing*, 22(1):3–13.

[Dickens and Winbow, 1997] Dickens, T. and Winbow, G. (1997). *J. Acoust. Soc. Am.*, 101(1):77–86.

[Dines and Lytle, 1979] Dines, K. and Lytle, J. (1979). *Proc. IEEE*, 67:1065–1073.

[Dixon, 1991] Dixon, T. (1991). *Rev. Geophys.*, 29:249–276.

[Duchene et al., 1985] Duchene, B., Lesselier, D., and Tabbara, W. (1985). *J. Opt. Soc. Am.*, 2(11):1943–1953.

[Faddeev, 1976] Faddeev, L. (1976). The inverse problem in the quantum theory of scattering II. In *J. Soviet Math.*, 5; 334–396. Consultants Bureau, New York.

[Fadeev, 1956] Fadeev, L. (1956). *Vestnik LGU. Seriya phiz. i khim.*, 7:126–130.

[Fante, 1975] Fante, R. (1975). Electromagnetic beam propagation in turbulent media. *Proc. IEEE*, 63(12):1669–1692.

[Fedoryuk, 1987] Fedoryuk, M. (1987). *Asimptotika. Integrali i Ryadi.* Nauka, Moskva.

[Fehmers, 1994] Fehmers, G. (1994). A new algorithm for ionospheric tomography. In Kersley, L. and other, editors, *Proc. Int. Beacon Satellite Symp.*, pp. 52–55, Aberystwyth, UK.

[Fehmers, 1996] Fehmers, G. (1996). *Tomography of the Ionosphere.* Ph.D. thesis, Technische Universiteit Eindhoven.

[Fehmers et al., 1998] Fehmers, G., Kamp, L., Sluijter, F., and others (1998). *Radio Sci.*, 33(1):149–163.

[Fejer, 1981] Fejer, B. (1981). *J. Atmos. Terr. Phys.*, 43:377–386.

[Fischer, 1976] Fischer, L. (1976). Probability and statistics. In Pearson, C., editor, *Handbook of Applied Mathematics.* Van Nostrand Reinhold, New York.

[Fjeldbo and Eshleman, 1965] Fjeldbo, G. and Eshleman, V. (1965). *J. Geophys. Res.*, 70(13):3217–3225.

[Fjeldbo et al., 1975] Fjeldbo, G., Seidel, B., Sweetnam, D., and Howard, T. (1975). *J. Atmos. Sci.*, 32(6):1232–1239.

[Fok, 1957] Fok, V. (1957). *Raboti po kvantovoy teorii polya.* LGU, Leningrad.

[Fok, 1970] Fok, V. (1970). *Problemi difraktziyi i rasprostraneniya elyektromagnitnikh voln.* Sovetskoye Radio, Moskva.

[Foster et al., 1994] Foster, J., Buonsanto, M., Holt, J., Klobuchar, J., Fougere, P., Pakula, W., Raymund, T., Kunitsyn, V., Andreeva, E., Tereshchenko, E., and Khudukon, B. (1994). *Int. J. Imaging Syst. Technol.*, 5:148–159.

[Foster et al., 1998] Foster, J., Cummer, S., and Inan, U. (1998). *J. Geophys. Res.*, 103(11):26359–26366.

[Foster and Rich, 1998] Foster, J. and Rich, F. (1998). *J. Geophys. Res.*, 103(11):26367–26372.

[Foster et al., 1992] Foster, J., Tetenbaum, D., del Pozo, C., and others (1992). *J. Geophys. Res.*, 97:8601–8617.

[Fösterling, 1942] Fösterling, K. (1942). *Hochfrequenz-technik und Elektroakustik*, 59:11–22.

[Fougere, 1995] Fougere, P. (1995). *Radio Sci.*, 30:429–444.

[Fougere, 1997] Fougere, P. (1997). *Radio Sci.*, 32:1623–1634.
[Franke et al., 2003] Franke, S., Yeh, K., Andreeva, E., and Kunitsyn, V. (2003). *Radio Sci.*, 38 (in press).
[Fremouw et al., 1994a] Fremouw, E., Howe, B., Secan, J., and Bussey, R. (1994a). *Int. J. Imaging Syst. Technol.*, 5:97–105.
[Fremouw and Secan, 1984] Fremouw, E. and Secan, J. (1984). *Radio Sci.*, 19(3):687–694.
[Fremouw et al., 1994b] Fremouw, E., Secan, J., Bussey, R., and Howe, B. (1994b). Simulation results from a tomographic processor employing the WDLS formulation of discrete inverse theory. In Kersley, L. and others, editors, *Proc. Int. Beacon Satellite Symp.*, pp. 172–175, Aberystwyth, UK.
[Fremouw et al., 1992] Fremouw, E., Secan, J., and Howe, B. (1992). *Radio Sci.*, 27:721–732.
[Fremouw et al., 1997] Fremouw, E., Secan, J., and Zhou, C. (1997). *Acta Geod. Geophys. Hung.*, 32:365–377.
[Fuks, 1963] Fuks, B. (1963). *Theory of Analytic Functions of Several Complex Variables.* Amer. Math. Soc., Providence, RI.
[Ganguly et al., 2001] Ganguly, S., Brown, A., DasGupta, A., and Ray, S. (2001). *Radio Sci.*, 36(4):789–800.
[Getmansev and Eroukhimov, 1967] Getmansev, G. and Eroukhimov, L. (1967). *Ann. ISQY*, 5:229–259.
[Ginzburg, 1961] Ginzburg, V. (1961). *Propagation of Electromagnetic Waves in Plasma.* Gordon and Breach, New York.
[Ginzburg and Rukhadze, 1975] Ginzburg, V. and Rukhadze, A. (1975). *Volni v magnitoaktivnoy plazme.* Nauka, Moskva.
[Golley and Rossiter, 1970] Golley, M. and Rossiter, D. (1970). *J. Atmos. Terr. Phys.*, 32(7):1215–1233.
[Grib et al., 1980] Grib, A., Mamayev, S., and Mostepanenko, V. (1980). *Kvantoviye effekti v intensivnikh vneshnikh polyakh.* Atomizdat, Moskva.
[Gusev and Kunitsyn, 1982] Gusev, V. and Kunitsyn, V. (1982). *Radiotekhnika i elyektronika*, 27(3):409–415.
[Gusev and Kunitsyn, 1986] Gusev, V. and Kunitsyn, V. (1986). *Radiotekhika i elyektronika*, 31(10):1894–1902.
[Gustavsson et al., 1986] Gustavsson, M., Ivansson, S., and Moren, P. (1986). *Proc. IEEE*, 74:339–346.
[Hagfors, 1977] Hagfors, T. (1977). In Brekke, A., editor, *Radar Probing of the Auroral Plasma*, pp. 15–28. Universitetsforlaget, Tromsø.
[Hagfors et al., 1968] Hagfors, T., Nanni, B., and Stone, K. (1968). *Radio Sci.*, 3:491–509.
[Hagfors and Tereshchenko, 1991] Hagfors, T. and Tereshchenko, E. (1991). *Radio Sci.*, 26:1199–1203.
[Hajj et al., 1994] Hajj, G., Ibañez-Meier, R., Kursinski, E., and Romans, L. (1994). *Int. J. Imaging Syst. Technol.*, 5:174–184.
[Hajj and Romans, 1998] Hajj, G. and Romans, L. (1998). *Radio Sci.*, 33:175–190.
[Heaton et al., 1996] Heaton, J., Jones, G., and Kersley, L. (1996). *Antarct. Sci.*, 8:297–302.
[Heaton et al., 1993] Heaton, J., Kersley, L., Pryse, S., Raymond, T., Rice, D., Sheen, P., Walker, I., and Willson, C. (1993). *Inst. Elect. Eng.*, 37:677–680.
[Heaton et al., 1995] Heaton, J., Pryse, S., and Kersley, L. (1995). *Ann. Geophys.*, 13:1297–1302.
[Hernández-Pajares et al., 2000] Hernández-Pajares, M., Huan, J., and Sanz, J. (2000). *Geophys. Res. Lett.*, 27(13):2009–2012.

[Hoeg et al., 1995] Hoeg, P., Syndergaard, S., Hauchecorne, A., and other (1995). *Derivation of Atmospheric Properties Using a Radio Occultation Technique.* ESA, Copenhagen.

[Hoenders, 1978] Hoenders, B. (1978). *Inverse Source Problems in Optics*, pp. 41–82. Springer-Verlag, New York. Ed. H.Baltes.

[Hörmander, 1986] Hörmander, L. (1986). *Analiz lineynikh differentzial'nikh operatorov s chastnimi proizvodnimi*, Vol. 1. Mir, Moskva.

[Howe et al., 1998] Howe, B., Runciman, K., and Secan, J. (1998). *Radio Sci.*, 33(1):109–128.

[Huang et al., 1997] Huang, C., Liu, C., Yeh, H., and Tsai, W. (1997). *J. Atmos. Terr. Phys.*, 59:1553–1567.

[Huang et al., 1999] Huang, C., Liu, C., Yeh, K., Lin, K., Tsai, W., Yeh, H., and Liu, J. (1999). *J. Geophys. Res.*, 104(A1):79–94.

[Hudson and Tyler, 1983] Hudson, D. and Tyler, G. (1983). *Icarus*, 54(2):337.

[Hunsucker, 1991] Hunsucker, R. (1991). *Radio Techniques for Probing the Terrestrial Ionosphere.* Springer-Verlag, Berlin, Heidelberg, New York.

[Idenden et al., 1998] Idenden, D., Moffett, R., Williams, M., Spencer, P., and Kersley, L. (1998). *Ann. Geophys.*, 16:969–973.

[Isakov, 1998] Isakov, V. (1998). *Inverse Problems for Partial Differential Equations.* Springer-Verlag, New York.

[Ishimaru, 1978] Ishimaru, A. (1978). *Wave Propagation and Scattering in Random Media.* Academic Press, New York.

[Ivansson, 1986] Ivansson, S. (1986). *Proc. IEEE*, 74:328–338.

[Jenkins and Watts, 1969] Jenkins, G. and Watts, D. (1969). *Spectral Analysis and Its Applications.* Holden-Day, San Francisco, Cambridge.

[Johnson and Tracy, 1983] Johnson, S. and Tracy, M. (1983). *Ultrasonic Imaging*, 5(6):361–392.

[Johnson et al., 1984] Johnson, S., Zhou, Y., Tracy, M., and other (1984). *Ultrasonic Imaging*, 6:103–106.

[Jones et al., 1997] Jones, D., Walker, I., and Kersley, L. (1997). *Ann. Geophys.*, 15:740–746.

[Kamalabadi et al., 1999] Kamalabadi, F., Karl, W., Semeter, J., Cotton, D., Cook, T., and Chakrabarti, S. (1999). *Radio Sci.*, 34(2):437–447.

[Kersley and E.Pryse, 1994] Kersley, L. and Pryse, S.E. (1994). *Int. J. Imaging Syst. Technol.*, 5:141–147.

[Kersley et al., 1993] Kersley, L., Heaton, J., Pryse, S., and Raymund, T. (1993). *Ann. Geophys.*, 11:1064–1074.

[Kersley et al., 1997] Kersley, L., Pryse, S., Walker, I., Heaton, J., Mitchell, C., Williams, M., and Willson, C. (1997). *Radio Sci.*, 32:1607–1621.

[Keskinen and Ossakow, 1983] Keskinen, M. and Ossakow, S. (1983). *Radio Sci.*, 18:1077–1091.

[Khurgin and Yakovlev, 1971] Khurgin, Y. and Yakovlev, V. (1971). *Finitniye funktzii v nauke i tekhnike.* Nauka, Moskva.

[Kliore et al., 1969] Kliore, A., Fjeldbo, G., Seidel, B., and Rasool, S. (1969). *Science*, 166(3911):373.

[Kliore et al., 1976] Kliore, A., Woiceshyn, P., and Hubbard, W. (1976). *Geophys. Res. Lett.*, 3(3):113.

[Klobuchar et al., 1994] Klobuchar, J., Doherty, P., Bailey, G., and Davies, K. (1994). Limitations in determining absolute total electron content from dual-frequency GPS group delay measurements. In Kersley, L. and others, editors, *Proc. Int. Beacon Satellite Symp.*, pp. 1–4, Aberystwyth, UK.

[Klyatzkin, 1980] Klyatzkin, V. (1980). *Stokhasticheskiye uravneniya i volni v sluchayno-neodnorodnikh sredakh.* Nauka, Moskva.

[Knipp and others, 1998] Knipp, D. and others (1998). *J. Geophys. Res.*, 103(11):26197–26220.

[Kolosov et al., 1979] Kolosov, M., Yakovlev, O., Efimov, A., and others (1979). *Radio Sci.*, 14(1):163–167.

[Krall and Trivelpiece, 1973] Krall, N. and Trivelpiece, A. (1973). *Principles of Plasma Physics*. McGraw–Hill, New York.

[Krasnushkin, 1981] Krasnushkin, P. (1981). *Geomagnetism and Aeronomy*, 21:1133–1135.

[Kravtsov et al., 2000] Kravtsov, Y., Kunitsyn, V., and Tereshchenko, E. (2000). *Phys. Chem. Earth (C)*, 25(1–2):59–62.

[Kreyn, 1964] Kreyn, S., editor (1964). *Funktzional'niy analiz.* Nauka, Moskva.

[Kronschnabl et al., 1995] Kronschnabl, G., Bust, G., Cook, J., and Vasicek, C. (1995). *Radio Sci.*, 30:105–108.

[Kuklinski, 1997] Kuklinski, W. (1997). *Radio Sci.*, 32(3):1037–1049.

[Kunitake et al., 1995] Kunitake, M., Maruyama, T., Morioka, A., Ohtaka, K., Tokumaru, M., and Watanabe, S. (1995). *Ann. Geophys.*, 13:1303–1310.

[Kunitsyn, 1985] Kunitsyn, V. (1985). *Moscow Univ. Phys. Bull.*, 26(6):33–40.

[Kunitsyn, 1986a] Kunitsyn, V. (1986a). *Geomagnetism and Aeronomy*, 26(1):58–62.

[Kunitsyn, 1986b] Kunitsyn, V. (1986b). *Moscow Univ. Phys. Bull.*, 41(2):43–48.

[Kunitsyn, 1992] Kunitsyn, V. (1992). *Proc. SPIE*, 1843:172–182.

[Kunitsyn et al., 1995a] Kunitsyn, V., Andreeva, E., Popov, A., and Razinkov, O. (1995a). *Ann. Geophys.*, 13(12):1421–1428.

[Kunitsyn et al., 1997] Kunitsyn, V., Andreeva, E., and Razinkov, O. (1997). *Radio Sci.*, 32(5):1953–1963.

[Kunitsyn et al., 1989] Kunitsyn, V., Preobrazhenskiy, N., and Tereshchenko, E. (1989). *Doklady Akademii Nauk SSSR*, 306(3):19–22.

[Kunitsyn and Smorodinov, 1986] Kunitsyn, V. and Smorodinov, V. (1986). *Geomagnetism and Aeronomy*, 26(1):123–125.

[Kunitsyn and Tereshchenko, 1990] Kunitsyn, V. and Tereshchenko, E. (1990). *The reconstruction of the ionosphere irregularities structure*, pp. 1–60. Preprint Polar Geophys. Inst. 90-01-69, Apatity.

[Kunitsyn and Tereshchenko, 1991] Kunitsyn, V. and Tereshchenko, E. (1991). *Tomography of the Ionosphere.* Nauka, Moscow (in Russian).

[Kunitsyn and Tereshchenko, 1992] Kunitsyn, V. and Tereshchenko, E. (1992). *IEEE Antennas Propagation Magn.*, 34:22–32.

[Kunitsyn et al., 1994a] Kunitsyn, V., Tereshchenko, E., Andreeva, E., Khudukon, B., and Nygrén, T. (1994a). *Int. J. Imaging Syst. Technol.*, 5:112–127.

[Kunitsyn et al., 1990] Kunitsyn, V., Tereshchenko, E., Andreeva, E., and others (1990). Radiotomography of global ionospheric structures (in Russian).

[Kunitsyn et al., 1995b] Kunitsyn, V., Tereshchenko, E., Andreeva, E., and others (1995b). *Ann. Geophys.*, 13(12):1351–1359.

[Kunitsyn et al., 1994b] Kunitsyn, V., Tereshchenko, E., Andreeva, E., and Razinkov, O. (1994b). *Int. J. Imaging Syst. Technol.*, 5:128–140.

[Kursinski et al., 1996] Kursinski, E., Hajj, G., Beritger, W., and others (1996). *Science*, 271:1107.

[Kursinski et al., 1995] Kursinski, E., Hajj, G., Hardy, K., Romans, L., and Schofield, J. (1995). *Geophys. Res. Lett.*, 22(17):2365.

[Kursinski et al., 1997] Kursinski, E., Hajj, G., Schofield, J., Linfield, R., and Hardy, K. (1997). *J. Geophys. Res.*, 102(D19):23429–23465.

[Lam et al., 1976] Lam, D., Schmidt-Weinmar, H., and Wouk, A. (1976). *Can. J. Phys.*, 54:1925–1926.

[Landau and Pollak, 1961] Landau, H. and Pollak, H. (1961). *Bell Syst. Tech. J.*, 40; 41:65–84; 1295–1336.

[Landau and Lifshitz, 1984] Landau, L. and Lifshitz, E. (1984). Course of theoretical physics. In *Electrodynamics of Continuous Media*, Vol. 8. Pergamon Press, Oxford, New York.

[Langeberg, 1987] Langeberg, K. (1987). Applied inverse problems for acoustic, electromagnetic and elastic wave scattering. In Sabatier, P., editor, *Basic Methods of Tomography and Inverse Problems*, pp. 127–467. Adam Hilger, Bristol.

[Lavrentiev et al., 1986] Lavrentiev, M., Romanov, V., and Shishatskij, S. (1986). *Ill-Posed Problems of Mathematical Physics and Analysis.* Amer. Math. Soc., Providence, RI.

[Leitinger, 1996] Leitinger, R. (1996). Tomography. In Kohl, H., Rüster, R., and Schlegel, K., editors, *Modern Ionospheric Science*, pp. 346–370. European Geophysical Society, Katlenburg-Lindau, Germany.

[Leitinger, 1999] Leitinger, R. (1999). *Rev. Radio Sci. 1996–1999*, 581–623.

[Leitinger et al., 1984] Leitinger, R., Hartman, G., Lohman, F., and other (1984). *Radio Sci.*, 19:789–797.

[Leitinger et al., 1975] Leitinger, R., Schmidt, G., and Taurianen, A. (1975). *J. Geophys. Res.*, 41:201–213.

[Leone et al., 1999] Leone, G., Brancaccio, A., and Pierri, R. (1999). *J. Opt. Soc. Am.*, A 15(12):2887–2895.

[Lesselier et al., 1985] Lesselier, D., Vuillet-Laurent, D., Jouvie, F., and others (1985). *Electromagnetics*, 5:147–189.

[Lindal, 1992] Lindal, G. (1992). *Astronom. J.*, 103(3):967.

[Lindal et al., 1987] Lindal, G., Lyons, I., Sweetnam, D., and others (1987). *J. Geophys. Res.*, 92(A13):14987.

[Lindal et al., 1981] Lindal, G., Wood, G., Levy, G., and others (1981). *J. Geophys. Res.*, 86(A10):8721.

[Louis, 1989] Louis, A. (1989). *Inverse and Ill-Posed Problems.* Teubner, Stuttgart.

[Lunt et al., 1999a] Lunt, N., Kersley, L., and Bailey, G. (1999a). *Radio Sci.*, 34(3):725–732.

[Lunt et al., 1999b] Lunt, N., Kersley, L., and Bishop, G. (1999b). *Radio Sci.*, 34(5):1261–1280.

[Markkanen et al., 1995] Markkanen, M., Lehtinen, M., Nygrén, T., Pirttila, J., Henelius P., Vilenius, E., Tereshchenko, E., and Khudukon, B. (1995). *Ann. Geophys.*, 13:1277–1287.

[Marple Jr., 1987] Marple Jr., S. (1987). *Digital Spectral Analysis.* Prentice-Hall, Englewood Cliffs, New Jersey.

[Mast, 1999] Mast, T. (1999). *J. Acoust. Soc. Am.*, 106(6):3061–3071.

[Mast et al., 1997] Mast, T., Nachman, A., and Waag, R. (1997). *J. Acoust. Soc. Am.*, 102(2):715–725. Part 1.

[McDade et al., 1991] McDade, I., Leoyd, N., and Llewellyn, E. (1991). *Planetary Space Sci.*, 39:895–906.

[McDade and Llewellyn, 1991] McDade, I. and Llewellyn, E. (1991). *Can. J. Phys.*, 69:1059.

[Menke, 1990] Menke, V. (1990). *Geophysical Data Analysis: Discrete Inverse Theory.* Academic Press, Orlando, Florida.

[Minerbo et al., 1980] Minerbo, G., Sanderson, J., van Hulsteyn, D., and others (1980). *Appl. Opt.*, 19(10):1723–1728.

[Mitchell et al., 1995] Mitchell, C., Jones, D., Kersley, L., Pryse, S., and Walker, I. (1995). *Ann. Geophys.*, 13:1311–1319.

[Mitchell et al., 1997a] Mitchell, C., Kersley, L., Heaton, J., and Pryse, S. (1997a). *Ann. Geophys.*, 15:747–752.

[Mitchell et al., 1997b] Mitchell, C., Kersley, L., and Pryse, S. (1997b). *J. Atmos. Terr. Phys.*, 59:1411–1415.

[Mitchell et al., 1997c] Mitchell, C., Pryse, S., Kersley, L., and Walker, I. (1997c). *J. Atmos. Terr. Phys.*, 59:2077–2087.

[Mitchell et al., 1998] Mitchell, C., Walker, I., Pryse, S., Kersley, L., McCrea, I., and Jones, T. (1998). *Ann. Geophys.*, 16:1519–1522.

[Moen et al., 1998] Moen, J., Berry, S., Kersley, L., and Lybekk, B. (1998). *Ann. Geophys.*, 16:574–582.

[Mudrov and Kushko, 1976] Mudrov, V. and Kushko, V. (1976). *Metodi obrabotki izmyeryeniy.* Sovyetskoye Radio, Moskva.

[Mueller et al., 1979] Mueller, R., Kaveh, M., and Wade, G. (1979). *Proc. IEEE*, 67(4):567–587.

[Munk and Wunsch, 1983] Munk, W. and Wunsch, C. (1983). *Rev. Geophys. Space Phys.*, 21:777–793.

[Na et al., 1995] Na, H., Hall, B., and Sutton, E. (1995). *Ann. Geophys.*, 13:1288–1296.

[Na and Lee, 1991] Na, H. and Lee, H. (1991). *Int. J. Imaging Syst. Technol.*, 3:354–365.

[Na and Sutton, 1994] Na, H. and Sutton, E. (1994). *Int. J. Imaging Syst. Technol.*, 5:169–173.

[Namgaladze et al., 1988] Namgaladze, A., Korenkov, Y., Klimenko, V., Karpov, I., Bessarab, F., Surotkin, V., Glushchenko, T., and Naumova, N. (1988). *Pure Appl. Geophys.*, 127(2/3):219–254.

[Namgaladze et al., 1996] Namgaladze, A., Martynenko, O., Namgaladze, A., Volkov, M., Korenkov, Y., Klimenko, V., Karpov, I., and Bessarab, F. (1996). *J. Atmos. Terr. Phys.*, 58(1-4):297–306.

[Natterer, 1986] Natterer, F. (1986). *The Mathematics of Computerized Tomography.* B.G. Teubner, Stuttgart, and J. Wiley and Sons.

[Nayfe, 1976] Nayfe, A. (1976). *Metodi vozmushcheniy.* Mir, Moskva.

[Newton, 1990] Newton, R. (1990). *Inverse Schrödinger Scattering in Three Dimensions.* Springer-Verlag, New York.

[Ney et al., 1984] Ney, M., Smith, A., and Stuchly, S. (1984). *IEEE Trans. Medical Imaging*, 3:155–162.

[Nolet, 1987] Nolet, G. (1987). Reidel, Dordrecht.

[Nygrén et al., 1997] Nygrén, T., Kaila, K., Markkanen, M., and Lehtinen, M. (1997). *Adv. Space Res.*, 19:639.

[Nygrén et al., 1997] Nygrén, T., Markannen, M., Lehtinen, M., Tereshchenko, E., and Khudukon, B. (1997). *Radio Sci.*, A(32):2359–2372.

[Nygrén et al., 1996a] Nygrén, T., Markkanen, M., Lehtinen, M., and Kaila, K. (1996a). *Ann. Geophys.*, 14:1124–1133.

[Nygrén et al., 1996b] Nygrén, T., Markkanen, M., Lehtinen, M., Tereshchenko, E., Khudukon, B., Evstafiev, O., and Pollari, P. (1996b). *Ann. Geophys.*, 14:1422–1428.

[Nygrén et al., 1997] Nygrén, T., Tereshchenko, E., Khudukon, B., Evstafiev, O., Lehtinen, M., and Markkanen, M. (1997). *Acta Geod. Geophys. Hung.*, 32:395–405.

[Nygrén et al., 2000] Nygrén, T., Tereshchenko, E., Khudukon, B., Evstafiev, O., Lehtinen, M., and Markkanen, M. (2000). *Adv. Space Res.*, 26(6):939–942.

[Oraevsky et al., 1995] Oraevsky, V., Kunitsyn, V., Ruzhin, Y., Andreeva, E., Depueva, A., Kozlov, E., Razinkov, O., and Shagimuratov, I. (1995). *Geomagnetism and Aeronomy*, 35:117–122.

[Oraevsky et al., 1993] Oraevsky, V., Kunitsyn, V., Ruzhin, Y., Razinkov, O., Andreeva, E., Depueva, A., Kozlov, E., and Shagimuratov, I. (1993). Radiotomographic sections of subauroral ionosphere along Moscow-Arkhangelsk trace. Preprint IZMIRAN 100, p. 1047 (in Russian).

[Orlov, 1975a] Orlov, S. (1975a). *Kristallographiya*, 20(3):511–515.

[Orlov, 1975b] Orlov, S. (1975b). *Kristallographiya*, 20(4):701–709.

[Pakula et al., 1995] Pakula, W., Fougere, P., Klobuchar, J., Keunzler, H., Buonsanto, M., Roth, J., Foster, J., and Sheehan, R. (1995). *Radio Sci.*, 30:89–103.

[Palamodov, 1990] Palamodov, V. (1990). Nekotoriye singulyarniye zadachi tomographii. In: *Voprosi kibernetiki. Matematicheskiye problemi tomographii.* Nauka, Moscow.

[Park et al., 1996] Park, C., Park, S., and Ra, J. (1996). *Radio Sci.*, 31(6):1877–1886.

[Pavlenko, 1976] Pavlenko, Y. (1976). *Gamil'tonovi metodi v elyektrodinamike i v kvantovoy mekhanike.* Nauka, Moskva.

[Philips and Spenser, 1955] Philips, G. and Spenser, M. (1955). *Proc. Phys. Soc.*, B68:481–497.

[Phinney and Anderson, 1968] Phinney, R. and Anderson, D. (1968). *J. Geophys. Res.*, 73(5):1819–1827.

[Pikalov and Preobrazhenskiy, 1987b] Pikalov, V. and Preobrazhenskiy, N. (1987b). *Rekonstruktivnaya tomographiya v gazodinamike i fizike plazmi.* Nauka, Novosibirsk.

[Pikalov and Preobrazhenskiy, 1987a] Pikalov, V. and Preobrazhenskiy, N. (1987a). *Rekonstruktivnaya tomographiya v gazodinamike i phizike plazmi.* Nauka, Novosibirsk.

[Porter, 1970] Porter, R. (1970). *J. Opt. Soc. Am.*, 60(1051–1059).

[Porter and Devaney, 1982a] Porter, R. and Devaney, A. (1982a). *J. Opt. Soc. Am.*, 72(3):327–330.

[Porter and Devaney, 1982b] Porter, R. and Devaney, A. (1982b). *J. Opt. Soc. Am.*, 72(12):1707–1713.

[Preobrazhenskiy and Sedel'nikov, 1984] Preobrazhenskiy, N. and Sedel'nikov, A. (1984). *Izvestiya AN SSSR. Seriya phiz.*, 48(4):810–814.

[Prett, 1982] Prett, U. (1982). *Tzifrovaya obrabotka izobrazheniy.* Mir, Moskva.

[Prosser, 1969] Prosser, R. (1969). *J. Math. Phys.*, 10(10):1819–1822.

[Prosser, 1976] Prosser, R. (1976). *J. Math. Phys.*, 17(10):1775–1779.

[Pryse and Kersley, 1992] Pryse, S. and Kersley, L. (1992). *J. Atmos. Terr. Phys.*, 54:1007–1012.

[Pryse et al., 1998a] Pryse, S., Kersley, L., Mitchell, C., Spencer, P., and Williams, M. (1998a). *Radio Sci.*, 33:1767–1779.

[Pryse et al., 1998b] Pryse, S., Kersley, L., Mitchell, C., Spenser, P., and Williams, M. (1998b). *Radio Sci.*, 33(6):1767–1779.

[Pryse et al., 1993] Pryse, S., Kersley, L., Rice, D., and others (1993). *Ann. Geophys.*, 11:144–149.

[Pryse et al., 1998c] Pryse, S., Kersley, L., Williams, M., and Walker, I. (1998c). *Ann. Geophys.*, 16:1169–1179.

[Pryse et al., 1997] Pryse, S., Kersley, L., Williams, M., Walker, I., and Willson, C. (1997). *J. Atmos. Terr. Phys.*, 59:1953–1959.

[Pryse et al., 1995] Pryse, S., Mitchell, C., Heaton, J., and Kersley, L. (1995). *Ann. Geophys.*, 13:1325–1342.

[Pryse et al., 1999] Pryse, S., Smith, A., Moen, J., and Lorentzen, D. (1999). *Geophys. Res. Lett.*, 26:25–28.

[Ra and Cho, 1981] Ra, J. and Cho, Z. (1981). *Proc. IEEE*, 69:668–670.

[Radon, 1917] Radon, J. (1917). *Sitzungsberichte der Preussischen Akademie der Wissenschaften*, 69(of Math.-Phys. Klasse):262–277.

[Ramm, 1992] Ramm, A. (1992). *Multidimensional Inverse Scattering Theory.* John Wiley & Sons, New York.

[Rao and Carin, 2000] Rao, B. and Carin, L. (2000). *Radio Sci.*, 35(2):315–329.

[Ratcliffe, 1959] Ratcliffe, J. (1959). *The Magnetoionic Theory and Its Applications to the Ionosphere.* Cambridge University Press, London.

[Ratcliffe, 1972] Ratcliffe, J. (1972). *An Introduction to the Ionosphere and Magnetosphere.* Cambridge University Press, London.

[Raymund, 1994] Raymund, T. (1994). *Int. J. Imaging Syst. Technol.*, 5(2):75–85.

[Raymund, 1995] Raymund, T. (1995). *Ann. Geophys.*, 13:1254–1262.

[Raymund et al., 1990] Raymund, T., Austen, J., Franke, S., Liu, C., Klobuchar, J., and Stalker, J. (1990). *Radio Sci.*, 25:771–789.

[Raymund et al., 1994a] Raymund, T., Bresler, Y., Anderson, D., and Daniell, R. (1994a). *Radio Sci.*, 29:1493–1512.

[Raymund et al., 1994b] Raymund, T., Franke, S., and Yeh, K. (1994b). *J. Atmos. Terr. Phys.*, 56:637–657.

[Raymund et al., 1993] Raymund, T., Pryse, S., Kersley, L., and Heaton, J. (1993). *Radio Sci.*, 28:811–817.

[Reed and Simon, 1979] Reed, M. and Simon, B. (1972–1979). *Methods of Modern Mathematical Physics*, Vol. I–IV. Academic Press, New York.

[Reinish, 1996] Reinish, B. (1996). Modern ionosondes. In Kohl, H., Ruster, R., and Shlegel, K., editors, *Modern Ionospheric Science.* EGS, Katlenburg-Lindau.

[Rishbeth and Garriott, 1969] Rishbeth, H. and Garriott, O. (1969). *Introduction to Ionospheric Physics.* Academic Press, New York.

[Rius et al., 1997] Rius, A., Ruffini, G., and Cucurull, L. (1997). *Geophys. Res. Lett.*, 24:2291–2294.

[Rocken et al., 1997] Rocken, C., Anthes, R., Exner, M., Hunt, D., Sokolovsky, S., Ware, R., Gorbunov, M., and Svhreiner, W. (1997). *J. Geophys. Res.*, 102(D25):29849–29866.

[Rogers, 1956] Rogers, G. (1956). *Nature*, 177:613.

[Romanov, 1987] Romanov, V. (1987). *Inverse Problems of Mathematical Physics.* VNU Science Press, Utrecht.

[Ruffini et al., 1998] Ruffini, G., Flores, A., and Rius, A. (1998). *IEEE Trans. Geoscience Remote Sensing*, 36(1):143–153.

[Rytov et al., 1989] Rytov, D., Kravtsov, Y., and Tatarskii, V. (1989). Introduction to statistical radiophysics. In *Wave Propagation through Random Media.* Springer-Verlag, New York.

[Saenko et al., 1991] Saenko, Y., Shagimuratov, I., Namagaladze, A., Yakimova, G., Natsvalian, H., Biriukov, O., and Kuzin, V. (1991). *Geomagnetism and Aeronomy*, 31:558–561.

[Saksman et al., 1997] Saksman, E., Nygrén, T., and Markkanen, M. (1997). *Radio Sci.*, 32(2):605–616.

[Schmidt, 1972] Schmidt, G. (1972). *Zs. Geophys.*, 38:891–913.

[Schmidt and Taurianen, 1975] Schmidt, G. and Taurianen, A. (1975). *J. Geophys. Res.*, 80(31):4313–4324.

[Schmidt-Weinmar et al., 1975] Schmidt-Weinmar, H., Lam, D., and Wouk, A. (1975). *J. Opt. Soc. Am.*, 65(10):1188–1189.

[Schuster, 1889] Schuster, A. (1889). The diurnal variation of the terrestrial magnetism. *Philos. Trans. R. Soc. London*, A(180):467–518.

[Schwinger, 1951] Schwinger, J. (1951). *Phys. Rev.*, 82:664–679.

[Secan et al., 1997] Secan, J., Bussey, R., Fremouw, E., and Basu, S. (1997). *Radio Sci.*, 32:1567–1574.

[Shabat, 1976] Shabat, B. (1976). *Vvedeniye v kompleksniy analiz.* Nauka, Moskva.

[Sheffield, 1975] Sheffield, J. (1975). *Plasma Scattering of Electromagneic Radiation.* Academic Press, New York.

[Shuleykin, 1923] Shuleykin, M. (1923). *Rasprostraneniye elyektromagnitnoy energiyi.* Pervoye russkoye radiobyuro, Moskva.

[Slaney and Kak, 1983] Slaney, M. and Kak, A. (1983). *Proc. SPIE Int. Soc. Opt. Eng.*, 413:2–19.

[Slepyan, 1976] Slepyan, D. (1976). *TIIER*, 64(3):4–14.

[Snow and Romanowski, 1994] Snow, W. and Romanowski, P. (1994). A comparison of ionosphere total electron content measurements with code and codeless GPS receivers. In Kersley, L. and other, editors, *Proc. Int. Beacon Satellite Symp.*, pp. 17–20, Aberystwyth, UK.

[Solodovnikov et al., 1988] Solodovnikov, G., Sinelnikov, V., and Krokhmalnikov, E. (1988). *Remote Sounding of the Earth Ionosphere by Satellite Beacons.* Nauka, Moscow (in Russian).

[Solomon et al., 1985] Solomon, S., Hays, P., and Abreu, V. (1985). *Appl. Opt.*, 24:4134–4140.

[Spenser et al., 1998] Spenser, P., Kersley, L., and Pryse, S. (1998). *Radio Sci.*, 33(3):607–616.

[Spindel, 1982] Spindel, R. (1982). *Oceanus*, 25:12–21.

[Stewart and Hogan, 1973] Stewart, R. and Hogan, I. (1973). *Radio Sci.*, 8(2):109–115.

[Stone, 1976a] Stone, W. (1976a). *J. Atmos. Terr. Phys.*, 38(6):583–592.

[Stone, 1976b] Stone, W. (1976b). *J. Atmos. Terr. Phys.*, 38(6):583–592.

[Stone, 1987] Stone, W. (1987). *Radio Sci.*, 22(6):1026–1030.

[Strohben, 1968] Strohben, J. (1968). *Proc. IEEE*, 56:1301–1318.

[Suni et al., 1989] Suni, A., Tereshchenko, V., Tereshchenko, E., and Khudukon, B. (1989). *Nekogerentnoye rasseyaniye radiovoln v visokoshirotnoy ionosphere.* Izd. KNC RAN, Apatity.

[Sutton and Na, 1994] Sutton, E. and Na, H. (1994). *Int. J. Imaging Syst. Technol.*, 5:106–111.

[Sutton and Na, 1995] Sutton, E. and Na, H. (1995). *Radio Sci.*, 30(1):115–125.

[Sutton and Na, 1996a] Sutton, E. and Na, H. (1996a). *Radio Sci.*, 31:489–496.

[Sutton and Na, 1996b] Sutton, E. and Na, H. (1996b). *Int. J. Imaging Syst. Technol.*, 7(3):238–245.

[Tabbara et al., 1988] Tabbara, W., Duchene, B., Pichot, C., and others (1988). *Inverse Problems*, 4:305–331.

[Tatarskii, 1968] Tatarskii, V. (1968). *Izvestiya AN SSSR, PhAO*, 4(8):811–818. (in Russian).

[Tatarskii, 1971] Tatarskii, V. (1971). *The Effects of the Turbulent Atmosphere on Wave Propagation.* Keter, Jerusalem.

[Tereshchenko, 1984] Tereshchenko, E. (1984). *Geomagnetizm i aeronomiya*, 24(6):1016–1018.

[Tereshchenko, 1987] Tereshchenko, E. (1987). *Radiogolograficheskiy metod issledovaniya ionosphernich neodnorodnostey.* Izd-vo Kol'skogo filiala AN SSSR, Apatity.

[Tereshchenko and Hagfors, 1997] Tereshchenko, E. and Hagfors, T. (1997). *Radio Sci.*, 32(4):1469–1475.

[Tereshchenko and Khudukon, 1981] Tereshchenko, E. and Khudukon, B. (1981). *Geomagnetizm i aeronomiya*, 21(2):255–258.

[Tereshchenko et al., 2000a] Tereshchenko, E., Khudukon, B., Kozlova, M., Evstafjev, O., Nygrén, T., Rietveld, M., and Brekke, A. (2000a). *Ann. Geophys.*, 18(8):918–927.

[Tereshchenko et al., 1998] Tereshchenko, E., Kozlova, M., and Evstafiev, O. (1998). *Geomagnetizm i aeronomiya*, 38(4).

[Tereshchenko et al., 2000b] Tereshchenko, E., Kozlova, M., Evstafjev, O., Khudukon, B., Nygrén, T., Rietveld, M., and Brekke, A. (2000b). *Ann. Geophys.*, 18(9):1197–1209.

[Tereshchenko et al., 2002] Tereshchenko, E., Kozlova, M., Kunitsyn, V., and Andreeva, E. (2002). Statistical tomography of subkilometer irregularities in the high latitude ionosphere. *Proc. Ionospheric Effects Symp.*, 6A5, pp. 1–8, Alexandria.

[Tereshchenko et al., 1982] Tereshchenko, E., Popov, A., Tereshchenko, A., and others (1982). *Issledovaniye visikoshirotnoy ionosphri i magnitospheri Zemli*, pp. 90–101. Nauka, Leningrad.

[Tereshchenko et al., 1983] Tereshchenko, E., Romanov, V., and Khudukon, B. (1983). *Geomagnetizm i aeronomiya*, 23(2):223–229.

[Testorf and Fiddy, 1999] Testorf, M. and Fiddy, M. (1999). *J. Opt. Soc. Am.*, A 16(7):1806–1813.

[Tikhonov, 1963] Tikhonov, A. (1963). *Sov. Math. Doklady*, 4:1035–1038.

[Tikhonov and Arsenin, 1977] Tikhonov, A. and Arsenin, V. (1977). *Methods for the Solution of Ill-Posed Problems.* Wiley & Sons, New York.

[Tikhonov et al., 1995] Tikhonov, A., Goncharsky, A., Stepanov, V., and Yagola, A. (1995). *Numerical Methods for the Solution of Ill-Posed Problems.* Kluwer, Dordrecht/Boston/London.

[Tønsager and La Hoz, 1999] Tønsager, H. and La Hoz, C. (1999). Simulations of range deconvolution using two-dimensional chirp mapping. In *Abstracts of 9th International EISCAT Workshop, Wernigerode*, p. 47. Max-Planck-Institut für Aeronomie.

[Tsihrintzis and Devaney, 2000] Tsihrintzis, G. and Devaney, A. (2000). *IEEE Trans. Inf. Theory*, 46(5):1748–1761.

[Tsunoda, 1988] Tsunoda, R. (1988). *Rev. Geophys.*, 26:719–760.

[Tyler et al., 1981] Tyler, G., Eshleman, V., and Anderson, I. (1981). *Science*, 212(4491):201.

[Vallance et al., 1991] Vallance, J., Gattinger, K., Creutzberg, F., Harris, F., McNamara, A., Yau, A., Llewellyn, E., Lumerzheim, D., Ress, M., McDade, I., and Marget, J. (1991). *Planetary Space Sci.*, 39:1677–1705.

[Vasicek and Kronschnabl, 1995] Vasicek, C. and Kronschnabl, G. (1995). *J. Atmos. Terr. Phys.*, 57:875–888.

[Vaynshteyn, 1988] Vaynshteyn, L. (1988). *Elyektromagnitniye volni.* Radio i svyaz', Moskva.

[Vladimirov, 1988] Vladimirov, V. (1988). *Uravneniya matematicheskoy phiziki.* Nauka, Moskva.

[Vorob'ev and Krasilnikova, 1993] Vorob'ev, V. and Krasilnikova, T. (1993). *Izvestiya AN SSSR. Seriya phiz.*, 29(5):626–633.

[Walker et al., 1998] Walker, I., Moen, J., Mitchell, C., Kersley, L., and Sandholt, P. (1998). *Geophys. Res. Lett.*, 25:293–296.

[Walker et al., 1996] Walker, K., Heaton, J., Kersley, L., Mitchell, C., Pryse, S., and Williams, M. (1996). *Ann. Geophys.*, 14:1413–1421.

[Wanniger et al., 1994] Wanniger, L., Sardon, E., and Warnat, R. (1994). Determination of the total ionospheric electron content with GPS – difficulties and their solution. In Kersley, L. and other, editors, *Proc. Int. Beacon Satellite Symp.*, pp. 172–175, Aberystwyth, UK.

[Ware et al., 1996] Ware, R., Exner, M., Gorbunov, M., Sokolovskiy, S., and others (1996). *Bull. Am. Meteorol. Soc.*, 77(1):19.

[Wells et al., 1986] Wells, D., Beck, N., and Delikaraoglou, D. (1986). *Guide to GPS Positioning.* Canadian GPS Associates, Fredericton, N.B.

[Wolf, 1969] Wolf, E. (1969). *Opt. Commun.*, 1(4):153–156.

[Wood and Perry, 1980] Wood, C. and Perry, G. (1980). *Philos. Trans. R. Soc. London*, A(294):307–315.

[Wyatt, 1968] Wyatt, P. (1968). *Appl. Opt.*, 7(10):1879–1896.

[Xiaochuan et al., 1999] Xiaochuan, Pan, and other (1999). *J. Opt. Soc. Am.*, 19(12):2896–2903.

[Yakovlev et al., 1995] Yakovlev, O., Matyugov, S., and Vilkov, I. (1995). *Radio Sci.*, 30(3):591.

[Yaroslavskiy, 1987] Yaroslavskiy, L. (1987). *Tzifrovaya obrabotka signalov v optike i golographiyi.* Radio i svyaz', Moskva.

[Yeh et al., 2001] Yeh, K., Franke, S., Andreeva, E., and Kunitsyn, V. (2001). *Geophys. Res. Lett.*, 28(24):4517–4520.

[Yeh and Raymund, 1991] Yeh, K. and Raymund, T. (1991). *Radio Sci.*, 26:1361–1380.

[Yen and Swenson, 1964] Yen, K. and Swenson, G. (1964). *J. Res. Nat. Bur. Stand. Radio Sci.*, 68(D):881–894.

[Yura, 1972] Yura, H. (1972). *Appl. Opt.*, 11(6):1399.

[Zakharov and Kunitsyn, 1998] Zakharov, V. and Kunitsyn, V. (1998). *Moscow Univ. Phys. Bull.*, 4:45–48.

[Zakharov and Kunitsyn, 1999] Zakharov, V. and Kunitsyn, V. (1999). *Moscow Univ. Phys. Bull.*, (4):42–46.

[Zhou et al., 1999] Zhou, C., Fremouw, E., and Sahr, J. (1999). *Radio Sci.*, 34(1):155–162.

[Zuev et al. 1988] Zuev, V., Banakh, V., and Pokasov, V. (1988). *Optika turbulentnoy atmospheri.* Gidrometeoizdat, Moskva.

Index

Printing: Druckhaus Beltz, Hemsbach
Binding: Buchbinderei Schäffer, Grünstadt